D1087567

Springer Handbook of Auditory Research

Series Editors: Richard R. Fay and Arthur N. Popper

For other titles published in this series, go to
www.springer.com/series/2506

Ray Meddis · Enrique A. Lopez-Poveda
Richard R. Fay · Arthur N. Popper
Editors

Computational Models of the Auditory System

Editors
Ray Meddis
University of Essex
Colchester CO4 3SQ
UK
rmeddis@essex.ac.uk

Enrique A. Lopez-Poveda
Neuroscience Institute of Castilla y León
University of Salamanca
37007 Salamanca, Spain
ealopezpoveda@usal.es

Richard R. Fay
Loyola University of Chicago
Chicago IL 60626
USA
rfay@luc.edu

Arthur N. Popper
University of Maryland
College Park, MD 20742
USA
apopper@umd.edu

ISBN 978-1-4419-1370-8 e-ISBN 978-1-4419-5934-8
DOI 10.1007/978-1-4419-5934-8
Springer New York Dordrecht Heidelberg London

Library of Congress Control Number: 2010921204

Printed on acid-free paper

Springer is part of Springer Science+Business Media (www.springer.com)

Series Preface

The *Springer Handbook of Auditory Research* presents a series of comprehensive and synthetic reviews of the fundamental topics in modern auditory research. The volumes are aimed at all individuals with interests in hearing research including advanced graduate students, post-doctoral researchers, and clinical investigators. The volumes are intended to introduce new investigators to important aspects of hearing science and to help established investigators to better understand the fundamental theories and data in fields of hearing that they may not normally follow closely.

Each volume presents a particular topic comprehensively, and each serves as a synthetic overview and guide to the literature. As such, the chapters present neither exhaustive data reviews nor original research that has not yet appeared in peer-reviewed journals. The volumes focus on topics that have developed a solid data and conceptual foundation rather than on those for which a literature is only beginning to develop. New research areas will be covered on a timely basis in the series as they begin to mature.

Each volume in the series consists of a few substantial chapters on a particular topic. In some cases, the topics will be ones of traditional interest for which there is a substantial body of data and theory, such as auditory neuroanatomy (Vol. 1) and neurophysiology (Vol. 2). Other volumes in the series deal with topics that have begun to mature more recently, such as development, plasticity, and computational models of neural processing. In many cases, the series editors are joined by a co-editor having special expertise in the topic of the volume.

Richard R. Fay, Chicago, IL
Arthur N. Popper, College Park, MD

Volume Preface

Models have always been a special feature of hearing research. The particular models described in this book are special because they seek to bridge the gap between physiology and psychophysics and ask how the psychology of hearing can be understood in terms of what we already know about the anatomy and physiology of the auditory system. However, although we now have a great deal of detailed information about the outer, middle, and inner ear as well as an abundance of new facts concerning individual components of the auditory brainstem and cortex, models of individual anatomically defined components cannot, in themselves, explain hearing. Instead, it is necessary to model the system as a whole if we are to understand how man and animals extract useful information from the auditory environment. A general theory of hearing that integrates all relevant physiological and psychophysical knowledge is not yet available but it is the goal to which all of the authors of this volume are contributing.

The volume starts with the auditory periphery by Meddis and Lopez-Poveda (Chapter 2) which is fundamental to the whole modeling exercise. The next level in the auditory system is the cochlear nucleus. In Chapter 3, Voigt and Zheng attempt to simulate accurately the responses of individual cell types and show how the connectivity among the different cell types determines the auditory processing that occurs in each subdivision.

Output from the cochlear nucleus has two main targets, the superior olivary complex and the inferior colliculus. The superior olivary complex is considered first in Chapter 4 by Jennings and Colburn because its output also passes through the inferior colliculus, which is discussed in Chapter 6 by Davis, Hancock, and Delgutte, who draws explicit links between the modeling work and psychophysics. Much less is known about the thalamus and cortex, and Chapter 5 by Eggermont sets out what has been achieved so far in understanding these brain regions and what the possibilities are for the future.

Four more chapters conclude this volume by looking at the potential of modeling to contribute to the solution of practical problems. Chapter 7 by Heinz addresses the issue of how hearing impairment can be understood in modeling terms. In Chapter 8, Brown considers hearing in connection with automatic speech recognition and reviews the problem from a biological perspective, including recent progress that has been made. In Chapter 9, Wilson, Lopez-Poveda, and Schatzer look more closely at cochlear implants and consider whether models can help to improve the coding strategies that are used. Finally, in Chapter 10, van Schaik, Hamilton, and Jin address these issues and show how models can be incorporated into very large scale integrated devices known more popularly as "silicon chips."

As is the case with volumes in the Springer Handbook of Auditory Research, previous volumes have chapters relevant to the material in newer volumes. This is clearly the case in this volume. Most notably, the advances in the field can be easily seen when comparing the wealth of new and updated information since the publication of Vol. 6, *Auditory Computation*. As pointed out in this Preface, and throughout this volume, the models discussed rest upon a thorough understanding of the anatomy and physiology of the auditory periphery and the central nervous system. Auditory anatomy was the topic of first volume in the series (*The Mammalian Auditory Pathway: Neuroanatomy*) and physiology in the second (*The Mammalian Auditory Pathway: Physiology*). These topics were brought up to date and integrated in the more recent Vol. 15 (*Integrative Functions in the Mammalian Auditory Pathway*). There are also chapters in several other volumes that are germane to the topic in this one, including chapters in *Cochlear Implants* (Vol. 20), *The Cochlea* (Vol. 8), and *Vertebrate Hair Cells* (Vol. 27).

Ray Meddis, Colchester, UK
Enrique A. Lopez-Poveda, Salamanca, Spain
Richard R. Fay, Chicago, IL
Arthur N. Popper, College Park, MD

Contents

Contributors

Guy J. Brown
Speech and Hearing Research Group, Department of Computer Science, University of Sheffield, Sheffield S1 4DP, UK, g.brown@dcs.shef.ac.uk

H. Steven Colburn
Department of Biomedical Engineering, Boston University, Boston, MA 02215, USA, colburn@bu.edu

Kevin A. Davis
Departments of Biomedical Engineering and Neurobiology and Anatomy, University of Rochester, Rochester, NY 14642, USA, kevin_davis@urmc.rochester.edu

Bertrand Delgutte
Eaton-Peabody Laboratory, Massachusetts Eye and Ear Infirmary, Boston, MA 02114, USA, Bertrand_Delgutte@meei.harvard.edu

Jos J. Eggermont
Department of Psychology, University of Calgary, Calgary, AB, Canada T2N 1N4, eggermon@ucalgary.ca

Tara Julia Hamilton
School of Electrical Engineering and Telecommunications, The University of New South Wales, NSW 2052, Sydney, Australia, tara@itee.uq.edu.au

Kenneth E. Hancock
Eaton-Peabody Laboratory, Massachusetts Eye and Ear Infirmary, Boston, MA 02114, USA, Ken_Hancock@meei.harvard.edu

Michael G. Heinz
Department of Speech, Language, and Hearing Sciences & Weldon School of Biomedical Engineering, Purdue University, West Lafayette, IN 47907, USA, mheinz@purdue.edu

Todd R. Jennings
Department of Biomedical Engineering, Boston University, Boston, MA 02215, USA, toddj@bu.edu

Craig Jin
School of Electrical and Information Engineering, The University of Sydney, Sydney, NSW 2006, Australia, craig@ee.usyd.edu.au

Enrique A. Lopez-Poveda
Instituto de Neurociencias de Castilla y León, University of Salamanca, 37007 Salamanca, Spain, ealopezpoveda@usal.es

Ray Meddis
Hearing Research Laboratory, Department of Psychology, University of Essex, Colchester CO4 3SQ, UK, rmeddis@essex.ac.uk

Reinhold Schatzer
C. Doppler Laboratory for Active Implantable Systems, Institute of Ion Physics and Applied Physics, University of Innsbruck, 6020 Innsbruck, Austria, reinhold.schatzer@uibk.ac.at

André van Schaik
School of Electrical and Information Engineering, The University of Sydney, Sydney, NSW 2006, Australia, andre@ee.usyd.edu.au

Herbert F. Voigt
Department of Biomedical Engineering, Boston University, Boston, MA 02215, USA, hfv@enga.bu.edu

Blake S. Wilson
Duke Hearing Center, Duke University Medical Center, Durham, NC 27710, USA; Division of Otolaryngology, Head and Neck Surgery, Department of Surgery, Duke University Medical Center, Durham, NC 27710, USA; MED-EL Medical Electronics GmbH, 6020 Innsbruck, Austria, blake.wilson@duke.edu

Xiaohan Zheng
Biomedical Engineering Department, Boston University, Boston, MA 02215, USA, xhzheng@bu.edu

Chapter 1
Overview

Ray Meddis and Enrique A. Lopez-Poveda

Models have always been a special feature of hearing research. Von Helmholtz (1954) likened the ear to a piano, an array of resonances each tuned to a different frequency. In modern psychophysics, the dominant models are often drawn from radio or radar technology and feature filters, amplifiers, oscillators, detectors, integrators, etc. In physiology, there have been many models of the individual components along the auditory pathway such as the Davis (1965) battery theory of cochlear transduction and Hodgkin and Huxley (1952) models of the initiation of spike activity in nerve fibers. These models are attractive to researchers because they are explicit and quantitative.

The particular models described in this book are special because they seek to bridge the gap between physiology and psychophysics. They ask how the psychology of hearing can be understood in terms of what we already know about the anatomy and physiology of the auditory system. Rapid recent progress in anatomy and physiology means that we now have a great deal of detailed information about the outer, middle, and inner ear as well as an abundance of new facts concerning individual components of the auditory brain stem and cortex. However, models of individual anatomically defined components cannot, in themselves, explain hearing. Instead, it is necessary to model the system as a whole if we are to understand how humans and animals extract useful information from the auditory environment.

Although a general theory of hearing that integrates all relevant physiological and psychophysical knowledge is not yet available, it is the goal to which all of the authors of this volume are contributing. Despite the considerable complexity implied by a general theory of hearing, this goal looks to be achievable now that computers are available. Computers provide the ability to represent complexity by adopting a systems approach wherein models of individual components are combined

R. Meddis (✉)
Hearing Research Laboratory, Department of Psychology, University of Essex,
Colchester CO4 3SQ, UK
e-mail: rmeddis@essex.ac.uk

R. Meddis et al. (eds.), *Computational Models of the Auditory System*,
Springer Handbook of Auditory Research 35, DOI 10.1007/978-1-4419-5934-8_1,

to form larger collections of interacting elements. Each element of the model can be developed independently but with the prospect of integrating it into the larger whole at a later stage.

Computational models are attractive because they are explicit, public, quantitative, and they work. They are explicit in that a computer program is a rigorous, internally consistent and unambiguous definition of a theory. They are public in that computer programs are portable and can be studied in anyone's laboratory so long as a general-purpose computer is available. Anyone can check whether the model really does what the author says that it does by obtaining and running the program. A good modeler can, and should, make the computer code available for public scrutiny. Computer models are quantitative in that all parameters of the model must be specified before the program can be run. They also "work" in the sense that a good computational model should be a practical tool for those designing better hearing aids, cochlear implant coding strategies, automatic speech recognizers, and robotic vehicles.

A computer model is a flexible working hypothesis shared among researchers. It is a research tool and will always be a work in progress. At this level of complexity, no modeler can hope to own the "final" model that is correct in every detail but he or she can hope to make a contribution to one or more components of the model. Perhaps he or she can design a better model of the basilar membrane response, or a primary-like unit in the cochlear nucleus. Even better, he or she might show how some puzzle in psychophysics is explained by how a particular structure processes acoustic information. In other words, auditory models can provide a shared framework within which research becomes a cooperative enterprise where individual contributions are more obviously seen to fit together to form a whole that is indeed greater than its parts.

These general considerations have shaped the design of this book and produced four major requirements for individual authors to consider. First, a general model of hearing must be based on the anatomy and physiology of the auditory system from the pinna up to the cortex. Second, each location along the auditory processing pathway presents its own unique problems demanding its own solution. Each subcomponent is a model in itself and the design of connections within and between these components are all substantial challenges. Third, the models need to be made relevant, as far as possible, to the psychology of hearing and provide explanations for psychophysical phenomena. Finally, the practical requirements of clinicians and engineers must be acknowledged. They will use the models to conceptualize auditory processing and adapt them to solve practical design problems. Each chapter addresses these issues differently depending on the amount of relevant progress in different areas, but it is hoped that all of these issues have been addressed in the volume as a whole.

The auditory periphery by Meddis and Lopez-Poveda (Chapter 2) is fundamental to the whole modeling exercise. It is the gateway to the system, and model mistakes at this level will propagate throughout. This stage has often been characterized as a bank of linear filters followed by half-wave rectification but, in reality, it is much more complex. Processing in the auditory periphery is nonlinear with respect to level at almost every stage, including the basilar membrane response, the receptor potential, and the generation of action potential in auditory nerve (AN) fibers. Adaptation of firing rates during exposure to sound and the slow recovery from adaptation mean that the system is also nonlinear with respect to time. For example, the effect of an acoustic event depends on what other acoustic events have occurred in the recent past.

Fortunately, the auditory periphery has received a great deal of attention from physiologists despite being the least accessible part of the auditory system because it is buried inside the temporal bone. As a result, there is a great deal of information concerning the properties of the individual peripheral components, including the stapes, outer hair cells, basilar membrane, inner hair cells, and the auditory nerve itself. Moreover, one element is very much like another along the cochlear partition. As a consequence, it is relatively easy to check whether models of the individual component processing stages are working as required.

In humans, there are about 30,000 afferent AN fibers, each responding with up to 300 action potentials per second. The cochlear nucleus (CN, Chapter 3) receives all of the AN output and has the function of processing this information before passing it on to other nuclei for further processing. The anatomy of the cochlear nucleus is surprisingly complex, with a number of subdivisions each receiving its own copy of the AN input. Detailed analysis shows that each subdivision contains different types of nerve cells each with its own electrical properties. Some are inhibitory and some excitatory, and all respond in a unique way to acoustic stimulation. Patterns of interconnections between the cells are also complex. It is the job of the modeler to simulate accurately the responses of individual cell types and show how the connectivity among the different cell types determines the auditory processing that occurs in each subdivision. Models of the CN could occupy a whole volume by itself. Here we can only give a flavor of what has been achieved and what lies ahead.

Output from the cochlear nucleus has two main targets, the superior olivary complex (SOC) and the inferior colliculus (IC). The SOC is considered first (Chapter 4) because its output also passes to the IC. Like the CN it is also complex and adds to the impression that a great deal of auditory processing is carried out at a very early stage in the passage of signals toward the cortex. As in the CN, there are different cell types and suggestive interconnections between them. However, we also see the beginning of a story that links the physiological with the psychophysical.

The SOC has long been associated with the localization of sounds in the popular Jeffress (1948) model that uses interaural time differences (ITDs) to identify where a sound is coming from. The reader will find in Chapter 4 that the modern story is more subtle and differentiated than the Jeffress model would suggest. Recent modeling efforts are an excellent example of how the computational approach can deal simultaneously with the details of different cell types, their inhibitory or excitatory nature, and how they are interconnected. In so doing, they lay the foundation for an understanding of how sounds are localized.

All outputs from both the CN and the SOC find their way to the central nucleus of the IC, which is an obligatory relay station en route to the cortex. In comparison with the CN or the SOC, it is much less complex, with fewer cell types and a more homogeneous anatomical structure. The authors of Chapter 6 (Davis, Hancock, and Delgutte) use Aitkin's (1986) characterization of the IC as a "shunting yard of acoustical information processing" but then go on to show that its significance is much greater than this. For the first time, in this volume, explicit links are drawn between the modeling work and psychophysics. Localization of sounds, the precedence effect, sensitivity to amplitude modulation, and the extraction of pitch of harmonic tones are all dealt with here. The full potential of computer models to explain auditory processing is most evident in this chapter.

The next stage is the thalamus, where information from the IC is collected and passed to the cortex. Unfortunately, relatively little is known about what this stage contributes to auditory processing. It does, however, have strong reciprocal links with the cortex, with information passing back and forth between them, and it may be best to view the thalamus and cortex as a joint system. Undoubtedly, the most sophisticated analyses of acoustic input occur in this region, and Chapter 5 (Eggermont) sets out what has been achieved so far and what the possibilities are for the future. Pitch, speech, language, music, and animal vocalization are all analyzed here and are affected when the cortex is damaged. Theories are beginning to emerge as to how this processing is structured and detailed computational models such as the spectrotemporal receptive fields are already being tested and subjected to critical analysis. Nevertheless, considerable effort will be required before it is possible to have detailed working models of the cortical processing of speech and music.

Four more chapters conclude this volume by looking at the potential of modeling to contribute to the solution of practical problems. Chapter 7 by Heinz addresses the issue of how hearing impairment can be understood in modeling terms. Aging, genetic heritage, noise damage, accidents, and pharmaceuticals all affect hearing, but the underlying mechanisms remain unclear. Many of these questions need to be addressed by empirical studies but modeling has a role to play in understanding why damage to a particular part of the system has the particular effect that it does. Hearing loss is not simply a case of the world becoming a quieter place. Patients complain variously that it can be too noisy, that their problems occur only when two or more people are speaking simultaneously, that their hearing is "distorted," or that they hear noises (tinnitus) that bear no relationship to events in the real world. Hearing loss is complex. Modeling has the potential to help make sense of the relationship between the underlying pathology and the psychological experience. It should also contribute to the design of better hearing prostheses.

Computer scientists have a long-standing interest in hearing in connection with automatic speech recognition (ASR). Considerable progress has been made using the techniques of spectral and temporal analysis of speech signals in an engineering tradition. However, there has always been a minority interest in building models that mimic human hearing. This interest has become more pressing as the limitations of the engineering approach have become evident. One of these limitations concerns how to separate speech from a noisy background before identification. However, this is only one aspect of the general problem of how to segregate sounds from different sources, a problem more generally known as "auditory scene analysis." In Chapter 8, Brown reviews the problem from a biological perspective and reviews recent progress. This is the highest level of auditory modeling and the chapter addresses the very high-level issue of the focus of attention. These are all issues of interest to psychologists, computer scientists, and philosophers alike.

In Chapter 9, Wilson, Lopez-Poveda, and Schatzer look more closely at cochlear implants and consider whether models can help to improve the coding strategies that they use. It is remarkable just how much progress has been made in the design and fitting of these devices and the enormous benefit that many patients have received. Nevertheless, the benefits vary considerably from patient to patient, and some types of acoustic stimulation benefit more than others. For example, implants work better with speech than with music. It is natural to want to push this technology to its limits, and one way forward is to explore the possibility of simulating natural hearing as closely as possible and incorporating these natural models into new coding strategies. Work has already begun but there is much more to be done.

For some, the greatest justification of auditory modeling will come from the useful artefacts that will ultimately result from the modeling efforts. These are hearing devices that can be embedded in many applications in everyday life. Such devices will need to operate in "real time," consume little power, and be inexpensive to manufacture. The final chapter in this book, by van Schaik, addresses these issues and shows how models can be incorporated into VLSI (very large scale integrated) devices known more popularly as "silicon chips." It is in the nature of these efforts that they will need to wait until individual models have been produced and tested. Even then the technical challenges are formidable. Nevertheless, considerable progress has already been made and working devices have been designed and built. It is likely that they will be the medium by which auditory modeling has its greatest impact on the welfare of the general public.

Taken together, these chapters reveal a mountain of achievement and show a field of intellectual endeavour on the verge of maturity. We do not yet have a complete working model of the auditory system and it is true that most modeling research projects are concentrated on small islands along the pathway between the periphery and the cortex. Nevertheless, it is increasingly clear that computer models will one day link up these islands to form a major theoretical causeway directing our understanding of how the auditory system does what it does for those fortunate enough to have normal hearing. Where hearing is imperfect as a result of genetics, damage, or simply aging, computer models of hearing offer the fascinating possibility of new explanations and new prostheses. While science atomizes hearing by focusing on ever smaller details, computer models have the power to resynthesize the hard-won findings of anatomists, physiologists, psychophysicists, and clinicians into a coherent and useful structure.

References

Aitkin LM (1986) The Auditory Midbrain: Structure and Function of the Central Auditory Pathway. Clifton, NJ: Humana.
Davis HA (1965) A model for transducer action in the cochlea. Cold Spring Harb Symp Quant Biol 30:81–189.
Helmholtz HLF (1954) On the Sensations of Tone as a Physiological Basis for the Theory of Music. New York: Dover. English translation of 1863 (German) edition.
Hodgkin A, Huxley A (1952) A quantitative description of membrane current and its application to conduction and excitation in nerve. J Physiol 117:500–544.
Jeffress LA (1948) A place theory of sound localization. J Comp Physiol Psychol 41:35–39.

Chapter 2
Auditory Periphery: From Pinna to Auditory Nerve

Ray Meddis and Enrique A. Lopez-Poveda

Abbreviations and Acronyms

AC	Alternating current
AN	Auditory nerve
BF	Best frequency
BM	Basilar membrane
BW	Bandwidth
CF	Characteristic frequency
dB	Decibel
DC	Direct current
DP	Distortion product
DRNL	Dual-resonance nonlinear
f_c	Center frequency
FFT	Fast Fourier transform
FIR	Finite impulse response
HRIR	Head-related impulse response
HRTF	Head-related transfer function
HSR	High-spontaneous rate
IHC	Inner hair cell
IIR	Infinite impulse response
kHz	KiloHertz
LSR	Low-spontaneous rate
MBPNL	Multiple bandpass nonlinear
ms	Milliseconds
OHC	Outer hair cell
SPL	Sound pressure level

R. Meddis (✉)
Hearing Research Laboratory, Department of Psychology, University of Essex,
Colchester CO4 3SQ, UK
e-mail: rmeddis@essex.ac.uk

R. Meddis et al. (eds.), *Computational Models of the Auditory System*,
Springer Handbook of Auditory Research 35, DOI 10.1007/978-1-4419-5934-8_2,

2.1 Introduction

The auditory periphery begins at the point where the pressure wave meets the ear and it ends at the auditory nerve (AN). The physical distance is short but the sound is transformed almost beyond recognition before it reaches the end of its journey. The process presents a formidable challenge to modelers, but considerable progress has been made over recent decades.

The sequence starts as a pressure wave in the auditory meatus, where it causes vibration of the eardrum. These vibrations are transmitted to the stapes in the middle ear and then passed on to the cochlear fluid. Inside the cochlea, the basilar membrane (BM) responds with tuned vibrations that are further modified by neighboring outer hair cells (OHCs). This motion is detected by inner hair cells (IHCs) that transduce it into fluctuations of an electrical receptor potential that control indirectly the release of transmitter substance into the AN synaptic cleft. Finally, action potentials are generated in the tens of thousands of auditory nerve fibers that carry the auditory message to the brain stem. Each of these successive transformations contributes to the quality of hearing, and none can be ignored in a computer model of auditory peripheral processing.

This combined activity of processing stages is much too complex to be understood in an intuitive way, and computer models have been developed to help us visualize the succession of changes between the eardrum and the AN. The earliest models used analogies with electrical tuned systems such as radio or radar, and these continue to influence our thinking. However, the most recent trend is to simulate as closely as possible the individual physiological processes that occur in the cochlea. Model makers are guided by the extensive observations of anatomists and physiologists who have mapped the cochlea and measured the changes that occur in response to sound. Their measurements are made at a number of places along the route and include the vibration patterns of the eardrum, stapes, and BM; the electrical potentials of the OHCs and IHCs; and, finally, the action potentials in the AN fibers. These places mark "way points" for modelers who try to reproduce the physiological measurements at each point. Successful simulation of the physiological observations at each point is the main method for verifying their models. As a consequence, most models consist of a cascade of "stages" with the physiological measurement points marking the boundary between one stage and another. The freedom to model one stage at a time has greatly simplified what would otherwise be an impossibly complex problem.

Figure 2.1 illustrates a cascade model based on the work conducted by the authors. The signal is passed from one stage to another, and each stage produces a unique transformation to simulate the corresponding physiological processes. Two models are shown. On the left is a model of the response at a single point along the BM showing how the stapes displacement is transformed first into BM displacement, then into the IHC receptor potential, and then into a probability that a vesicle of transmitter will be released onto the IHC/AN synaptic cleft (if one is available). The bottom panel shows the spiking activity of a number of auditory

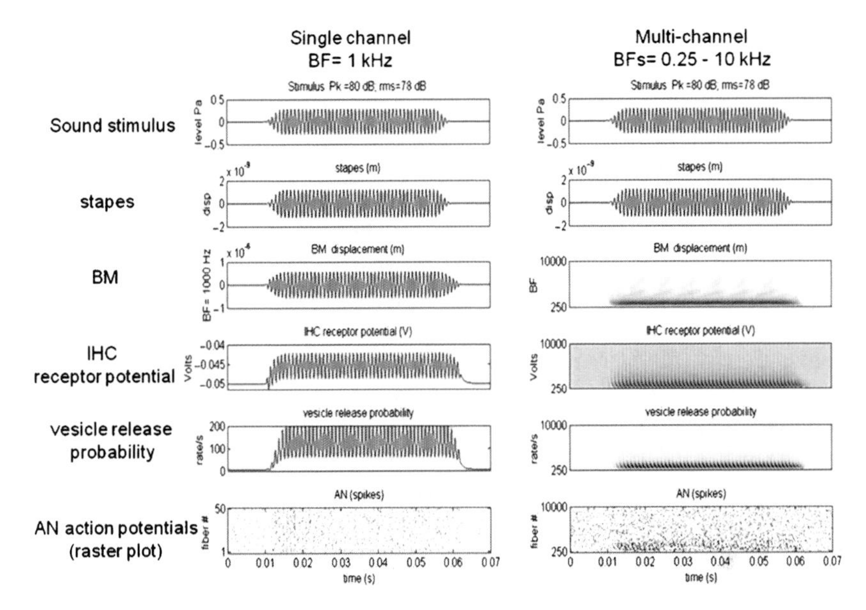

Fig. 2.1 The response of a multistage computer model of the auditory periphery is illustrated using a 1-kHz pure tone presented for 50 ms at 80 dB SPL. Each *panel* represents the output of the model at a different stage between the stapes and the auditory nerve. The *left-hand panels* show a single channel model (BF = 1 kHz) representing the response at a single point along the basilar membrane. Each plot shows the response in terms of physical units: stapes (displacement in meters), the BM (displacement in meters), the IHC receptor potential (volts), and vesicle release (probability). The *right-hand panels* show surface plots representing the response of a 40-channel model with BFs ranging between 250 Hz and 10 kHz. Channels are arranged across the *y*-axis (high BFs at the *top*) with time along the *x*-axis. *Darker shading* indicates more activity. Note that high-BF channels are only weakly affected by the 1-kHz pure tone and most activity is concentrated in the low-BF channels. The *bottom panel* of both models is the final output of the model. It shows the spiking activity of a number of AN fibers represented as a raster plot where each *row of dots* is the activity of a single fiber and each *dot* is a spike. The *x*-axis is time. In the single-channel model (*left*), all fibers have the same BF (1 kHz). In the multichannel model (*right*), the fibers are arranged with high-BF fibers at the *top*. Note that all fibers show spontaneous activity and the response to the tone is indicated only by an increase in the firing rate, particularly at the beginning of the tone. In the multichannel model, the *dots* can be seen to be more closely packed in the low-BF fibers during the tone presentation

nerve fibers presented as a raster plot where each dot represents a spike in a nerve fiber. On the right, a more complex model is shown. This represents the activity at 40 different sites along the cochlear partition each with a different best-frequency (BF). Basal sites (high BFs) are shown at the top of each panel and apical sites (low BF) at the bottom with time along the x-axis. Darker shades indicate more intense activity.

The input to the model is a 1-kHz ramped tone presented for 50 ms at a level of 80 dB SPL. The multichannel model shows frequency selectivity in that only some channels are strongly affected by the stimulus. It is also important to note that the AN fibers are all spontaneously active, and this can be seen most clearly before the tone begins to play. The single-channel model (left) shows most frequent firing soon after the onset of the tone, and this is indicated by more closely packed dots in the raster plot. When the tone is switched off, the spontaneous firing is less than before the tone, as a consequence of the depletion of IHC presynaptic transmitter substance that has occurred during the presentation of the tone. The multichannel model (right) shows a substantial increase of AN fiber firing only in the apical channels (low-BFs at the bottom of the plot). Only a small number of fibers are shown in the figure to illustrate the basic principles. A full model will represent the activity of thousands of fibers.

Models serve many different purposes, and it is important to match the level of detail to the purpose in hand. For example, psychophysical models such as the loudness model of Moore et al. (1997) are based only loosely on physiology including a preemphasis stage (outer–middle ear), as well as frequency tuning and compression (BM). When compared with the model in Fig. 2.1, it is lacking in physiological detail. Nevertheless, it serves an important purpose in making useful predictions of how loud sounds will appear to the listener. When fitting hearing aids, for example, this is very useful and the model is fit for its purpose. By contrast, the more detailed simulations of the auditory periphery (discussed in this chapter) cannot at present make loudness predictions.

A more detailed model such as that offered by Derleth et al. (2001) includes peripheral filtering and a simulation of physiological adaptation without going so far as to model the individual anatomical components. This has proved useful in simulating human sensitivity to amplitude modulation. It may yet prove to be the right level of detail for low-power hardware implementations such as hearing aids because the necessary computing power is not available in a hearing aid to model all the details of a full physiological model. Different degrees of detail are required for different purposes. Nevertheless, in this chapter, emphasis is placed on computer models that simulate the anatomy and physiology as closely as possible because these are the only models that can be verified via actual physiological measurements.

Auditory models can be used in many different ways. From a purely scientific point of view, the model represents a theory of how the auditory periphery works. It becomes a focus of arguments among researchers with competing views of the underlying "truth." In this respect, computer models have the advantage of being quantitatively specified because their equations make quantitative predictions that can be checked against the physiological data. However, models also have the potential for practical applications. Computer scientists can use a peripheral model as an input to an automatic speech recognition device in the hope that it will be better than traditional signal-processing methods. Such attempts have had mixed success so far but some studies have found this input to be more robust (Kleinschmidt et al. 1999). Another application involves their use in the design of algorithms for generating the signals used in cochlear implants or hearing aids (e.g., Chapter 9; Chapter 7). Indeed, any problem involving the analysis of acoustic signals might benefit from the use of auditory models, but many of these applications lie in the future.

Before examining the individual stages of peripheral auditory models, some preliminary remarks are necessary concerning the nature of compression or "nonlinearity" because it plays an important role in many of these stages. In a linear system, an increase in the input signal results in a similar-size increase at the output; in other words, the level of the output can be predicted as the level of the input multiplied by a constant. It is natural to think of the auditory system in these terms. After all, a sound is perceived as louder when it becomes more intense. However, most auditory processing stages respond in a nonlinear way. The vibrations of the BM, the receptor potential in the IHC, the release of transmitter at the IHC synapse, and the auditory nerve firing rate are all nonlinear functions of their inputs. The final output of the system is the result of a cascade of nonlinearities. Such systems are very difficult to intuit or to analyze using mathematics. This is why computer models are needed. This is the only method to specify objectively and test how the system works.

The auditory consequences of this compression are important. They determine the logarithmic relationship between the intensity of a pure tone and its perceived intensity. It is for this reason that it is important to describe intensity using decibels rather than Pascals when discussing human hearing. Further, when two tones are presented at the same time they can give rise to the perception of mysterious additional tones called "combination tones" (Goldstein 1966; Plomp 1976). The rate of firing of an auditory nerve in response to a tone can sometimes be reduced by the addition of a second tone, known as two-tone suppression (Sachs and Kiang 1968). The width of an AN "tuning curve" is often narrow when evaluated near threshold but becomes wider when tested at high signal levels. These effects are all the emergent properties of a complex nonlinear system. Only computer models can simulate the consequences of nonlinearity, especially when complex broadband sounds such as speech and music are being studied.

The system is also nonlinear in time. The same sound produces a different response at different times. A brief tone that is audible when presented in silence may not be audible when it is presented after another, more intense tone, even though a silent gap may separate the two. The reduction in sensitivity along with the process of gradual recovery is known as the phenomenon of "adaptation" and it is important to an understanding of hearing in general. Once again, this nonlinearity can be studied effectively only by using computer simulation.

This chapter proceeds, like a peripheral model, by examining each individual processing stage separately and ending with the observation that the cascade of stages is complicated by the presence of feedback loops in the form of the efferent system that has only recently began to be studied. Finally, some examples of the output of a computer model of the auditory periphery are evaluated.

2.2 Outer Ear

The first stage of a model of the auditory periphery is the response of the middle ear, but it must be remembered that sounds are modified by the head and body of the listeners before they enter the ear canal. In a free-field situation, the spectrum of a sound is first altered by the filtering action of the body (Shaw 1966; Lopez-Poveda 1996). The acoustic transfer function of the body in the frequency domain is commonly referred to as the head-related transfer function (HRTF) to stress that the principal filtering contributions come from the head and the external ear (Shaw 1975; Algazi et al. 2001). In the time domain, the transfer function is referred to as the head-related impulse response (HRIR). The HRIR is usually measured as the click response recorded by either a miniature microphone placed in the vicinity of the eardrum (Wightman and Kistler 1989) or by the microphone of an acoustic manikin (Burkhard and Sachs 1975). The filtering operation of the body is linear; thus a Fourier transform serves to obtain the HRTF from its corresponding HRIR.

The spectral content of an HRTF reflects diffraction, reflection, scattering, resonance, and interference phenomena that affect the incoming sound before it reaches the eardrum (Shaw 1966; Lopez-Poveda and Meddis 1996). These phenomena depend strongly on the location of the sound source relative to the ear's entrance, as well as on the size and shape of the listener's torso, head, pinnae, and ear canal. As a result, HRTFs, particularly their spectral characteristics above 4 kHz, are different for different sound source locations and for different individuals (Carlile and Pralong 1994). Further, for any given source location and individual, the HRTFs for the left and the right ear are generally different as a result of the two ears being slightly dissimilar in shape (Searle et al. 1975). The location-dependent spectral content of HRTFs is a useful cue for sound localization, and for this reason HRTFs have been widely studied (Carlile et al. 2005).

2.2.1 Approaches to Modeling the Head-Related Transfer Function

All of the aforementioned considerations should give an idea of the enormous complexity involved in producing a computational model of HRTFs. Nevertheless, the problem has been attempted from several angles. There exists one class of models that try to reproduce the main features of the HRTFs by mathematically formulating the physical interaction of the sound waves with the individual anatomical elements of the body. For example, Lopez-Poveda and Meddis (1996) reproduced the elevation-dependent spectral notches of the HRTFs considering that the sound is diffracted at the concha aperture and then reflected on the concha back wall before reaching the ear canal entrance. The total pressure at the ear canal entrance would be the sum of the direct sound plus the diffracted/reflected sound. Similar physical models have been developed by Duda and Martens (1998) to model the response of a spherical head, by Algazi et al. (2001) to model the combined contributions of a spherical head and a spherical torso, and by Walsh et al. (2004) to model the combined contribution of the head and the external ear.

One of the main advantages of physical models is that they help elucidate the contributions of the individual anatomical elements to the HRTFs. Another advantage is that they allow approximate HRTFs to be computed for (theoretically) arbitrary body geometries, given the coordinates of the sound source(s). In practice, however, they are usually evaluated for simplified geometrical shapes (an exception is the model of Walsh et al. 2004) and are computationally very expensive. Another disadvantage is that, almost always, these models are developed in the frequency domain, although the HRIR can be obtained from the model HRTF by means of an inverse Fourier transform (Algazi et al. 2001). For these reasons, physical models of HRTFs are of limited practical use as part of composite models of spectral processing by the peripheral auditory system.

An alternative method is to reproduce specific HRTFs by means of finite- (FIR) or infinite-impulse response (IIR) digital filters. An immediately obvious way to approach it is to treat the sample values of the experimental digital HRIRs as the coefficients of an FIR filter (Kulkarni and Colburn 2004). Alternatively, such coefficients may be obtained by an inverse Fourier transform of the amplitude HRTF (e.g., Lopez-Poveda and Meddis 2001), although this method does not preserve the phase spectra of HRIRs that may be perceptually important (Kulkarni et al. 1999).

A more challenging problem, however, is to develop computationally efficient digital filter implementations of HRIRs, that is, digital filters of the lowest possible order that preserve the main amplitude and phase characteristics of the HRTFs. This is important to obtain HRIRs that can be computed in real time. The problem is twofold. First, it is necessary to identify the main spectral characteristics of HRTFs that are common to all individuals and provide important sound localization information (Kistler and Wightman 1992). Second, it is necessary to reproduce those features using low-order IIR filters, as they are more efficient than FIR filters. Kulkarni and Colburn (2004) have recently reported a reasonable solution to the problem by demonstrating that stimuli rendered through a 6-pole, 6-zero IIR-filter model of the HRTF had inaudible differences from stimuli rendered through the actual HRTF.

The main advantages of these digital-filter-type models is that they can process time-varying signals in real or quasi-real time. Their disadvantages are that they shed no light on the physical origin or the anatomical elements responsible for the characteristic spectral features of the HRTFs. Further, they require that the HRTFs of interest be measured beforehand (several publicly available databases already exist). Nevertheless, this type of model is more frequently adopted in composite models of signal processing by the peripheral auditory system.

2.3 Middle Ear

The middle ear transmits the acoustic energy from the tympanic membrane to the cochlea through a chain of three ossicles: the malleus, in contact with the eardrum, the incus, and the stapes, which contacts the cochlea at the oval window. The middle ear serves to adapt the low acoustic impedance of air to that of the cochlear perilymphatic fluid, which is approximately 4,000 times higher (von Helmholtz 1877; Rosowski 1996). For frequencies below approximately 2 kHz, this impedance transformation is accomplished mainly by the piston-like functioning of the middle ear (Voss et al. 2000) that results from the surface area of the eardrum being much larger than that of the stapes footplate. The lever ratio of the ossicles also contributes to the impedance transformation for frequencies above approximately 1 kHz (Goode et al. 1994).

In signal processing terms, the middle ear may be considered as a linear system whose input is a time-varying pressure signal near the tympanic membrane, and whose corresponding output is a time-varying pressure signal in the scala vestibuli of the cochlea, next to the stapes footplate. Therefore, its transfer function is expressed as the ratio (in decibels) of the output to the input pressures as a function of frequency (Nedzelnitsky 1980; Aibara et al. 2001). The intracochlear pressure relates directly to the force exerted by the stapes footplate, which in turn relates to the displacement of the stapes with respect to its resting position. For pure tone signals, stapes velocity (v) and stapes displacement (d) are related as follows: $v = 2\pi fd$, where f is the stimulus frequency in Hertz. For this reason, it is also common to express the frequency transfer function of the middle ear as stapes displacement or stapes velocity vs. frequency for a given sound level (Goode et al. 1994).

The middle ear is said to act as a linear system over a wide range of sound levels (<130 dB SPL) for two reasons. First, the intracochlear peak pressure at the oval window (Nedzelnitsky 1980), the stapes peak displacement (Guinan and Peake 1966), or the stapes peak velocity (Voss et al. 2000) is proportional to the peak pressure at the eardrum. The second reason is that sinusoidal pressure variations at the tympanic membrane produce purely sinusoidal pressure variations at the oval window (Nedzelnitsky 1980). In other words, the middle ear does not introduce distortion for sound levels below approximately 130 dB SPL.

The middle ear shapes the sound spectrum because it acts like a filter. However, a debate has been recently opened on the type of filter. Recent reports (Ruggero and Temchin 2002, 2003) suggest that the middle ear is a wide-band pressure transformer with a flat velocity-response function rather than a bandpass pressure transformer tuned to a frequency between 700 and 1,200 Hz, as previously thought (Rosowski 1996). The debate is still open.

2.3.1 *Approaches to Modeling the Middle Ear Transfer Function*

The function of the middle ear has been classically modeled by means of analog electrical circuits (Møller 1961; Zwislocki 1962; Kringlebotn 1988; Goode et al. 1994; Pascal et al. 1998; Voss et al. 2000; reviewed by Rosowski 1996). These models regard the middle ear as a transmission line with lumped mechanical elements and, as such, its functioning is described in electrical terms thanks to the analogy between electrical and acoustic elements (this analogy is detailed in Table 2.2 of Rosowski 1996). These models commonly describe the middle ear as a linear filter, although the model of Pascal et al. (1998) includes the nonlinear effects induced by the middle-ear reflex that occur at very high levels (>100 dB SPL). Electrical analogues have also been developed to model the response of pathological (otosclerotic) middle ear function (Zwislocki 1962).

The function of the middle ear has also been modeled by means of biomechanical, finite element methods (e.g., Gan et al. 2002; Koike et al. 2002; reviewed by Sun et al. 2002). This approach requires reconstructing the middle ear geometry, generally from serial sections of frozen temporal bones. The reconstruction is then used to develop a finite-element mesh description of the middle ear mechanics. So far, the efforts have focused on obtaining realistic descriptions of healthy systems that include the effects of the attached ligaments and tendons. However, as noted by Gan et al. (2002), finite element models will be particularly useful to investigate the effects of some pathologies (e.g., tympanic perforations or otosclerosis) on middle ear transmission, as well as to design and develop better middle ear prostheses (Dornhoffer 1998). These models also allow detailed research on the different modes of vibration of the tympanic membrane (e.g., Koike et al. 2002), which influence middle ear transmission for frequencies above approximately 1 kHz (Rosowski 1996). The main drawback of finite element models is that they are computationally very expensive.

A third approach is that adopted by most signal processing models of the auditory periphery. It consists of simulating the middle ear function by a linear digital filter with an appropriate frequency response. As a first approximation, some studies (e.g., Lopez-Poveda 1996; Robert and Eriksson 1999; Tan and Carney 2003) have used a single IIR bandpass filter while others (Holmes et al. 2004; Sumner et al. 2002, 2003a, b) use a filter cascade in an attempt to achieve more realistic frequency response characteristics. In any case, the output signal must be multiplied by an appropriate scalar to achieve a realistic gain.

Some authors have suggested that the frequency response of the middle ear determines important characteristics of the basilar response, such as the asymmetry of the iso-intensity response curves (Cheatham and Dallos 2001; see later) or the characteristic frequency modulation of basilar membrane impulse responses, that is, the so-called "glide" (e.g., Tan and Carney 2003; Lopez-Najera et al. 2005). This constitutes a reasonable argument in favor of using more realistic middle ear filter functions as part of composite models of the auditory periphery. To produce such a filters, some authors (e.g., Lopez-Poveda and Meddis 2001) employ FIR digital filters whose coefficients are obtained as the inverse fast Fourier transform (FFT) of an experimental stapes frequency response curve, whereas others (e.g., Lopez-Najera et al. 2007) prefer to convolve the tympanic pressure waveform directly with an experimental stapes impulse response. The latter approach guarantees realistic amplitude and phase responses for the middle ear function in the model.

2.4 Basilar Membrane

The motion of the stapes footplate in response to sound creates a pressure gradient across the cochlear partition that sets the organ of Corti to move in its transverse direction. The characteristics of this motion are commonly described in terms of BM velocity or displacement with respect to its resting position.

The BM responds tonotopically to sound. The response of each BM site is strongest for a particular frequency (termed the best frequency or BF) and decreases gradually with moving the stimulus frequency away from it. For this reason, each BM site is conveniently described to function as a frequency filter and the whole BM as a bank of overlapping filters. Each BM site is identified by its characteristic frequency (CF), which is defined as the BF for sounds near threshold.

BM filters are nonlinear and asymmetric. They are asymmetric in that the magnitude of the BM response decreases faster for frequencies above the BF than for frequencies below it as the stimulus frequency moves away from the BF (e.g., Robles and Ruggero 2001). The asymmetry manifests also in that the impulse (or click) response of a given BM site is modulated in frequency. This phenomenon is sometimes referred to as the chirp or glide of BM impulse responses. For basal sites, the instantaneous frequency of the impulse response typically increases with increasing time (Recio et al. 1998). The direction of the chirp for apical sites is still controversial (e.g., Lopez-Poveda et al. 2007), but AN studies suggest it could happen in the direction opposite to that of basal sites (Carney et al. 1999).

Several phenomena demonstrate the nonlinear nature of BM responses (Robles and Ruggero 2001). First, BM responses show more gain at low than at high sound levels. As a result, the magnitude of the BM response grows compressively with increasing sound level (slope of ~0.2 dB/dB). BM responses are linear (slope of 1 dB/dB) for frequencies an octave or so below the CF. This frequency response pattern, however, is true for basal sites only. For apical sites (CFs below ~1 kHz), compressive responses appear to extend to a wider range of stimulus frequencies relative to the CF (Rhode and Cooper 1996; Lopez-Poveda et al. 2003).

BM responses are nonlinear also because the BF and the bandwidth of a given cochlear site change depending on the stimulus level. The BF of basal sites decreases with increasing sound level. There is still controversy on the direction of change of the BF of apical cochlear sites. AN studies suggest that it increases with increasing level (Carney et al. 1999), but psychophysical studies suggest a downward shift (Lopez-Poveda et al. 2007). The bandwidth is thought to increase always with increasing level.

Suppression and distortion are two other important phenomena pertaining to BM nonlinearity (reviewed in Lopez-Poveda 2005). Suppression occurs when the magnitude of BM response to a given sound, called the suppressee, decreases in the presence of a second sound, called the suppressor. It happens only for certain combinations of the frequency and level of the suppressor and the suppressee (Cooper 1996, 2004). Suppression leads to decreases in both the degree (i.e., the slope) and dynamic range of compression that can be observed in the BM response. The time course of the two-tone suppression appears to be instantaneous (Cooper 1996).

Distortion can occur for any stimulus but is more clearly seen when the BM is stimulated with pairs of tones of different frequencies (f_1 and f_2, $f_2>f_1$) referred to as primaries. In response to tone pairs, the BM excitation waveform contains distortion products (DPs) with frequencies f_2-f_1, $(n+1)f_1-nf_2$ and $(n+1)f_2-nf_1$ ($n=1, 2, 3,\ldots$) (Robles et al. 1991). These DPs are generated at cochlear sites with CFs equal to the primaries but can travel along the cochlea and excite remote BM regions with CFs equal to the DP frequencies (Robles et al. 1997). DPs can be heard as combination tones (Goldstein 1966) and are thought to be the source of distortion-product otoacoustic emissions.

The characteristics of BM responses are not steady. Instead, they change depending on the activation of the efferent cochlear system, which depends itself on the characteristics of the sound being presented in the ipsilateral and contralateral ears. Activation of the efferent system reduces the cochlear gain (Russell and Murugasu 1997).

BM responses depend critically on the physiological state of the cochlea. Some diseases or treatments with ototoxic drugs (furosemide, quinine, aminoglycosides) damage cochlear outer hair cells, reducing the gain and the tuning of BM responses. Responses are fully linear postmortem or in cochleae with total OHC damage (reviewed in Ruggero et al. 1990; Robles and Ruggero 2001). Consequently, BM responses are sometimes described as the sum of an active (nonlinear) component, present only in cochleae with remaining OHCs, and a passive (linear) component, which remains post-mortem.

The BM response characteristics described in the preceding text determine important physiological properties of the AN response as well as perceptual properties in normal-hearing listeners and in those with cochlear hearing loss (Moore 2007). To a first approximation they determine, for instance, the frequency tuning of AN fibers near threshold (Narayan et al. 1998), the dynamic range of hearing (reviewed in Bacon 2004), our ability (to a limited extent) to resolve the frequency components of complex sounds (reviewed in Moore 2007), and even our perception of combination tones not present in the acoustic stimulus (Goldstein 1966). In addition, suppression is thought to facilitate the perception of speech immersed in certain kinds of noise (Deng and Geisler 1987; Chapter 9). Therefore, it is fundamental that composite AN models and models of auditory perception include a good BM nonlinear model.

2.4.1 Phenomenological BM Models

BM models aim at simulating BM excitation (velocity or displacement) in response to stapes motion. Many attempts have been made to achieve this with models of different nature. We review only a small a selection of phenomenological, signal-processing models. These types of models attempt to account for BM responses using signal-processing elements (e.g., digital filters). The advantage of this approach is that the resulting models can be implemented and evaluated easily for digital, time-varying signals. Models of a different kind are reviewed elsewhere: a succinct review of transmission line models is provided by Duifhuis (2004) and van Schaik (Chapter 10); mechanical cochlear models are reviewed by de Boer (1996). A broader selection of phenomenological models is reviewed in Lopez-Poveda (2005).

2.4.1.1 The MBPNL Model

The Multiple BandPass NonLinear (MBPNL) model of Goldstein (1988, 1990, 1993, 1995) was developed in an attempt to provide a unified account of complex BM nonlinear phenomena such as compression, suppression, distortion, and simple-tone interference (the latter phenomenon is described later). It simulates the filtering function of a given cochlear partition (a given CF) by cascading a narrowly tuned bandpass filter followed by a compressive memoryless nonlinear gain, followed by another more broadly tuned bandpass filter (Fig. 2.2a). This structure is similar to

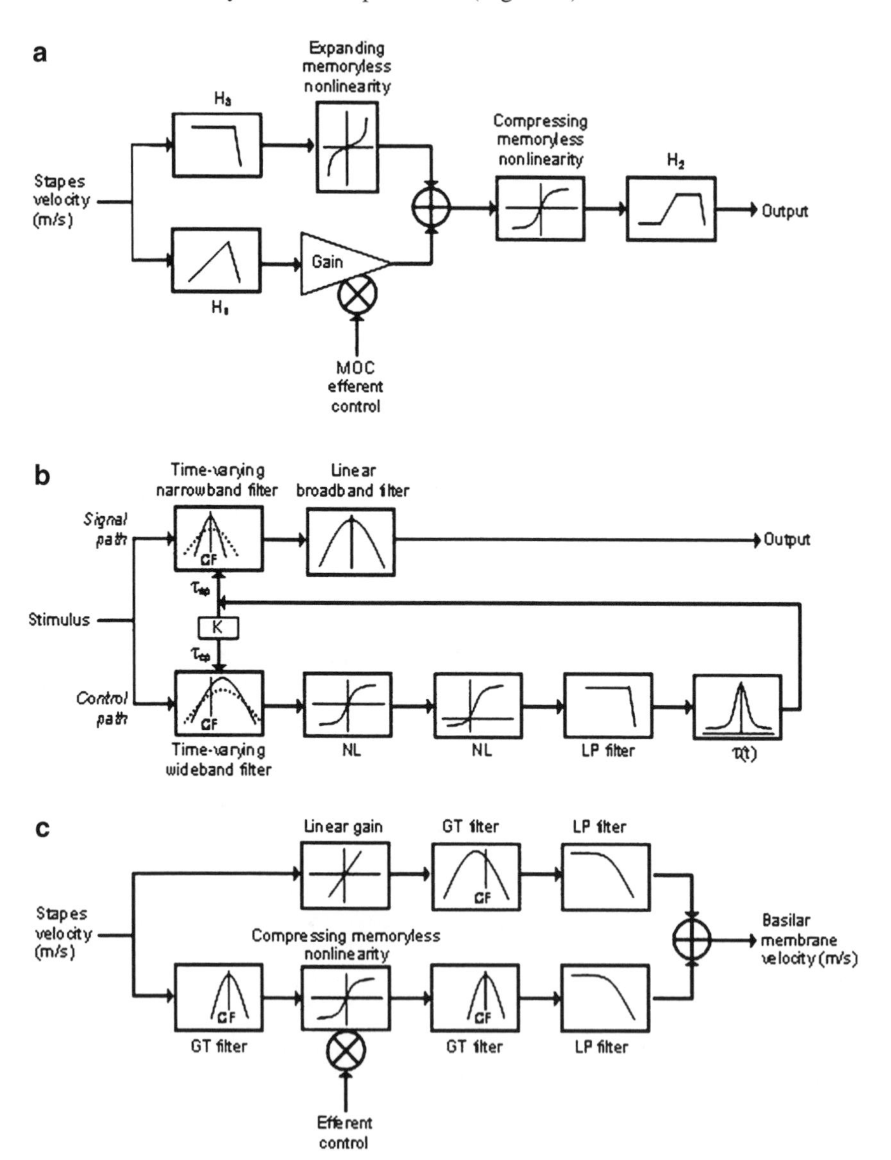

Fig. 2.2 Comparative architecture of three phenomenological nonlinear BM models. (**a**) The multiple bandpass nonlinear filter of Goldstein (adapted from Goldstein 1990). (**b**) The model of Zhang et al. (adapted from Zhang et al. 2001). (**c**) The dual-resonance nonlinear filter of Meddis et al. (adapted from Lopez-Poveda and Meddis 2001). See text for details. *GT* gammatone; *LP* low-pass; *NL* nonlinearity; *MOC* medio-olivocochlear

the bandpass nonlinear filter of Pfeiffer (1970) and Duifhuis (1976). The narrow and broad filters account for BM tuning at low and high levels, respectively. By carefully choosing their shapes and the gain of the compressive gain, the model reproduces level-dependent tuning and BF shifts (Goldstein 1990).

The model was specifically designed to reproduce the nonlinear cyclic interactions between a moderate-level tone at CF and another highly intense tone with a very low frequency, a phenomenon usually referred to as "simple-tone interaction" (or simple-tone interference; Patuzzi et al. 1984). This required incorporating an expanding nonlinearity (inverse in form to the compressing nonlinearity) whose role in the model is to enhance the low frequencies before they interact with on-CF tones at the compressive stage (Fig. 2.2a). With this expanding nonlinearity, the model reproduces detailed aspects of BM suppression and combination tones (Goldstein 1995). However, propagation of combination tones is lacking in the model, although it appears necessary to account for the experimental data regarding the perception of the $2f_1-f_2$ combination tone (Goldstein 1995).

The MBPNL model was further developed into a version capable of reproducing the response of the whole cochlear partition by means of a bank of interacting MBPNL filters (Goldstein 1993). This newer version gave the model the ability to account for propagating combination tones. However, to date systematic tests have not been reported on this MBPNL filterbank.

2.4.1.2 The Gammatone Filter

It is not possible to understand many of the current signal-processing cochlear models without first understanding the characteristics of their predecessor: the gammatone filter. The gammatone filter was developed to simulate the impulse response of AN fibers as estimated by reverse correlation techniques (Flanagan 1960; de Boer 1975; de Boer and de Jongh 1978; Aertsen and Johannesma 1980). The impulse response of the gammatone filter basically consists of the product of two components: a carrier tone of a frequency equal to the BF of the fiber and a statistical gamma-distribution function that determines the shape of the impulse response envelope. One of the advantages of the gammatone filter is that its digital, time-domain implementation is relatively simple and computationally efficient (Slaney 1993), and for this reason it has been largely used to model both physiological and psychophysical data pertaining to auditory frequency selectivity. It has also been used to simulate the excitation pattern of the whole cochlear partition by approximating the functioning of the BM to that of a bank of parallel gammatone filters with overlapping passbands, a filterbank (e.g., Patterson et al. 1992).

On the other hand, the gammatone filter is linear, thus level independent, and it has a symmetric frequency response. Therefore, it is inadequate to model asymmetric BM responses. Several attempts have been made to design more physiological versions of the gammatone filter. For instance, Lyon (1997) proposed an all-pole digital version of the filter with an asymmetric frequency response. This all-pole version also has the advantage of being simpler than the conventional gammatone filter in terms of parameters, as its gain at center frequency and its bandwidth are both controlled by a single parameter, namely, the quality factor (Q) of the filter (the quality factor of a filter is defined as the ratio of the filter center frequency, f_C, to the filter bandwidth, BW, measured at a certain number of decibels below the maximum gain, $Q=f_C/\text{BW}$).

2.4.1.3 The Gammachirp Filter

The gammachirp filter of Irino and Patterson (1997), like the all-pole gammatone filter, was designed to produce an asymmetric gammatone-like filter. This was achieved by making the carrier-tone term of the analytic impulse response of the gammatone filter modulated in frequency, thus the suffix chirp. This property was inspired by the fact that the impulse responses of the BM and of AN fibers are also frequency modulated (Recio et al. 1998; Carney et al. 1999).

In its original form, the gammachirp filter was level independent (linear), hence inadequate to simulate the nonlinear, compressive growth of BM response with level. Further refinements of the filter led to a compressive gammachirp filter with a level-independent chirp (Irino and Patterson 2001), hence more consistent with the physiology. The compressive gammachirp filter can be viewed as a cascade of three fundamental filter elements: a gammatone filter followed by a low-pass filter, followed by a high-pass filter with a level-dependent corner frequency. Combined, the first two filters produce an asymmetric gammatone-like filter, which can be approximated to represent the "passive" response of the BM. Because of its asymmetric frequency response, the associated impulse response of this "passive" filter shows a chirp.

The third element in the cascade, the high-pass filter, is responsible for the level dependent gain and tuning characteristics of the compressive gammachirp filter. It is designed to affect only frequencies near the center frequency of the gammatone filter in a level-dependent manner. At low levels, its corner frequency is configured to compensate for the effect of the low-pass filter, thus making the frequency response of the global gammachirp filter symmetric. At high levels, by contrast, its corner frequency is set so that the frequency response of the "passive" filter is almost unaffected and thus asymmetric. The chirping properties of the gammachirp filter are largely determined by those of its "passive" asymmetric filter at all levels, and have been shown to fit well those of AN fibers (Irino and Patterson 2001).

The compressive gammachirp filter has proved adequate to design filterbanks that reproduce psychophysically estimated human auditory filters over a wide range of center frequencies and levels (Patterson et al. 2003). It could probably be used to simulate physiological BM iso-intensity responses directly, although no studies have been reported to date aimed at testing the filter in this regard. Its BF shifts with level as do BM and AN iso-intensity curves, but the trends shown by Irino and Patterson (2001) are not consistent with the physiological data (Tan and Carney 2003). More importantly, we still lack detailed studies aimed at examining the ability of this filter to account for other nonlinear phenomena such as level-dependent phase responses, combination tones, or two-tone suppression. Some authors have suggested that it cannot reproduce two-tone suppression because it is not a "true" nonlinear filter, but rather a "quasilinear" filter whose shape changes with level (Plack et al. 2002). Recently, a dynamic (time-domain) version of the compressive gammachirp filter adequate for processing time-varying signals has become available (Irino and Patterson 2006).

2.4.1.4 The Model of Carney and Colleagues

Carney and colleagues (Heinz et al. 2001; Zhang et al. 2001) have proposed an improved version of Carney's (1993) composite phenomenological model of the AN response that reproduces a large number of nonlinear AN response characteristics. A version of this model (Tan and Carney 2003) also reproduces level-independent frequency glides (the term "frequency glide" is synonymous with the term "chirp" and both refer to the frequency-modulated character of BM and AN impulse responses).

An important stage of this composite AN model is designed to account for the nonlinear response of a single BM cochlear site (Fig. 2.2b). In essence, it consists of a gammatone filter whose gain and bandwidth vary dynamically in time depending on the level of the input signal (this filter is referred to in the original reports as "the signal path"). For a gammatone filter, both these properties, gain and bandwidth, depend on the filter's time constant, τ (see Eq. (2) of Zhang et al. 2001). In the model, the value of this time constant varies dynamically in time depending on the amplitude of the output signal from a feed-forward control path, which itself depends on the level of the input signal. As the level of the input signal to the control path increases, then the value of τ decreases, thus increasing the filter's bandwidth and decreasing its gain. The structure of the control path is carefully designed to reflect the "active" cochlear process of the corresponding local basilar-membrane site as well as that of neighboring sites. It consists of a cascade of a wideband filter followed by a saturating nonlinearity. This saturating nonlinearity can be understood to represent the transduction properties of outer hair cells and is responsible for the compressive character of the model input/output response. Finally, the bandwidth of the control-path filter also varies dynamically with time, but it is always set to a value greater than that of the signal-path filter. This is necessary to account for two-tone suppression, as it allows for frequency components outside the pass-band of the signal-path filter to reduce its gain and thus the net output amplitude.

This model uses symmetric gammatone filters and, therefore, does not produce asymmetric BM frequency responses or click responses showing frequency glides. The model version of Tan and Carney (2003) solves these shortcomings by using asymmetrical digital filters that are designed in the complex plane (i.e., by positioning their poles and zeros) to have the appropriate glide (or "chirp"). Further, by making the relative position of these poles and zeros in the complex plane independent of level, the model can also account for level-independent frequency glides, consistent with the physiology (de Boer and Nuttall 1997; Recio et al. 1998; Carney et al. 1999).

2.4.1.5 The DRNL Filter of Meddis and Colleagues

The Dual-Resonance NonLinear (DRNL) filter model of Meddis and co-workers (Lopez-Poveda and Meddis 2001; Meddis et al. 2001; Lopez-Poveda 2003) simulates the velocity of vibration of a given site on the BM (Fig. 2.2c). This filter is inspired by Goldstein's MBPNL model and its predecessors (see earlier), although the structure of the DRNL filter is itself unique. The input signal to the filter is processed through two asymmetric bandpass filters arranged in parallel: one linear and broadly tuned, and one nonlinear and narrowly tuned. Gammatone filters are employed that are made asymmetric by filtering their output through a low-pass filter. A compressing memoryless (i.e., instantaneous) gain is applied to the narrow filter that produces linear responses at low levels but compressive responses for moderate levels. The output from the DRNL filter is the sum of the output signals from both paths. Level-dependent tuning is achieved by setting the relative gain of the two filter paths so that the output from the narrow and broad filters dominate the total filter response at low and high levels, respectively. Level-dependent BF shifts are accounted for by setting the center frequency of the broad filter to be different from that of the narrow filter.

The model reproduces suppression because the narrow nonlinear path is actually a cascade of a gammatone filter followed by the compressive nonlinearity, followed by another gammatone filter (Fig. 2.2c). For a two-tone suppression stimulus, the first gammatone filter passes both the suppressor and the probe tone, which are then compressed together by the nonlinear gain. Because the probe tone is compressed with the suppressor, its level at the output of the second filter is less than it would be if it were presented alone. Some versions of the DRNL filter assume that the two gammatone filters in this pathway are identical (Lopez-Poveda and Meddis 2001; Meddis et al. 2001; Sumner et al. 2002), while others (e.g., Plack et al. 2002) allow for the two filters to have different center frequencies and bandwidths to account for suppression phenomena more realistically (specifically, it can be assumed that the first filter is broader and has a higher center frequency than the second filter). On the other hand, the characteristics of the first gammatone filter in this nonlinear pathway determine the range of primary frequencies for which combination tones occur, while the second gammatone filter determines the amplitude of the generated combination tones.

The DRNL filter has proved adequate to reproduce frequency- and level-dependent BM amplitude responses for a wide range of CFs (Meddis et al. 2001; Lopez-Najera et al. 2007). It also reproduces local combination tones (i.e., combination tones that originate at BM regions near the measurement site) and some aspects of two-tone suppression (Meddis et al. 2001; Plack et al. 2002). Its impulse response resembles that of the BM and it shows frequency glides (Meddis et al. 2001; Lopez-Najera et al. 2005). These characteristics, however, appear very sensitive to the values of the model parameters, particularly to the total order of the filters in both paths and to the frequency response of the middle-ear filter used in the model (Lopez-Najera et al. 2005).

Filterbank versions of the DRNL filter have been proposed for human (Lopez-Poveda and Meddis 2001), guinea pig (Sumner et al. 2003b), and chinchilla (Lopez-Najera et al. 2007) based on corresponding experimental data. These filterbanks do not consider interaction between neighboring filters or propagation of combination tones. The parameters of the DRNL filter may be simply adjusted to model BM responses in cochleae with OHC loss (Lopez-Poveda and Meddis 2001). A version of the DRNL exists designed to account for effect of efferent activation on BM responses (Ferry and Meddis 2007).

This filter has been successfully employed for predicting the AN representation of stimuli with complex spectra, such as HRTF (Lopez-Poveda 1996), speech (Holmes et al. 2004), harmonic complexes (Gockel et al. 2003; Wiegrebe and Meddis 2004), or amplitude-modulated stimuli (Meddis et al. 2002). The model has also been used to drive models of brain stem units (Wiegrebe and Meddis 2004). It has also been used as the basis to build a biologically inspired speech processor for cochlear implants (Wilson et al. 2005, 2006; see also Chapter 9).

2.5 Inner Hair Cells

IHCs are responsible for the mechanoelectrical transduction in the organ of Corti of the mammalian cochlea. Deflection of their stereocilia toward the tallest cilium in the bundle increases the inward flow of ions and thus depolarizes the cell. Stereocilia deflection in the opposite direction closes transducer channels and prevents the inward flow of ions to the cell. This asymmetric gating of transducer channels has led to the well-known description of the IHC as a half-wave rectifier. Potassium (K^+) is the major carrier of the transducer current. The "excess" of intracellular potassium that may result from bundle deflections is eliminated through K^+ channels found in the IHC basolateral membrane, whose conductance depends on the IHC basolateral transmembrane potential (Kros and Crawford 1990). Therefore, the intracellular voltage variations produced by transducer currents may be modulated also by currents flowing through these voltage-dependent basolateral K^+ conductances. The intracellular voltage is further determined by the capacitive effect of the IHC membrane and by the homeostasis of the organ of Corti.

The in vivo IHC inherent input/output response characteristics are hard to assess because in vivo measurements reflect a complex combination of the response characteristics of the middle ear, the BM, and the IHC itself (Cheatham and Dallos 2001). Inherent IHC input/output functions have been inferred from measurements of the growth of the AC or DC components of the receptor potential with increasing sound level for stimulus frequencies an octave or more below the characteristic frequency of the IHC. The BM responds linearly to these frequencies (at least in basal regions). Therefore, any sign of nonlinearity is attributed to inherent IHC processing characteristics (Patuzzi and Sellick 1983). These measurements show that the dc component of the receptor potential grows expansively (slope of 2 dB/dB) with increasing sound level for sound levels near threshold and that the AC and DC components of the receptor potential grow compressively (slope <1 dB/dB) for moderate to high sound levels (Patuzzi and Sellick 1983). These nonlinear transfer characteristics reflect the combination of nonlinear activation of transducer and basolateral K^+ currents (described by Lopez-Poveda and Eustaquio-Martín 2006).

The in vivo IHC inherent frequency response is also difficult to assess (Cheatham and Dallos 2001). Some authors have estimated it as the ratio of the AC to the DC components of the in vivo receptor potential (AC/DC ratio) on the assumption that this ratio is normalized for constant input to the cell (Sellick and Russell 1980). The AC/DC ratio decreases with increasing stimulus frequency (Russel and Sellick 1978). This low-pass filter effect is attributed to the resistor-capacitance properties of the IHC membrane. To a first approximation, this is independent of the driving force to the cell (Russel and Sellick 1978) and of the cell's membrane potential (cf. Kros and Crawford 1990; Lopez-Poveda and Eustaquio-Martín 2006). Therefore, it is considered that the low-pass filter behavior is independent of sound level (Russel and Sellick 1978). This low-pass filter effect is thought to be responsible for the rapid roll-off of AN phase-locking with increasing frequency above approximately 1.5–2 kHz (Palmer and Russell 1986) and has led to the common description of the IHC as a low-pass filter.

It is worth mentioning that while the AC/DC ratio shows a low-pass frequency response, the AC component alone shows a bandpass response tuned at a frequency of approximately 500 Hz (Sellick and Russell 1980) or 1 kHz (Dallos 1984, 1985) for low sound levels. This result is important because it is for a basal IHC in response to low-frequency stimuli. The excitation of basal BM sites is linear and untuned in response to low-frequency tones. Therefore, the result of Sellick and Russell (1980) constitutes direct evidence for bandpass AC responses without substantial contributions from BM tuning. They argued that the rising slope of the response indicates that the IHC receptor potential responds to BM velocity for frequencies below approximately 200 Hz and to BM displacement above that frequency (see also Shamma et al. 1986).

The IHC responds nonlinearly also in time. The time-dependent activation of basolateral K^+ channels induces a nonlinear, time-dependent adaptation of the receptor potential (Kros and Crawford 1990) that could contribute to adaptation as observed in the AN (Kros 1996). This in vitro result, however, is awaiting confirmation in vivo, but computational modeling studies support this suggestion (Zeddies and Siegel 2004; Lopez-Poveda and Eustaquio-Martín 2006).

2.5.1 Approaches to Modeling the IHC Transfer Function

IHC models aim to simulate the cell's intracellular potential in response to BM excitation because the latter determines the release of neurotransmitter from within the IHC to the synaptic cleft. It is common to model the function of the IHC using either biophysical analogs or signal-processing analogs. The latter consider the IHC as a cascade of an asymmetric, saturating nonlinear gain, which accounts for the activation of the transducer currents, followed by a low-pass filter, which accounts for the resistor-capacitor filtering of the IHC membrane. The order and cutoff frequency of this filter are chosen so as to mimic as closely as possible the physiological low-pass characteristics of the IHC.

These signal-processing models are easy to implement, fast to evaluate, and require very few parameters. For these reasons, they are widely used in composite peripheral auditory models (e.g., Robert and Eriksson 1999; Zhang et al. 2001). However, they neglect important aspects of IHC processing and are limited in scope. For instance, IHCs are modeled as a low-pass filter regardless of whether the input to the IHC model stage is BM velocity or displacement. As discussed in the preceding section, this is almost certainly inappropriate for sounds with frequencies below 0.2–1 kHz. In addition, these models do not account for the time-activation of basolateral K^+ currents, which could be significant, particularly for brief and intense sounds (Kros 1996). Another shortcoming is that their parameters do not represent physiological variables; hence they do not allow modeling some forms of hearing loss associated to IHC function without changing the actual transducer and/or filter function (see Chapter 7).

An alternative approach is to model the IHC using biophysical models (an early review is provided by Mountain and Hubbard 1996). Typically these are electrical-circuit analogs of the full organ of Corti. The model of Lopez-Poveda and Eustaquio-Martín (2006) is an example. It consists of several elements that describe the electrical properties of the apical and basal portions of the IHC and its surrounding fluids. The model assumes that the intracellular space is equipotential and thus can be represented by a single node. It assumes that the IHC intracellular potential is primarily controlled by the interplay of a transducer, variable (inward) K^+ current that results from stereocilia deflections and a basolateral (outward) K^+ current that eliminates the excess of intracellular K^+ from within the IHC. The magnitude of the transducer current is calculated from stereocilia displacement using a Boltzmann function that describes the gating of transducer channels. The excess of intracellular K^+ is eliminated through two voltage- and time-dependent nonlinear activating basolateral conductances, one with fast and one with slow-activation kinetics. The activation of these two conductances is modeled using a Hodgkin–Huxley approach. The reversal potential of each of the currents involved is accounted for by a shunt battery. The capacitive effects of the IHC membrane are modeled with a single capacitor. The flow of transducer current depends also on the endocochlear potential, which is simulated with a battery.

This relatively simple electrical circuit accounts for a wide range of well reported in vitro and in vivo IHC response characteristics without a need for readjusting its parameters across data sets. Model simulations support the idea that the basolateral K^+ conductances effectively reduce the rate of growth of IHC potential with increasing stereocilia displacement by more than a factor of two for displacements above approximately 5 nm. Such compression affects the DC component of the cell's potential in a similar way for all stimulation frequencies. The AC component is equally affected but only for stimulation frequencies below 800 Hz. The simulations further suggest that the nonlinear gating of the transducer current produces an expansive growth of the DC potential with increasing sound level (slope of 2 dB/dB) at low sound pressure levels (Lopez-Poveda and Eustaquio-Martín 2006).

The model of Shamma et al. (1986) is similar and simpler in that it considers voltage- and time-independent basolateral K^+ currents. A more sophisticated version of the model of Lopez-Poveda and Eustaquio-Martín (2006) exists that incorporates the role of transmembrane cloring and sodium currents and pumps in shaping the IHC intracellular potential (Zeddies and Siegel 2004).

Biophysical IHC models have been used successfully in composite models of the peripheral auditory system (e.g., Sumner et al. 2002, 2003a, b). In these cases, a high-pass filter is used to couple BM displacement to stereocilia displacement.

2.6 Auditory Nerve Synapse

AN activity is provoked by the release of transmitter substance (glutamate) into the synaptic cleft between the AN dendrites and the IHC. The rate of release of this transmitter is regulated by two factors, the IHC receptor potential and the availability of transmitter in the presynaptic area. These two processes can be modeled separately.

Researchers generally agree that vesicles of transmitter substance are held inside the cell in a local store close to the synaptic site from which the vesicles are released into the postsynaptic cleft between the cell and a dendrite of an AN fiber. As the electrical potential inside the cell increases, the probability of release of one or more vesicles also increases. The number of vesicles available for release is relatively small and a series of release events will result in a depletion of the available vesicle store. When this happens, the rate of release of vesicles falls even though the receptor potential is unchanged. The rate will remain depressed until the presynaptic store can be replenished (Smith and Zwislocki 1975; Smith et al. 1985). It is important to distinguish between the probability that a vesicle will be released (if it is available) and the number of vesicles available for release. The vesicle release rate is the product of these two values. If no transmitter is available for release, then none will be released even if the probability of release is high. In Fig. 2.1, the "release probability" in the second from bottom panel is the first of these two quantities.

The reduction of AN spike rate after stimulation is known as "adaptation." The speed of recovery from adaptation is thought to reflect the rate at which the available store can be replenished. While there is considerable uncertainty concerning the details of this process, it nevertheless remains an important goal for the modeler to generate an accurate representation of this process. This is because it is reflected in many aspects of psychophysics where sounds are presented in rapid succession, each influencing the response of later sounds as a function of the resulting depletion of the available pool of transmitter vesicles.

2.6.1 Calcium Control of Transmitter Release

Most early models of the transmitter release and recovery proposed a simple relationship between the receptor potential level and rate of release of transmitter

(Siebert 1965; Weiss 1966; Eggermont 1973; Schroeder and Hall 1974; Oono and Sujaku 1975; Nilsson 1975; Geisler et al. 1979; Ross 1982; Schwid and Geisler 1982; Smith and Brachman 1982). In so doing, they ignored the complex nature of the relationship. This was because research has only recently unraveled the details (see, e.g., Augustine et al. 1985). It is now known that the release of transmitter is only indirectly controlled by the internal voltage of the cell. Instead, the voltage controls the rate of flow of calcium into the cell and it is this calcium that promotes the release of available transmitter into the synaptic cleft.

While it might be thought that this is one complication too many, there are indications that it is an essential part of an understanding of the signal processing that occurs at this stage. For example, Kidd and Weiss (1990) have suggested that delays associated with the movement of calcium contribute to the reduction of AN phase-locking at high frequencies. Phase-locking is already limited by the IHC membrane capacitance (see earlier) but they suggest that the rate of accumulation of presynaptic calcium further limits this effect. To some extent this is inevitable and much depends on an exact knowledge of the rate of accumulation.

More recently, it has been suggested that the accumulation of presynaptic calcium might be the physiological basis for some aspects of psychophysical thresholds (Heil and Neubauer 2003). Sumner et al. (2003a) and Meddis (2006) have also suggested that differences in the rate of accumulation and dissipation of calcium might control the rate/level function of the fiber attached to the synapse, particularly the difference between low and high spontaneous rate (LSR, HSR) fibers. The synapse is very inaccessible and difficult to study. As a consequence, these ideas must remain speculative but they do justify the inclusion of the calcium control stage in recent models of transmitter release.

Calcium enters the cell through voltage-gated calcium ion channels located close to the synapse. The number of open calcium channels is determined by the receptor potential; as the voltage rises, more gates open. Calcium ions enter the cell and accumulate in the region of the synapse. The density of ions close to the synapse determines the probability that a transmitter vesicle will be released into the cleft. However, the calcium dissipates rapidly or is chemically inactivated by a process known as buffering and the calcium concentration falls rapidly if the receptor potential falls again. The opening and closing of these ion channels as well as calcium accumulation and dissipation can be modeled using equations that are generally agreed upon among physiologists (Meddis 2006).

2.6.2 Transmitter Release

Transmitter release is an important feature of auditory models because it is the basis for explaining adaptation in the AN. From the beginning, all models of the auditory periphery have included a stage that simulates this process of depletion and recovery. All assume that there is a reservoir of transmitter that releases its contents into the synaptic cleft at a rate proportional to the stimulus intensity.

Although this is a satisfactory model for many purposes, the data suggest that the situation is more complex. If only one reservoir is involved, we might expect only one time constant of adaptation when a stimulus is presented. However, the data indicate two or even three time constants (Smith and Brachman 1982). The same applies to the recovery process where the time course of recovery is complex (Harris and Dallos 1979). The most elegant solution to this problem was proposed by Westerman and Smith (1984, 1988), who suggested a cascade of reservoirs each with their own time constant (Fig. 2.3). When the reservoir closest to the synapse becomes depleted, it is slowly refilled by the reservoir immediately above it. The third reservoir refills the second and so on. In a cascade system, the time constants of all three reservoirs are reflected in the time course of release of transmitter from the pre-synaptic reservoir. Westerman's ideas have been adopted in the modeling of Carney (1993).

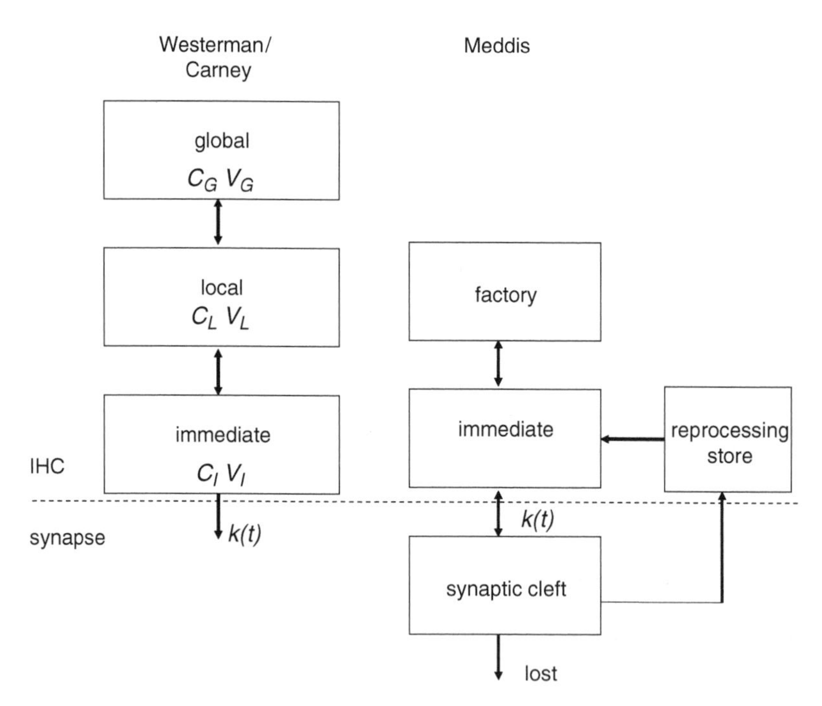

Fig. 2.3 Westerman/Carney and Meddis models of IHC/AN transmitter release. In both models $k(t)$ represents the rate at which transmitter substance is released into the synaptic cleft and this is indirectly controlled by the receptor potential of the IHC. In the Westerman/Carney model, C represents the concentration of transmitter in a reservoir and V represents its volume. P is the permeability of the path between two reservoirs. The *dashed line* indicates the IHC membrane that the transmitter must cross when released into the cleft. Equations controlling the model can be found in Zhang and Carney (2005). The Meddis model consists of reservoirs containing individual vesicles of transmitter (usually less than 20 vesicles). The equations controlling the probability that one vesicle is transferred from one reservoir to another can be found in Meddis (2006). The two models are arranged slightly differently but the behavior of the two systems is very similar

Meddis (1986, 1988) suggested an alternative system that also involved reservoirs of transmitter but used reuptake of transmitter from the synaptic cleft as the major source of replenishment of the presynaptic reservoir. Zhang and Carney (2005) have recently reevaluated both models and found that they are mathematically very similar. Recent studies of IHC physiology have confirmed that reuptake of transmitter does take place but on a much longer time scale than required by the Meddis model (see Griesinger et al. 2002).

Models of transmitter circulation are relatively straightforward and consist of a cascade of reservoirs with transmitter flowing between them. The flow of transmitter between reservoirs is determined by the relative concentrations of transmitter in the reservoirs as well as the permeability of the channels linking them. Details of the equations used to evaluate both models can be found in Zhang and Carney (2005) and Meddis (2006). The two models are illustrated in Fig. 2.3.

The most important reservoir is the "immediate" pool that releases transmitter into the synaptic cleft according to the level of the receptor potential. After stimulation, this pool becomes depleted and fewer vesicles are available for release, leading to adaptation of the response. It is important to note that the receptor potential is not affected during adaptation. The reduction in transmitter release is accounted for mainly by the reduction in available transmitter. Recovery takes place over time and as the result of replenishment either from transmitter reuptake (Meddis 1988) or a from a "global" reserve reservoir (Westerman and Smith 1988; Carney 1993).

2.7 Auditory Nerve Activity

The release of transmitter is generally agreed to be a stochastic process. The instantaneous probability of release is determined by the product of the concentration of presynaptic calcium and the number of available transmitter vesicles. However, the release event is itself a random outcome. Stochastic release of transmitter can be generated simply using random number generators to convert the release probabilities into binary release events. It is not known exactly how release events translate into AN spike events. Meddis (2006) makes the simplifying assumption that a single vesicle release event is enough to trigger an AN spike. This idea was based on some early observations of postsynaptic events by Siegel (1992). Goutman and Glowatzki (2007) offer some recent support for this view but the issue is the focus of continuing research. Certainly, the assumption of the model works well in practice.

Modelers often use the release rate as the final result of the modeling exercise. In the long run, the rate of release is a useful indication of the rate of firing of the AN fiber attached to the synapse. This is a quick and convenient representation if the model is to be used as the input to another computationally intensive application such as an automatic speech recognition device.

Modeling individual spike events in AN fibers is more time-consuming than computing probabilities alone but for many purposes it is essential, for example, when the next stage in the model consists of models of neurons in the brain stem.

Refractory effects should be included in the computation for greater accuracy. In common with other nerve cells, the AN fiber is limited in terms of how soon it can fire immediately after a previous spike. There is an absolute limit (~500 ms) on how soon a second spike can occur. The absolute refractory period is followed by a relative refractory period during which time the probability of an action potential recovers exponentially. Carney (1993) describes a useful method to simulate such effects.

2.8 Efferent Effects

So far we have considered the auditory periphery in terms of a one-way path, from the eardrum to the AN. In reality, many fibers travel in the other direction from the brain stem to the cochlea. Efferent feedback operates through two separate systems: lateral and medial (Guinan 2006). The lateral system acts directly on the dendrites of afferent auditory nerve fibers and is only poorly understood. The medial system acts by damping the response of the BM indirectly through the OHCs. This damping effect modifies the relationship between the stimulus level and the BM response. This reduced response also leads to less adaptation in the auditory nerve. It is widely believed that this latter effect is critical to the function of the medial efferent system by protecting the periphery from overstimulation.

The function of these efferent fibers is largely unknown and they rarely feature in computer models. A computer model has been developed (Ghitza et al. 2007; Messing et al. 2009) showing that efferent feedback can improve vowel discrimination against a background of noise. Ferry and Meddis (2007) have also shown that a model with efferent feedback can simulate physiological observations at the level of the BM and the AN.

2.9 Summary

It can be seen that a model of the auditory periphery is very complex. It is composed of many stages, each of which has its own associated scientific literature. Individual component stages are always compromises in terms of simulation accuracy. Part of the problem is the need to compute the result in a reasonable amount of time but it is also the case that researchers have not yet finally agreed on the details of any one processing stage. Models will need to change as new data and new insights are published. Nevertheless, models are already good enough to use them in a range of applications.

The nonlinear nature of the auditory periphery has many unexpected consequences, and it is important that the user of any model should appreciate from the outset that a computer model of the auditory periphery is not simply a biological way to generate a spectral analysis of the input sound. The ear appears to be doing something quite different. Figure 2.4 gives a simple example of a nonlinear effect that would not

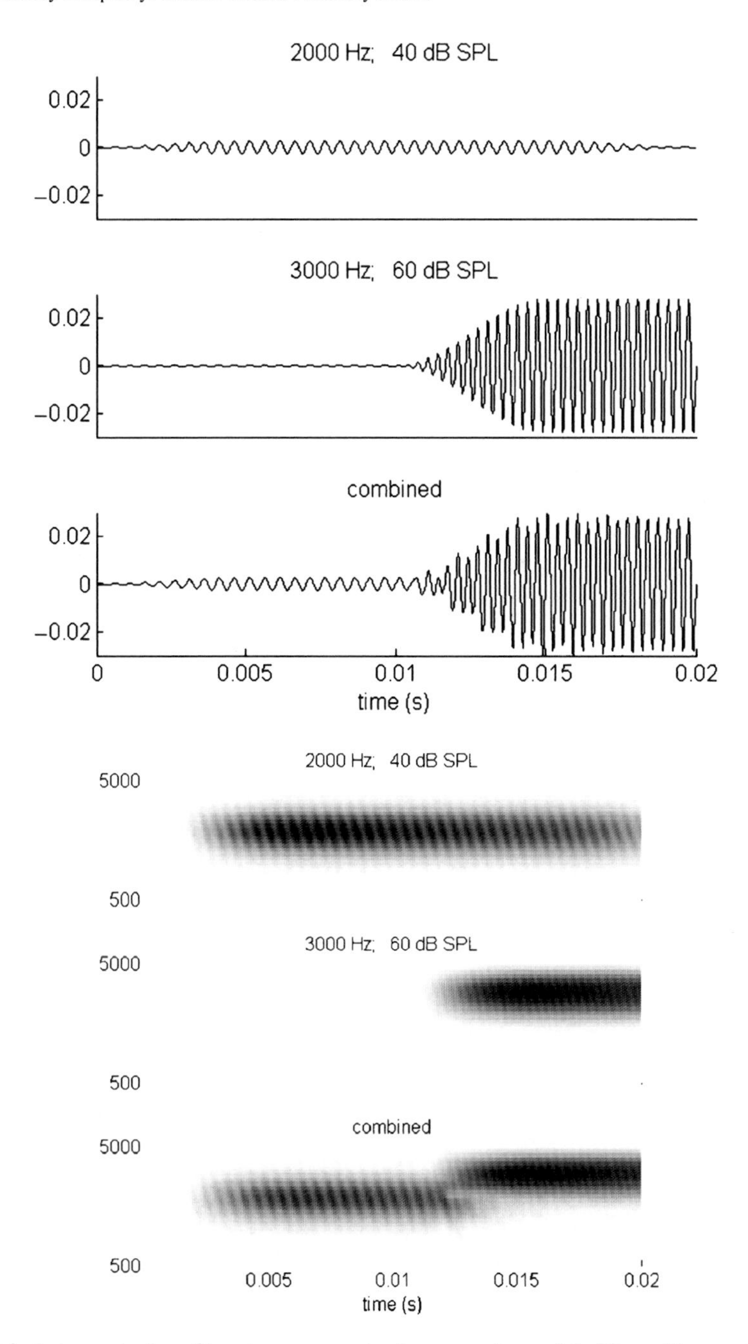

Fig. 2.4 A demonstration of two-tone suppression in a computer model of the auditory periphery. The model uses 30 channels with best frequencies distributed between 500 and 5 kHz. *Left*: Stimuli, all presented on the same scale. *Right*: Multichannel model showing probability of transmitter release. *Top panels*: 2-kHz, 20-ms tone (the probe) presented at 40 dB SPL. *Middle panels*: 3-kHz, 10-ms tone (the suppressor) presented at 60 dB SPL. *Bottom panels*: both tones presented together. The response to the probe tone is reduced when the suppressor begins

be seen in a discrete Fourier transform. The top panel shows the response to a single pure tone called the "probe." The second panel shows the response to a second pure tone called the "suppressor." Note that the suppressor is timed to start after the probe. The third panel shows what happens when the two tones are combined. When the suppressor tone starts, the response to the probe is substantially reduced. This is a consequence of the nonlinearities in the model and would never occur in a linear system. While this demonstration is very clear, it should not be assumed that all tones suppress all other tones. This effect occurs only with certain combinations of levels and tone frequencies. This example was found only after careful searching for an ideal combination.

Another difference from traditional signal processing can be seen with background firing rates in the auditory nerve. The majority of auditory nerve fibers are spontaneously active. They have spontaneous firing rates up to 100 spikes/s. When the fiber is driven by a steady high intensity tone, its firing rate will rarely exceed 300 spikes/s. Figure 2.5 shows the response of an auditory model to speech (the utterance "one-oh seven") at three speech levels. Two kinds of output are shown. The left-hand panels show the pattern of transmitter release rates while the right-hand panels show raster plots of spike activity in a single fiber per channel. Release rates are faster to compute and show a much clearer picture. The spiking activity is much less easy to interpret, but it must be remembered that a full model has thousands of

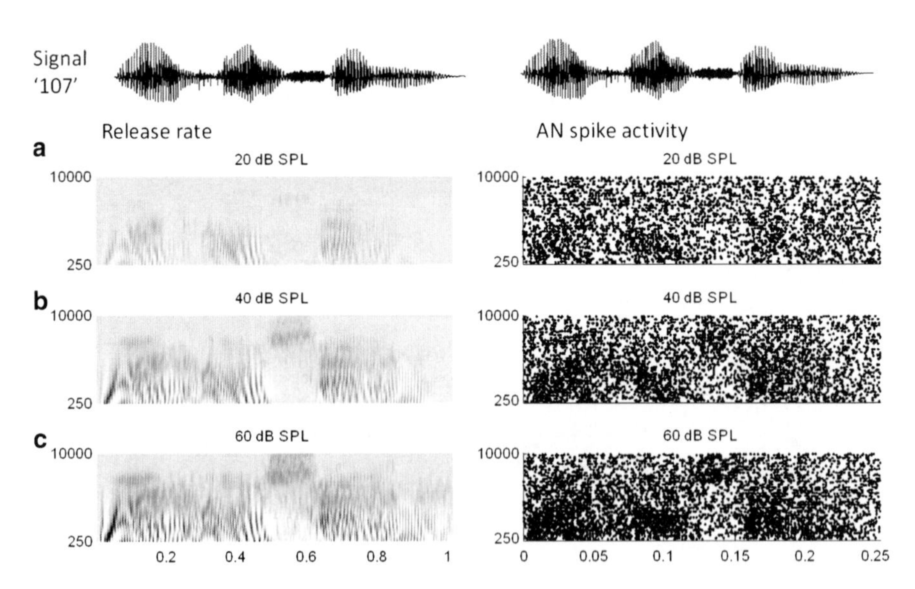

Fig. 2.5 Sixty-channel model AN response to the speech utterance "one oh seven" presented at three signal levels 20, 40, and 60 dB SPL. Channel best frequencies ranged between 250 Hz and 10 kHz. The model is based on equations in Meddis (2006). *top*: Transmitter vesicle release rate. *bottom*: Raster plot of individual AN fibers (1 per channel). The conventions used here are also explained in Fig. 2.1

fibers and the aggregate activity of all the fibers will follow the release rate pattern very closely (except for the refractory effects that are built into the fiber activity but not the transmitter release rates). The release rates are easier to interpret and link to the input signal but the spiking activity is shown to remind the reader that this is the true output of the model. This is what will be passed to later processing modules representing the activity in the cochlear nucleus. Clearly, the background activity of the fibers and the stochastic nature of the response present important challenges to the signal processing power of the brain stem neurons that receive AN input.

References

Aertsen AM, Johannesma PI (1980) Spectro-temporal receptive fields of auditory neurons in the grassfrog: I. Characterization of tonal and natural stimuli. Biol Cybern 38:223–234.

Aibara R, Welsch JT, Puria S, Goode RL (2001) Human middle-ear transfer function and cochlear input impedance. Hear Res 152:100–109.

Algazi VR, Duda RO, Morrison RP, Thompson DM (2001) Structural composition and decomposition of HRTFs. In: Proceedings of 2001 IEEE Workshop on Applications of Signal Processing to Audio and Acoustics. New Paltz, NY, pp. 103–106.

Augustine GJ, Charlton MP, Smith SJ (1985) Calcium entry into voltage-clamped pre-synaptic terminals of squid. J Physiol 367:143–162.

Bacon SP (2004) Overview of auditory compression. In: Bacon SP, Fay RR, Popper AN (eds), Compression: From Cochlea to Cochlear Implants. New York: Springer, pp. 1–17.

Burkhard MD, Sachs RM (1975) Anthropometric manikin for acoustic research. J Acoust Soc Am 58:214–222.

Carlile S, Pralong D (1994) The location-dependent nature of perceptually salient features of the human head-related transfer functions. J Acoust Soc Am 95:3445–3459.

Carlile S, Martin R, McAnally K (2005) Spectral information in sound localization. Int Rev Neurobiol 7:399–435.

Carney LH (1993) A model for the responses of low-frequency auditory-nerve fibers in cat. J Acoust Soc Am 93:402–417.

Carney LH, McDuffy MJ, Shekhter I (1999) Frequency glides in the impulse responses of auditory-nerve fibers. J Acoust Soc Am 105:2384–2391.

Cheatham MA, Dallos P (2001) Inner hair cell response patterns: implications for low-frequency hearing. J Acoust Soc Am 110:2034–2044.

Cooper NP (1996) Two-tone suppression in cochlear mechanics. J Acoust Soc Am 99:3087–3098.

Cooper NP (2004) Compression in the peripheral auditory system. In: Bacon SP, Fay RR, Popper AN (eds), Compression: From Cochlea to Cochlear Implants. New York: Springer, pp. 19–61.

Dallos P (1984) Some electrical circuit properties of the organ of Corti: II. Analysis including reactive elements. Hear Res 14:281–291.

Dallos P (1985) Response characteristics of mammalian cochlear hair cells. J Neurosci 5: 1591–1608.

de Boer E (1975) Synthetic whole-nerve action potentials for the cat. J Acoust Soc Am 58: 1030–1045.

de Boer E (1996) Mechanics of the cochlea: modeling efforts. In: Dallos P, Popper AN, Fay RR (eds), Auditory Computation. New York: Springer, pp. 258–317.

de Boer E, de Jongh HR (1978) On cochlear encoding: potentialities and limitations of the reverse correlation technique. J Acoust Soc Am 63:115–135.

de Boer E, Nuttall AL (1997) The mechanical waveform of the basilar membrane: I: Frequency modulation ("glides") in impulse responses and cross-correlation functions. J Acoust Soc Am 101:3583–3592.

Deng L, Geisler CD (1987) A composite auditory model for processing speech sounds. J Acoust Soc Am 82:2001–2012.
Derleth RP, Dau T, Kollmeier B (2001) Modeling temporal and compressive properties of the normal and impaired auditory system. Hear Res 159:132–149.
Dornhoffer JL (1998) Hearing results with the Dornhoffer ossicular replacement prostheses. Laryngoscope 108:531–536.
Duda RO, Martens WL (1998) Range dependence of the response of a spherical head model. J Acoust Soc Am 104:3048–3058.
Duifhuis H (1976) Cochlear nonlinearity and second filter: possible mechanism and implications. J Acoust Soc Am 59:408–423.
Duifhuis H (2004) Comments on "An approximate transfer function for the dual-resonance nonlinear filter model of auditory frequency selectivity." J Acoust Soc Am 115(5 Pt 1):1889–1990.
Eggermont JJ (1973) Analogue modeling of cochlea adaptation. Kybernetic 14:117–126.
Ferry RT, Meddis R (2007) A computer model of medial efferent suppression in the mammalian auditory system. J Acoust Soc Am 122:3519–3526.
Flanagan JL (1960) Models for approximating basilar membrane displacement. Bell Syst Technol J 39:1163–1191.
Gan RZ, Sun Q, Dyer RK, Chang K-H, Dormer KJ (2002) Three-dimensional modeling of middle ear biomechanics and its applications. Otol Neurotol 23:271–280.
Geisler CD, Le S, Schwid H (1979) Further studies on the Schroeder-hall hair-cell model. J Acoust Soc Am 65:985–990.
Ghitza O, Messing D, Delhorne L (2007) Towards predicting consonant confusions of degraded speech. In: Kollmeier B, Klump, G, Hohmann V, Langemann U, Mauermann M, Uppenkamp S, Verhey J (eds), Hearing: From Sensory Processing to Perception. New York: Springer, pp. 541–550.
Gockel H, Moore BCJ, Patterson RD, Meddis R (2003) Louder sounds can produce less forward masking effects: effects of component phase in complex tones. J Acoust Soc Am 114: 978–990.
Goldstein JL (1966) Auditory nonlinearity. J Acoust Soc Am 41:676–689.
Goldstein JL (1988) Updating cochlear driven models of auditory perception: a new model for nonlinear auditory frequency analysing filters. In: Elsendoorn BAG, Bouma H (eds), Working Models of Human Perception. London: Academic, pp. 19–58.
Goldstein JL (1990) Modeling rapid waveform compression on the basilar membrane as multiple-bandpass-nonlinearity filtering. Hear Res 49:39–60.
Goldstein JL (1993) Exploring new principles of cochlear operation: bandpass filtering by the organ of Corti and additive amplification by the basilar membrane. In: Duifhuis H, Horst JW, van Dijk P, van Netten SM (eds), Biophysics of Hair Cell Sensory Systems. Singapore: World Scientific, pp. 315–322.
Goldstein JL (1995) Relations among compression, suppression, and combination tones in mechanical responses of the basilar membrane: data and MBPNL model. Hear Res 89:52–68.
Goode RL, Killion M, Nakamura K, Nishihara S (1994) New knowledge about the function of the human middle ear: development of an improved analog model. Am J Otol 15:145–154.
Goutman JD, Glowatzki E (2007) Time course and calcium dependence of transmitter release at a single ribbon synapse. Proc Natl Acad Sci U S A 104:16341–16346.
Griesinger CB, Richards CD, Ashmore JF (2002) FM1-43 reveals membrane recycling in adult inner hair cells of the mammalian cochlea. J Neurosci 22:3939–3952.
Guinan JJ (2006) Olivocochlear efferents: anatomy, physiology, function, and the measurement of efferent effects in humans. Ear Hear 27:589–607.
Guinan JJ, Peake WT (1966) Middle-ear characteristics of anaesthetized cats. J Acoust Soc Am 41:1237–1261.
Harris DM, Dallos P (1979) Forward masking of auditory nerve fiber responses. J Neurophysiol 42:1083–1107.
Heil P, Neubauer H (2003) Unifying basis of auditory thresholds based on temporal summation. Proc Natl Acad Sci U S A 100:6151 6156.

Heinz MG, Zhang X, Bruce IC, Carney LH (2001) Auditory nerve model for predicting performance limits of normal and impaired listeners. Acoust Res Lett Online 2:91–96.

Holmes SD, Sumner CJ, O'Mard LPO, Meddis R (2004) The temporal representation of speech in a nonlinear model of the guinea pig cochlea. J Acoust Soc Am 116:3534–3545.

Irino T, Patterson RD (1997) A time-domain, level-dependent auditory filter: the gammachirp. J Acoust Soc Am 101:412–419.

Irino T, Patterson RD (2001) A compressive gammachirp auditory filter for both physiological and psychophysical data. J Acoust Soc Am 109:2008–2022.

Irino T, Patterson RD (2006) A dynamic, compressive gammachirp auditory filterbank. IEEE Audio Speech Lang Process 14:2222–2232.

Kidd RC, Weiss TF (1990) Mechanisms that degrade and timing information in the cochlea. Hear Res 49:181–208.

Kistler DJ, Wightman FL (1992) A model of head-related transfer functions based on principal components analysis and minimum-phase reconstruction. J Acoust Soc Am 91:1637–1647.

Kleinschmidt M, Tchorz J, Kollmeier B (1999) Combining speech enhancement and auditory feature extraction for robust speech recognition. Speech Commun 34:75–91.

Koike T, Wada H, Kobayashi T (2002) Modeling of the human middle ear using the finite-element method. J Acoust Soc Am 111:1306–1317.

Kringlebotn M (1988) Network model for the human middle ear. Scand Audiol 17:75–85.

Kros CJ (1996) Physiology of mammalian cochlear hair cells. In: Dallos P, Popper AN, Fay RR (eds), The Cochlea. New York: Springer, pp. 318–385.

Kros CJ, Crawford AC (1990) Potassium currents in inner hair cells isolated from the guinea-pig cochlea. J Physiol 421:263–291.

Kulkarni A, Colburn HS (2004) Infinite-impulse-response models of the head-related transfer function. J Acoust Soc Am 115:1714–1728.

Kulkarni A, Isabelle SK, Colburn HS (1999) Sensitivity of human subjects to head-related-transfer-function phase spectra. J Acoust Soc Am 105:2821–2840.

Lopez-Najera A, Meddis R, Lopez-Poveda EA (2005) A computational algorithm for computing non-linear auditory frequency selectivity: further studies. In: Pressnitzer, D, de Cheveigné A, McAdams S, Collet L (eds), Auditory Signal Processing: Physiology, Psychoacoustics, and Models. New York: Springer, pp. 14–20.

Lopez-Najera A, Lopez-Poveda EA, Meddis R (2007) Further studies on the dual-resonance non-linear filter model of cochlear frequency selectivity: responses to tones. J Acoust Soc Am 122:2124–2134.

Lopez-Poveda EA (1996) The physical origin and physiological coding of pinna-based spectral cues. PhD thesis, Loughborough University, UK.

Lopez-Poveda EA (2003) An approximate transfer function for the dual-resonance nonlinear filter model of auditory frequency selectivity. J Acoust Soc Am 114:2112–2117.

Lopez-Poveda EA (2005) Spectral processing by the peripheral auditory system: facts and models. Int Rev Neurobiol 70:7–48.

Lopez-Poveda EA, Eustaquio-Martín A (2006) A biophysical model of the inner hair cell: the contribution of potassium current to peripheral compression. J Assoc Res Otolaryngol 7:218–235.

Lopez-Poveda EA, Meddis R (1996) A physical model of sound diffraction and reflections in the human concha. J Acoust Soc Am 100:3248–3259.

Lopez-Poveda EA, Meddis R (2001) A human nonlinear cochlear filterbank. J Acoust Soc Am 10:3107–3118.

Lopez-Poveda EA, Plack CJ, Meddis R (2003) Cochlear nonlinearity between 500 and 8000 Hz in normal-hearing listeners. J Acoust Soc Am 113:951–960.

Lopez-Poveda EA, Barrios LF, Alves-Pinto A (2007) Psychophysical estimates of level-dependent best-frequency shifts in the apical region of the human basilar membrane. J Acoust Soc Am 121:3646–3654.

Lyon RF (1997) All-pole models of auditory filtering. In: Lewis ER, Lyon R, Long GR, Narins PM (eds), Diversity in Auditory Mechanics. Singapore: World Scientific, pp. 205–211.

Meddis R (1986) Simulation of mechanical to neural transduction in the auditory receptor. J Acoust Soc Am 79:702–711.
Meddis R (1988) Simulation of mechanical to neural transduction: further studies. J Acoust Soc Am 83:1056–1063.
Meddis R (2006) Auditory-nerve first-spike latency and auditory absolute threshold: a computer model. J Acoust Soc Am 119:406–417.
Meddis R, O'Mard LPO, Lopez-Poveda EA (2001) A computational algorithm for computing non-linear auditory frequency selectivity. J Acoust Soc Am 109:2852–2861.
Meddis R, Delahaye R, O'Mard LPO, Sumner C, Fantini DA, Winter I, Pressnitzer D (2002) A model of signal processing in the cochlear nucleus: comodulation masking release. Acta Acust/Acustica 88:387–398.
Messing DP, Delhorne L, Bruckert E, Braida LD, Ghitza O (2009) A non-linear efferent-inspired model of the auditory system; matching human confusion in stationary noise. Speech Commun 51:668–683.
Møller AR (1961) Network model of the middle ear. J Acoust Soc Am 33:168–176.
Moore BCJ (2007) Cochlear Hearing Loss. Physiological, Psychological and Technical Issues. Chichester: Wiley.
Moore BCJ, Glasberg BR, Baer T (1997) A model for the prediction of thresholds, loudness and partial loudness. J Audio Eng Soc 45:224–240.
Mountain DC, Hubbard AE (1996) Computational analysis of hair cell and auditory nerve processes. In: Hawkins HL, McMullen TA, Popper AN, Fay RR (eds), Auditory Computation. New York: Springer, pp. 121–156.
Narayan SS, Temchin AN, Recio A, Ruggero MA (1998) Frequency tuning of basilar membrane and auditory nerve fibers in the same cochleae. Science 282:1882–1884.
Nedzelnitsky V (1980) Sound pressures in the basal turn of the cat cochlea. J Acoust Soc Am 68:1676–1689.
Nilsson HG (1975) Model of discharge patterns of units in the cochlear nucleus in response to steady state and time-varying sounds. Biol Cybern 20:113–119.
Oono Y, Sujaku Y (1975) A model for automatic gain control observed in the firings of primary auditory neurons. Trans Inst Electron Comm Eng Jpn 58:352–358 (in Japanese) An abstract in English appears in Abstracts of the Trans Inst Elects on Comm Eng Jpn 58:61–62.
Palmer AR, Russell IJ (1986) Phase-locking in the cochlear nerve of the guinea-pig and its relation to the receptor potential of inner hair cells. Hear Res 24:1–15.
Pascal J, Bourgeade A, Lagier M, Legros C (1998) Linear and nonlinear model of the human middle ear. J Acoust Soc Am 104:1509–1516.
Patterson RD, Robinson K, Holdsworth J, McKeown D, Zhang C, Allerhand M (1992) Complex sounds and auditory images. In: Cazals Y, Horner K, Demany L (eds), Auditory Physiology and Perception, Oxford: Pergamon, pp. 429–443.
Patterson RD, Unoki M, Irino T (2003) Extending the domain of center frequencies for the compressive gammachirp auditory filter. J Acoust Soc Am 114:1529–1542.
Patuzzi R, Sellick PM (1983) A comparison between basilar membrane and inner hair cell receptor potential input-output functions in the guinea pig cochlea. J Acoust Soc Am 74:1734–1741.
Patuzzi R, Sellick PM, Johnstone BM (1984) The modulation of the sensitivity of the mammalian cochlea by low frequency tones: III. Basilar membrane motion. Hear Res 13:19–27.
Pfeiffer RR (1970) A model for two-tone inhibition of single cochlear-nerve fibers. J Acoust Soc Am 48:1373–1378.
Plack CJ, Oxenham AJ, Drga V (2002) Linear and nonlinear processes in temporal masking. Acta Acust/Acustica 88:348–358.
Plomp R (1976) Aspects of Tone Sensation: A Psychophysical Study. London: Academic.
Recio A, Rich NC, Narayan SS, Ruggero MA (1998) Basilar-membrane responses to clicks at the base of the chinchilla cochlea. J Acoust Soc Am 103:1972–1989.
Rhode WS, Cooper NP (1996) Nonlinear mechanics in the apical turn of the chinchilla cochlea in vivo. Audit Neurosci 3:101–121.

Robert A, Eriksson JL (1999) A composite model of the auditory periphery for simulating responses to complex sounds. J Acoust Soc Am 106:1852–1864.
Robles L, Ruggero MA (2001) Mechanics of the mammalian cochlea. Physiol Rev 81:1305–1352.
Robles L, Ruggero MA, Rich NC (1991) Two-tone distortion in the basilar membrane of the cochlea. Nature 349:413–414.
Robles L, Ruggero MA, Rich NC (1997) Two-tone distortion in the basilar membrane of the chinchilla cochlea. J Neurophysiol 77:2385–2399.
Rosowski JJ (1996) Models of external- and middle-ear function. In: Hawkins HL, McMullen TA, Popper AN, Fay RR (eds), Auditory Computation. New York: Springer, pp. 15–61.
Ross S (1982) A model of the hair cell-primary fiber complex. J Acoust Soc Am 71:926–941.
Ruggero MA, Temchin AN (2002) The roles of the external, middle, and inner ears in determining the bandwidth of hearing. Proc Natl Acad Sci U S A 99:13206–13210.
Ruggero MA, Temchin AN (2003) Middle-ear transmission in humans: wide-band, not frequency-tuned? Acoust Res Lett Online 4:53–58.
Ruggero MA, Rich NC, Robles L, Recio A (1990) The effects of acoustic trauma, other cochlear injury, and death on basilar-membrane responses to sound. In: Axelson A, Borchgrevink H, Hellström PA, Henderson D, Hamernik RP, Salvi RJ (eds), Scientific Basis of Noise-Induced Hearing Loss. New York: Thieme, pp. 23–35.
Russell IJ, Murugasu E (1997) Medial efferent inhibition suppresses basilar membrane responses to near characteristic frequency tones of moderate to high intensities. J Acoust Soc Am 102:1734–1738.
Russel IJ, Sellick PM (1978) Intracellular studies of hair cells in the mammalian cochlea. J Physiol 2:261–290.
Sachs MB, Kiang NY (1968) Two-tone inhibition in auditory nerve fibers. J Acoust Soc Am 43:1120–1128.
Schroeder MR, Hall JL (1974) Model for mechanical to neural transduction in the auditory receptor. J Acoust Soc Am 55:1055–1060.
Schwid HA, Geisler CD (1982) Multiple reservoir model of neurotransmitter release by a cochlear inner hair cell. J Acoust Soc Am 72:1435–1440.
Searle CL, Braida LD, Cuddy DR, Davis MF (1975) Binaural pinna disparity: another auditory localization cue. J Acoust Soc Am 57:448–455.
Sellick PM, Russell IJ (1980) The responses of inner hair cells to basilar membrane velocity during low frequency auditory stimulation in the guinea pig cochlea. Hear Res 2:439–445.
Shamma SA, Chadwick RS, Wilbur WJ, Morrish KA, Rinzel J (1986) A biophysical model of cochlear processing: intensity dependence of pure tone responses. J Acoust Soc Am 80:133–145.
Shaw EAG (1966) Earcanal pressure generated by a free sound field. J Acoust Soc Am 39: 465–470.
Shaw EAG (1975) The external ear. In: Keidel WD, Neff WD (eds), Handbook of Sensory Physiology. Berlin: Springer, pp. 455–490.
Siebert WM (1965) Some implications of the stochastic behavior of primary auditory neurons. Kybernetic 2:206–215.
Siegel JH (1992) Spontaneous synaptic potentials from afferent terminals in the guinea pig cochlea. Hear Res 59:85–92
Slaney M (1993) An efficient implementation of the Patterson-Holdsworth auditory filter bank. Apple Computer Technical Report #35. Apple Computer Inc.
Smith RL, Brachman ML (1982) Adaptation in auditory nerve fibers: a revised model. Biol Cybern 44:107–120.
Smith RL, Zwislocki JJ (1975) Short-term adaptation and incremental responses of single auditory-nerve fibers. Biol Cybern 17:169–182.
Smith RL, Brachman ML, Frisina RD (1985) Sensitivity of auditory-nerve fibers to changes in intensity: a dichotomy between decrements and increments. J Acoust Soc Am 78:1310–1316.
Sumner CJ, Lopez-Poveda EA, O'Mard LPO, Meddis R (2002) A revised model of the inner hair cell and auditory nerve complex. J Acoust Soc Am 111:2178–2188.

Sumner CJ, Lopez-Poveda EA, O'Mard LP, Meddis R (2003a) Adaptation in a revised inner-hair cell model. J Acoust Soc Am 113:893–901.
Sumner CJ, O'Mard LPO, Lopez-Poveda EA, Meddis R (2003b) A non-linear filter-bank model of the guinea-pig cochlear nerve. J Acoust Soc Am 113:3264–3274.
Sun Q, Gan RZ, Chang K-H, Dormer KJ (2002) Computer-integrated finite element modeling of human middle ear. Biomechan Model Mechanobiol 1:109–122.
Tan Q, Carney LH (2003) A phenomenological model for the responses of auditory-nerve fibers: II. Nonlinear tuning with a frequency glide. J Acoust Soc Am 114:2007–2020.
von Helmholtz HL (1877) The Sensation of tones. (Translated by AJ Ellis, 1954.) New York: Dover.
Voss SE, Rosowski JJ, Merchant SN, Peake WT (2000) Acoustic responses of the human middle ear. Hear Res 150:43–69.
Walsh T, Demkowicz L, Charles R (2004) Boundary element modelling of the external human auditory system. J Acoust Soc Am 115:1033–1043.
Weiss TF (1966) A model of the peripheral auditory system. Kybernetic 3:153–175.
Westerman LA, Smith RL (1984) Rapid and short term adaptation in auditory nerve responses. Hear Res 15:249–260.
Westerman LA, Smith RL (1988) A diffusion model of the transient response of the cochlear inner hair cell synapse. J Acoust Soc Am 83:2266–2276.
Wiegrebe L, Meddis R (2004) The representation of periodic sounds in simulated sustained chopper units of the ventral cochlear nucleus. J Acoust Soc Am 115:1207–1218.
Wightman FL, Kistler DJ (1989) Headphone simulation of free-field listening: I. Stimulus synthesis. J Acoust Soc Am 85:858–867.
Wilson BS, Schatzer R, Lopez-Poveda EA, Sun X, Lawson DT, Wolford RD (2005) Two new directions in speech processor design for cochlear implants. Ear Hear 26:73S–81S.
Wilson BS, Schatzer R, Lopez-Poveda EA (2006) Possibilities for a closer mimicking of normal auditory functions with cochlear implants. In: Waltzman SB, Roland JT (eds), Cochlear Implants. New York: Thieme, pp. 48–56.
Zeddies DG, Siegel JH (2004) A biophysical model of an inner-hair cell. J Acoust Soc Am 116:426–441.
Zhang X, Carney LH (2005) Analysis of models for the synapse between the inner hair cell and the auditory nerve. J Acoust Soc Am 118:1540–1553.
Zhang X, Heinz MG, Bruce IC, Carney LH (2001) A phenomenological model for the responses of auditory-nerve fibers: I. Nonlinear tuning with compression and suppression. J Acoust Soc Am 109:648–670.
Zwislocki J (1962) Analysis of the middle-ear function. Part I: Input impedance. J Acoust Soc Am 34:1514–1523.

Chapter 3
The Cochlear Nucleus: The New Frontier

Herbert F. Voigt and Xiaohan Zheng

Abbreviations and Acronyms

2D two-dimensional
AM amplitude-modulated
AN auditory nerve
AVCN anteroventral cochlear nucleus
b_K, B sensitivity to potassium conductance
BBN broad-band noise
BF best frequency
BW bandwidth
$BW_{A \to B}$ bandwidth of cell group A sending connections to cell B
c threshold sensitivity (0–1)
chop-S chopper-sustained
chop-T chopper-transient
$C_{A \to B}$ center frequency offset from cell group A to cell B
CN cochlear nucleus
dB SPL decibels sound pressure level
DCN dorsal cochlear nucleus
DSI depolarization-induced suppression of inhibition
E neuron transmembrane potential
E_{ex} excitatory reversal potential
E_{in} inhibitory reversal potential
E_K, EK potassium reversal potential
ES prior soma potential
$g_{A \to B}$ conductance of the synaptic connection from cell A to B
g_{ex} normalized excitatory synaptic conductance
g_{in} normalized inhibitory synaptic conductance

H.F. Voigt (✉)
Department of Biomedical Engineering, Boston University,
Boston, MA 02215, USA
e-mail: hfv@enga.bu.edu

R. Meddis et al. (eds.), *Computational Models of the Auditory System*,
Springer Handbook of Auditory Research 35, DOI 10.1007/978-1-4419-5934-8_3,

g_k	normalized potassium conductance
G	resting conductance
G_{ex}	excitatory synaptic conductance
G_{in}	inhibitory synaptic conductance
G_k, *GK*	potassium conductance
HRP	horseradish peroxidase
HRTF	head-related transfer function
I_h	a Ca^+-sensitive, hyperpolarization-activated inward rectifier
I–V	current-voltage
I2-cells	inhibitory interneurons with type II unit RMs
ISIH	interspike interval histogram
K_{LT}	low-threshold K^+
LTD	long-term depression
LTP	long-term potentiation
$N_{A\rightarrow B}$	number of cell A connections to cell B
On	Onset
On-C	onset-chopper
On-I	ideal onset
On-L	onset with late activity
P-cells	principal cells
PL	Primarylike
Pri-N	primarylike-with-notch
PS	potential in soma
PSTH	peristimulus time histogram
PVCN	posterioventral cochlear nucleus
R–C	resistance-capacitance
RM	response map
S	spiking variable indicating whether a cell has fired
S_A	input spike from cell A
SpAc	spontaneous activity
SC	magnitude of step current injected into a point neuron
SCN	the input current
SR	spontaneous rate
TGK	refractory time constant
TH	neuron threshold
*TH*0	initial neuron threshold
TMEM	membrane time constant
TTH	time constant for accommodation
VCN	ventral cochlear nucleus
V_m	membrane potential relative to rest
W-cells	wideband inhibitors
θ	spike threshold voltage
σ	conductance step
$\sigma_{A\rightarrow B}$	the connection strength from cell A to cell B
τ	time constant

$\tau_{A \to B}$ time constant of synaptic connection from cell A to cell B
τ_K potassium time constant
τ_m membrane time constant

3.1 Conceptual versus Computational Models of the Cochlear Nucleus

Paraphrasing Lord Kelvin, "If you can't make a model, you didn't understand." Conceptual models of the neuronal circuitry within the auditory brainstem have been around for a long time. With the advent of supercomputers and the ubiquity of laptops, computational modeling these days is relatively common.

In this chapter, please consider the differences between conceptual models and computational models, focusing specifically on the cochlear nucleus. Conceptual models are typically qualitative in nature. They deal in ideas about how system mechanisms (cellular, neural circuits, etc.) work. They deal in concepts. Computational models, conversely, are quantitative in nature and require identifying specific parameters, whose values must be specified. They are often expressed in differential or other kinds of equations, whose parameters are specific numbers or ranges of numbers. Conceptual models are generally first; these then become realized through mathematical equations and codified in computer languages that run today primarily on digital computers. In addition, there have been silicone-based implementations of some of these models.

Computational models of neurons and neural circuitry in the cochlear nucleus (CN) are briefly cited and reviewed. The literature is quite rich and space does not permit a detailed description of all of these computational models, but this chapter will allow the interested reader to find details in the original papers. An earlier treatment of this material may be found in Hawkins et al. (1996). A list of abbreviations and acronyms is provided at the beginning of the chapter.

3.2 Computational Models of the Cochlear Nucleus

The cochlear nucleus (CN) has been the subject of extensive anatomical and physiological experiments since the 20th century. The CN, which is divided into three subnuclei—the anteroventral (AVCN), posterioventral (PVCN), and dorsal (DCN) cochlear nuclei—displays a rich variety of anatomically distinct neurons that are non-uniformly distributed within the nucleus. Spatial gradients of neurons have been described in both the AVCN and DCN. For example, large spherical cells of the AVCN are highly concentrated in the rostral pole of the AVCN and become less so caudally. All three subnuclei are tonotopically organized and have their own output pathways for their principal neurons. The AVCN, PVCN, and DCN have been the subject of numerous computational modeling studies (see Table 3.1).

Table 3.1 Models of the cochlear nucleus

Implementation	AVCN	PVCN	DCN
Computational single neuron models	*Stellate Cells*	*Octopus Cells*	*Fusiform Cells*
	Arle and Kim 1991;	Levy and Kipke 1997;	Arle and Kim 1991;
	Banks and Sachs 1991;	Kipke and Levy 1997;	Hewitt and Meddis 1995
	Hewitt et al. 1992;	Cai et al. 1997, 2000	Reed and Blum 1995, 1997;
	Hewitt and Meddis 1993;	*On Cells*	Kanold and Manis 1999
	Bahmer and Langner 2006, 2008	Kalluri and Delgutte 2003a, b	
	Bushy Cells		
	Rothman et al. 1993;		
	Kim and D'Angelo 2000;		
	Rothman and Manis 2003		
	Marginal shell		
	Pathmanathan and Kim 2001		
	On Cells		
	Kalluri and Delgutte 2003a, b		
Computational Neural Network models	Eager et al. 2004		Pont and Damper 1991;
	Bahmer and Langner 2006, 2008		Davis and Voigt 1994, 1996;
			Nelken and Young 1994;
			Voigt and Davis 1996;
			Hancock et al. 1997;
			Reed and Blum 1995, 1997
			Hancock and Voigt 1999;
			Zheng and Voigt 2006a, b
Silicone-based	van Schaik et al. 1996		

Section 3.2 describes these efforts. For details about the rich anatomy and physiology of the cochlear nucleus, the reader is directed to Popper and Fay (1992) and Webster et al. (1992).

A large variety of computational models exist for the CN. Most of these model the electrophysiological properties of the neurons, such as current–voltage relations, spike generation, or more sophisticated features such as peristimulus time histograms (PSTHs), interspike interval histograms (ISIHs), rate-level curves, response maps (RMs), or other response-derived properties.

The model's input stimuli are sometimes simple, simulated electrical currents, which are commonly used in *in vitro* experiments, and acoustic sounds, which are typically applied in *in vivo* experiments. Data from *in vitro* experiments can provide more insight into the basic physiological properties of neurons and form the bases of single neuron modeling. *In vivo* experiments can provide higher abstract properties of the neuron and are essential for both single-neuron and neural network behaviors and models. With simulated acoustic inputs, an auditory periphery model is needed before neural modeling of the CN. Different acoustic stimuli are used to study the variety of electrophysiological responses of CN neurons. Pure tones and broadband noises are frequently used stimuli in the CN neuron model simulations. More natural stimuli are also frequently used, including notch noises, amplitude-modulated and frequency-modulated sine waves, and even species-specific sounds such as speech.

Given that the CN is the first stage in the central auditory pathway, few experiments focus on the correlations between behavior/consciousness and CN electrophysiological responses. Some neurons in CN, however, have complex analysis functions. For example, DCN fusiform cells appear to be specialized to detect spectral notches in head-related transfer functions (HRTFs) and thus may be involved in localization of sound sources in the median plane. DCN granule cells may be involved in detecting or relaying information regarding the animal's pinna movements. The study of the links between neurophysiological responses and behavior/consciousness are of high interest and challenge, and may provide better understanding of the neurons in the central auditory system.

3.2.1 Modeling Single Neurons in the VCN

3.2.1.1 Bushy and Stellate cells: I–V and PSTH Characteristics

Arle and Kim (1991) modeled both the bushy cells and stellate cells. This neuron model was a cell-based model composed as an R-C circuit with a branch containing the nonlinear voltage-dependent conductance added to the point neuron model from MacGregor (1987). (The MacGregor neuromime is described in detail below.) Arle and Kim's model was used to simulate neurons whose *I–V* characteristics were known.

Stellate cells have primarily linear current–voltage (*I–V*) characteristics; Bushy cells have highly nonlinear *I–V* characteristics. This nonlinearity was realized by implementing a nonlinear voltage-dependent conductance. Specifically, an exponentially increasing function of membrane voltage was used, instead of a stationary conductance that was used in the stellate cell model.

The model reproduced the salient features of the experimentally observed *I–V* characteristics of the cells and the spike discharge behavior to current steps injected intracellularly. Moreover, the chopper subtypes of stellate cells were related to the nature of the current at the cell soma during tonal stimuli. Steady average currents with small variability generated the chopper-sustained (chop-S) pattern, while more irregular current was required to generate chopper-transient (chop-T) patterns. Similar irregular current was required in generating onset-chopper (On-C) patterns. The accommodation term and threshold time constant (i.e., a voltage-dependent threshold change), however, needed some adjustment.

VCN (and Other) Chopper-Sustained (Chop-S) Units

Unlike the previous models, which were single-neuron models and relied exclusively on the membrane properties, Bahmer and Langner (2006, 2008) obtained precise properties of sustained chopper neurons by modeling them in a network. Their model was able to explain the oscillations of chopper neurons with periods of multiples of 0.4 ms, which was the time of synaptic delay, in responses to the acoustic amplitude-modulated (AM) sound. The simulated chop-S neurons were characterized by "sharp frequency tuning to pure tones and faithful periodicity coding over a large dynamic range." They received inputs from onset (On) neurons, which provided the appropriate responses to the wide dynamic frequency range of sound inputs. In addition, the network model showed properties of self-stabilization of spike intervals, which were in multiples of 0.4 ms as observed across different species and is consistent with anatomical data that neurons with chopper PSTHs build interconnected networks.

The smallest unit of the network consisted of two interconnected chopper neurons with a synaptic delay of 0.4 ms and an On neuron, which activated one of the chopper neurons. Both chopper neurons and the On neuron received auditory inputs, while the former received narrowband and the latter received wideband inputs. The two fast firing chopper neurons acted as "pacemakers" with spike intervals of 0.4 ms. Slower chopper neurons with spike intervals of multiples of 0.4 ms were obtained by receiving projections from these pacemakers, which resulted from their intrinsic property of longer refractory period and skipping the intervals shorter than this period. The auditory peripheral model was implemented by W. Hemmert, Infineon Technologies (Hemmert et al. 2003; Holmberg and Hemmert 2004). The On neuron model was a simplified version of the single electrical compartment model by Rothman and Manis (2003) and each On neuron received a broad spectral range of auditory nerve (AN) input. The neuromime consisted of a sodium, a low-threshold potassium, an excitatory synaptic and a leakage current channel, where the low-threshold potassium channel was responsible for the phasic discharge pattern observed in On bushy cells. The chopper neuron model was a leaky integrate-and-fire model. Sinusoidal amplitude-modulated sine waves were used as sound stimuli.

VCN Chopper-Transient (Chop-T) Units

Irregular current input was sufficient to generate chop-T behavior based on the model for neurons showing chop-S behavior, as verified by the model of Arle and Kim (1991). Physiologically, the irregularity might come from the inhibitory inputs of the cell. One way to explain this by the neuron circuitry is that perhaps the inhibitory inputs to stellate cells come from the efferents of the olivocochlear bundle (Fex 1962; Liberman and Brown 1986). This hypothesis was posed after observing the coincidence of the progressive irregularity of chop-T cells after a 10- to 15-ms tonal stimulus time course and the efferent activity latency of 10 to 15 ms in the olivocochlear bundle. Other possible inhibitory sources to stellate cells include the intranuclear axons from the deep DCN (Wickesberg and Oertel 1988) and from local stellate cell collaterals, leading to varying amounts of irregularity in the temporal responses.

Other neurons with nonlinear *I–V* characteristics could be modeled by the Arle and Kim model by modifying the nonlinear conductance, such as fusiform cells, which were modeled with an exponentially decreasing function of membrane voltage, or in a fashion other than exponential.

Hewitt and Meddis (1993), however, argued that the transient chopper pattern could be modeled without inhibitory input to the stellate cells. This is in contrast to Arle and Kim's (1991) model, where irregular step current input was required and hypothesized to arise from inhibitory inputs to the stellate cells. Other mechanisms in conjunction with inhibition might account for the transient response; for example, Smith and Rhode (1989) showed that some stellate cells exhibited irregular response patterns with high levels of sustained depolarization throughout the tone burst.

In the Hewitt and Meddis (1993) model, the spike trains generated by their AN model (Hewitt and Meddis 1991) was filtered by a first-order low-pass filter to simulate dendritic filtering before the stellate cell membrane model, which was a point neuron model of MacGregor (1987). A transient chopper response pattern could be generated from the model without the need for inhibitory inputs.

3.2.1.2 AVCN Bushy Cells

The model of Rothman et al. (1993) is a more realistic biophysical neuron model including physiologically existing ion channels of VCN bushy cells. Sufficient information about the ion channels present in bushy cells from whole-cell clamp studies (Manis and Marx 1991) allowed construction of a model of the postsynaptic events leading to action potential discharge in these cells.

This point neuron model consisted of a single compartment and included three voltage-sensitive ion channels: fast sodium, delayed-rectifier-like potassium, and low-threshold potassium. The model also contained a leakage channel, membrane capacitance, and synaptic inputs. The purpose of the model was to investigate the convergence of the AN fibers onto bushy cells. The number and synaptic strength of excitatory AN inputs were varied to investigate the relationship between input convergence parameters and response characteristics. The input AN spike trains were generated as a renewal point process that accurately modeled characteristics of AN fibers. Adaptation characteristics of both high and low spontaneous-rate AN fibers were simulated. Each spike produced brief (~0.5 ms) conductance increases.

This model accurately reproduced the membrane rectification seen in current clamp studies of bushy cells, as well as their current clamp responses. With AN inputs typical of high-SR AN fibers, the model reproduced primarylike (PL), primarylike-with-notch (Pri-N), and onset-L (On-L) responses of bushy cells. The regularity and phase-locking properties of bushy cells could also be reproduced, but only with constraints on the number and size of synaptic inputs. PL responses, however, could be reproduced only when the AN inputs have the adaptation characteristics of low-SR AN fibers, which suggested that neurons showing Pri-N and On-L responses required exclusively high-SR AN inputs. The model accounted well for the characteristics of spherical bushy cells, which receive a small number of large AN inputs, but failed to account for the globular bushy cells, which receive a large number of AN inputs and produce Pri-N or On-L responses. An inhibitory input, applied as a constant hyperpolarizing conductance, enabled the model to generate both Pri-N and On-L responses, which suggested that the inhibitory inputs known to exist on VCN bushy cells serve a role in regulating the cell's responses to the large synaptic inputs that they receive. The effects of inhibitory inputs to globular bushy cells, however, have not been evident in most single-unit studies (Rothman et al. 1993).

3.2.1.3 VCN On Cells

Kalluri and Delgutte (2003a, b) modeled neurons with On discharge patterns, which are recorded from all three major cell types in the VCN: stellate, bushy, and octopus cells. A deterministic, integrate-to-threshold, point-neuron model with a fixed refractory period was simulated to produce the On discharge patterns for a tonal stimulus with the auditory peripheral model originally from Carney (1993). The membrane time constant and the number and strength of the excitatory AN inputs were systematically varied to produce the characteristics of On cells. To accommodate both the entrainment of the On cells to low-frequency tones and the onset responses to high-frequency tones, the time constant should be short (~0.125 ms) and the On cell should receive a large number of weak AN fiber synapses. These two characteristics of On cells were contrasted when considering the interspike intervals of the cell responses. Entrainment to tones up to 1000 Hz required neurons to have short interspike intervals of 1 ms., while to produce onset response PSTHs, a neuron must prevent short interspike intervals during the steady-state of high-frequency tone bursts.

The model can produce On-C neuron response properties with the entrainment to low-frequency tones when fixed absolute refractory period was used in the spike generator. To obtain On-L and ideal onset (On-I) response properties, an input-dependent dynamic spike-blocking mechanism was introduced in the spike generator.

When the membrane voltage was over the threshold, a spike was generated and the dynamic spike-blocking mechanism prevented the model from reentering the integrate-and-fire routine until the membrane voltage fell below a transition voltage. This transition voltage had a value tightly constrained between the resting voltage and the threshold of discharge for both On-L and On-I response, however, with slightly different values for these two types.

3.2.1.4 PVCN Octopus Cells

Many aspects of octopus cells have been studied using a compartment model. The model of Levy and Kipke (1997) consists of an axon, a soma, and four dendrites. The model's parameters were obtained from many morphological and physiological experiments (Osen 1969; Kane 1973, 1977; Rhode et al. 1983; Oertel et al. 1990; Ostapoff et al. 1994). Octopus cells *in vivo* respond to high level tones and clicks with precise timing; their PSTHs are often onset responses. The octopus cell model was innervated by an AN model to study the octopus cell model's onset response to a tone burst (Levy and Kipke 1997). The model can generate both On-L and On-I units depending on the number of auditory-nerve inputs (60 and 120, respectively). Later sensitivity analysis on the parameters, however, suggests that the onset response to tone bursts depends on many cell parameters, rather than a single parameter (Kipke and Levy 1997). The impact of ion channels has also been studied extensively using a similar compartment model (Cai et al. 1997, 2000). Model K^+, Na^+, low-threshold K^+ (K_{LT}), and I_h channels have been imbedded into the previous model. The results show that K_{LT} (the low-threshold K^+ channel) is the most important channel for generating onset-like response. This result is supported by the importance of low membrane resistance from the physiological experiments (Golding et al. 1995, 1999; Ferragamo and Oertel 1998).

3.2.2 DCN Neural Circuitry

The neural circuitry of the DCN is the most complicated of the cochlear nucleus (Fig. 3.1). Its neurons are sensitive to the presence of anesthesia, and its cortical arrangement in most mammals suggests a structure suitable for integration of multiple inputs, widely believed to be from both auditory and nonauditory sources. Several groups have described computational models for individual fusiform cells, DCN projection neurons (Arle and Kim 1991; Hewitt and Meddis 1995; Kanold et al. 1999), and of the DCN neuronal circuitry (see DCN column of Table 3.1). These computational models involving fusiform cells are primarily models of the fusiform cell layer and the components of the neural circuitry basal to that layer, including the AN input. Much of the apical DCN circuitry has still not yet been addressed computationally.

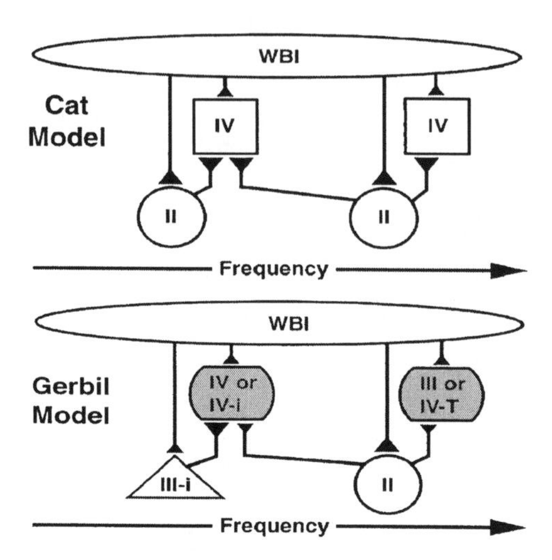

Fig. 3.1 Partial models of DCN circuitry in cat and gerbil showing that both type IV units in cat and type III units in gerbil receive input from wideband inhibitor interneurons (WBI) and type II units. All connections shown are inhibitory and the size of the connection symbol reflects the strength of connection. Cat type IV units receive strong input from type II units, whereas gerbil type III units receive weak input from type II units. The horizontal axis represents tonotopic excitatory input from the auditory nerve connections not shown, with the width of the cell symbol reflecting the frequency range of its auditory nerve input. (Modified from Davis and Voigt 1997.)

As an example of a single fusiform cell model consider Hewitt and Meddis (1995). They presented a fusiform cell model, modified from Hodgkin-Huxley (1952), to model the findings of Manis (1990). In Manis's study, chopper, pauser, and build-up PSTHs were obtained from DCN fusiform cells by manipulating the magnitudes of hyper- and depolarizing current pulse pairs that were injected into the fusiform cells in *in vitro* preparations. Hewitt and Meddis utilized a transient potassium channel, in addition to the sodium and delayed potassium channels in the standard Hodgkin–Huxley model, and generated results similar to Manis's intracellular recordings from fusiform cells. These results reinforce the conclusion that pre-hyperpolarizing input pulses could modulate a neuron's intrinsic membrane conductance and that the transient potassium conductance played an important role in these response profiles.

In the following section, one anatomically inspired model that can be used to understand the subtle differences in DCN neural response properties of fusiform cells, including RMs, from different species (cat and gerbil) is explored in some detail.

3.2.2.1 The Conceptual DCN Model

The conceptual model of the DCN neural circuitry can be traced back to the anatomical work of Lorente de Nó (1981), Kane (1974), Osen (1969, 1970), and Osen and Mugnaini (1981) and the physiological work of Young and his colleagues

(Young and Brownell 1976; Young 1980; Voigt and Young 1980, 1985, 1988; Young and Voigt 1982; Spirou and Young 1991; Nelken and Young 1994). In parallel to the physiological studies have been attempts to create anatomically and physiologically based computational models of the DCN's neural circuitry (see Section 3.2.2.2). Newer data from gerbil DCN studies (Davis et al. 1996; Davis and Voigt 1997), however, suggest that there may be species-specific differences in the response properties of the DCN's principal cells and this has caused a challenge to the conceptual model of DCN circuitry. Results of a DCN computational model that can account for data from both cat and gerbil without modification to the model's underlying neural architecture are shown in the next sections; only slight differences in synaptic efficacies are needed to account for the apparent inter- and intraspecies differences in response properties of DCN principal cells.

3.2.2.2 A Specific Computational DCN Model

The computational model of the DCN consists of five cell groups arranged into 800 isofrequency slices spaced 0.005 octaves apart and centered at 5 kHz. Figure 3.2a shows the connections among the model cells within a single frequency slice. AN-fibers represent auditory nerve fibers. I2-cells represent inhibitory interneurons with type II unit RMs. W-cells represent wideband inhibitors, which receive a wide range of AN-fiber inputs. P-cells represent principal cells. Briefly stated, AN-fibers excite the I2-cells, W-cells, and P-cells. W-cells inhibit P-cells and I2-cells. I2-cells

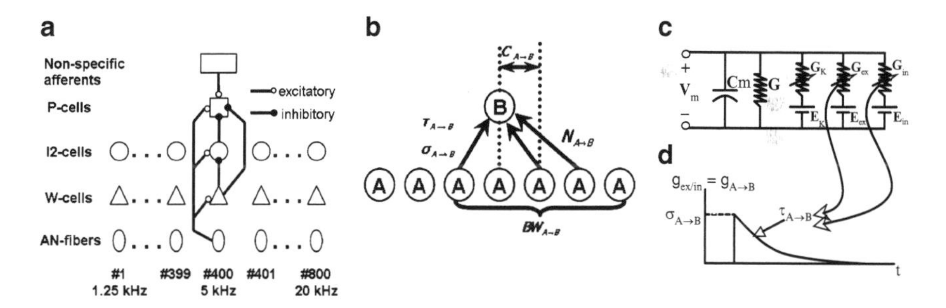

Fig. 3.2 (**a**) A patch of the DCN model. Five cell populations are organized in a tonotopic manner with frequency step equal to 0.005 octaves. The excitatory and inhibitory connections are shown in a single frequency slice. P cells represent type III units in gerbil and type IV units in cat; I2 cells represent type II units; W cells provide wide band inhibition; AN fibers, the auditory nerve fiber input. (Modified from Hancock and Voigt 1999.) (**b**) Group connection parameters. A cells are source cells projecting to B target cells. Parameters $\sigma_{A\to B}$ and $\tau_{A\to B}$ are the step conductance increase and time constant in response to the input spikes, as shown in **d**. The B cell receives *N* inputs from randomly chosen A cells of bandwidth *BW* whose center *BF* is offset from the B cell's *BF* by parameter *C*. (**c**) V_m is the membrane potential relative to rest; C_m is membrane capacitance; *G* is resting conductance; G_k and E_k are the variable conductance and reversal potential for potassium; $G_{ex/in}$ and $E_{ex/in}$ are excitatory/inhibitory synaptic conductance's and reversal potentials. (**d**) Expression of connection synapse conductance. Target cell will undergo a step increase and exponential decay in response to input spikes. (Modified from Zheng and Voigt 2006a.)

Table 3.2 Intrinsic parameters

Population	τ_m (ms)	θ *(mV)*	b_K	τ_K (ms)	E_K (mV)	E_{ex} (mV)	E_{in} (mV)
W-cells	5.0	4.0	2.00	1.0	–10	+70	–10
I2-cells	6.0	14.5	1.75	1.0	–10	+70	–10
P-cells	10.0	7.5	2.00	1.0	–10	+70	–10

Table 3.3 Connection parameters

Connection $A \rightarrow B$	$C_{A\rightarrow B}$ (octaves)	$BW_{A\rightarrow B}$ (octaves)	$N_{A\rightarrow B}$ (no. of synapses)	$\sigma_{A\rightarrow B}$ (normalized)	$\tau_{A\rightarrow B}$ (ms)
$AN \rightarrow W$	0.0	1.5/2.0/3.0	140	0.03/0.06/0.09	10
$AN \rightarrow I2$	0.0	0.2/0.4/0.6	48	0.30/0.55/0.80	10
$AN \rightarrow P$	0.0	0.2/0.4/0.6	48	0.1:1.0	10
$W \rightarrow I2$	0.0	0.1	15	1.40	10
$W \rightarrow P$	0.0	0.1	15	0.3/0.5/0.7	10
$I2 \rightarrow P$	0.0	0.3/0.6/0.9	21	0.05:2.75	1
$NSA \rightarrow P$	n/a	n/a	15	0.10/0.15/0.20	3

inhibit P-cells. P-cells are also excited by nonspecific afferents, which provide spontaneous activity (SpAc). The specific intrinsic parameters for these model neurons are shown in Table 3.2. The relationships among the model cells can be discussed in terms of the connection parameters as shown in Fig. 3.2b. For a given population of model cells (e.g., P-cells), the target cell "B" receives N inputs from the source cells "A," which are chosen from a band of "A" source cells with bandwidth BW, offset from the best frequency (BF) of B by C. The subscript A→B indicates the source unit, A, to the target unit, B. Table 3.3 shows the connection parameters of the DCN model.

3.2.2.3 The MacGregor Neuromime

MacGregor (1987) proposed several computational models for modeling the behavior of neurons. These are called neuromimes. The widely used single-point neuron model is a simplified approximation of a neural membrane patch. It is computationally efficient and can generate spike trains that are dependent on either direct input current or input spikes. The resulting spike trains are dependent on ascribed membrane properties. His two-point neuron model is capable of simulating the complex spiking properties that are seen, for example, in DCN cartwheel cells. More accurate single neuron models can also be found that consider the influence of dendritic trees and the propagation of action potentials. He also modeled individual neuronal assemblies as junctions and pools, networks of interconnected neuronal populations, and included the possible occurrence of Hebbian learning at prescribed junctions, based on the point neuron model and the neuron models with dendritic trees.

The equations for MacGregor's single-point neuron model are listed below:

$$\frac{dE}{dt} = \frac{-E + \{SCN + GK \times (EK - E)\}}{TMEM} \tag{3.1}$$

$$\frac{dTH}{dt} = \frac{-(TH - TH0) + c \times E}{TTH} \tag{3.2}$$

$$S = \begin{cases} 0(E < TH) \\ 1(E \geq TH) \end{cases} \tag{3.3}$$

$$\frac{dGK}{dt} = \frac{-GK + B \times S}{TGK} \tag{3.4}$$

$$PS = ES + S \times (50 - ES) \tag{3.5}$$

Figure 3.3 shows an example of how the point neuron reacts to the input stimuli. Equations (3.1) to (3.4) define the four state variables, *E*, *TH*, *S*, and *GK*, and the input current *SCN*. Equation (3.5) is the simulated potential of spikes in soma.

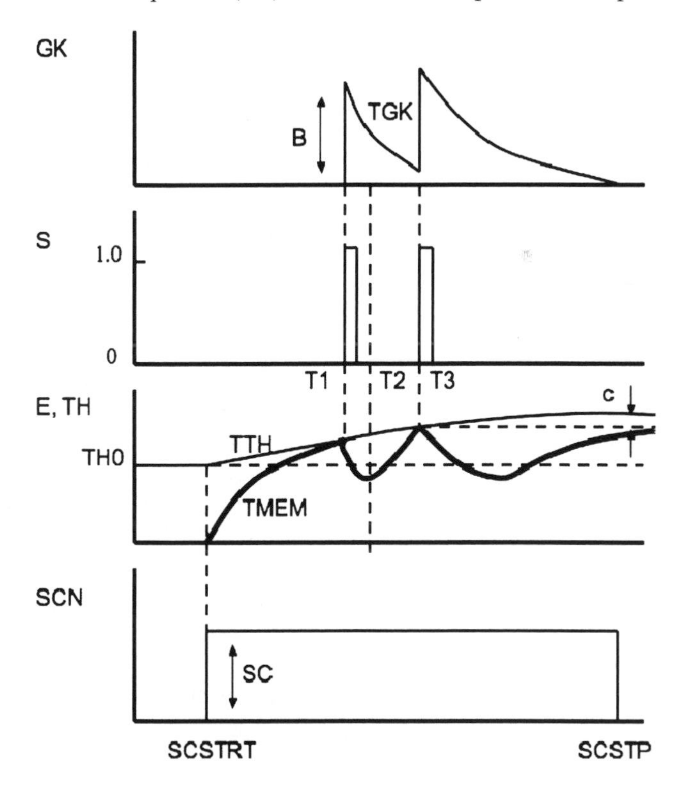

Fig. 3.3 Essential behavior of state variables in MacGregor's point neuron model PTNRN10. (Modified from Fig. 14.2 of MacGregor 1987.)

Here, the input current *SCN* is a step current with amplitude *SC* and excites the model neuron from time SCSTRT to SCSTP. The transmembrane potential *E* increases exponentially with positive *SCN* input, as shown in Eq. (3.1). The neuron threshold *TH* rises in proportion to the threshold sensitivity parameter *c*, whose value is from 0 to 1, as shown in Eq. (3.2). The speed of increase of *E* and *TH* depends on their time constants, *TMEM* and *TTH*, respectively. Obviously, the transmembrane potential *E* increases faster than the neuron's threshold *TH* and at time point *T*1, *E* reaches *TH* and the neuron fires, which is recorded by the spiking variable *S*, which changes value from 0 to 1, as shown in Eq. (3.3). The firing elevates the potassium conductance *GK* from 0 to a positive value proportional to *B*, and *GK* then declines exponentially with time constant *TGK*, as shown in Eq. (3.4). The influence of opening potassium channels on the membrane potential is also defined in Eq. (3.1), where *EK* denotes the potassium equilibrium potential with a constant value –10 mV. Thus, *GK* enlarges the negative term in the right side of differential Eq. (3.1), which means that the increase of *GK* draws the transmembrane potential *E* back toward its equilibrium potential. As the value of *GK* decreases, the excitatory effect of input current *SCN* compensates for the effect of potassium at time *T*2 and *E* starts increasing again and may reach *TH* again at time *T*3, which would incur another spike at time *T*3.

With *B* and *TGK* at relatively low levels, the potassium conductance *GK* has a smaller jump and decreases faster, resulting in the transmembrane potential *E* recovering sooner and therefore the neuron fires at a relatively higher rate. On the other hand, with *B* and *TGK* at relatively high levels, the neuron fires at a lower rate.

The model neuron is capable of tonic (steady) firing and phasic (transient) firing with different parameters *c*, *TTH* and *SC*, as shown in the *SC* vs. *c* plane in Fig. 3.4.

Without the potassium channel (i.e., $GK=0$) and when *SCN* is a constant value *SC*, the result for Eq. (3.1) is

$$E(t) = SC \times (1 - e^{-\frac{t}{TMEM}})$$

This means that the transmembrane potential *E*(*t*) will increase exponentially with time constant *TMEM* until it reaches *SC* as time goes to infinity.

If *E*(*t*) is fixed at *E*, we obtain the result for Eq. (3.2) as:

$$TH(t) = TH0 + c \times E \times (1 - e^{-\frac{t}{TTH}})$$

Thus, *TH*(*t*) increases from *TH*0 to $TH0 + c * SC$ as time goes to infinity.

So, if the upper limit of *TH*(*t*) is larger than the upper limit of *E*(*t*), even if the neuron fires [the neuron may not fire at all if the *TH*(t) is always larger than *E*(*t*)], it will stop firing at some point (phasic firing). In contrast, if the upper limit of *TH*(*t*) is smaller than or equal to the upper limit of *E*(*t*), the neuron may fire from some point and then keep firing (tonic firing). The boundary between these two behaviors for this neuron is $TH0 + c * SC = SC$ in the *SC* vs. *c* plane, which denotes the curve with *TTH* beside it in Fig. 3.4. Below the curve, the neuron gives phasic responses or does not fire at all; above the curve, the neuron gives tonic responses.

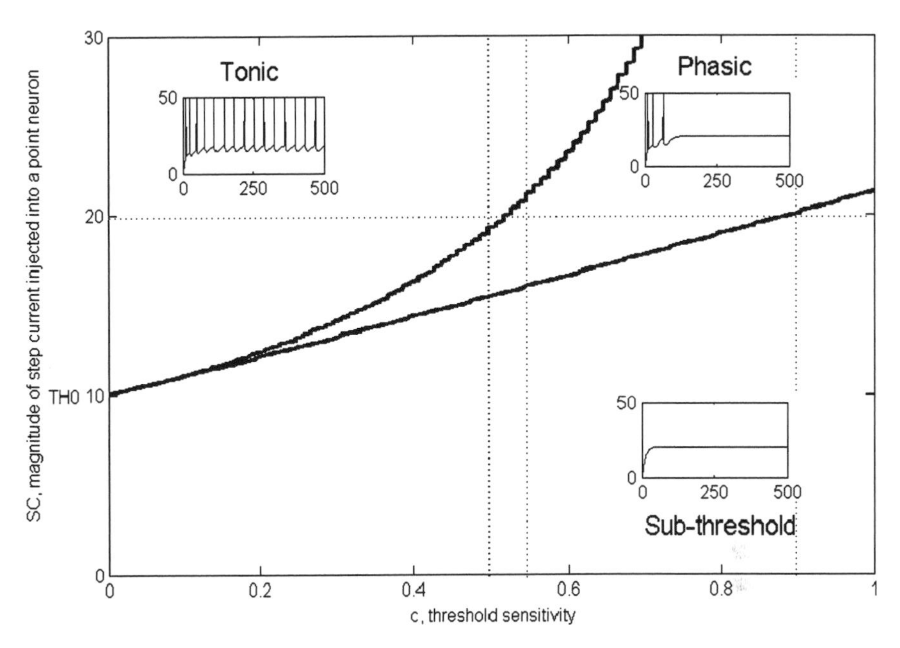

Fig. 3.4 Parameter space for MacGregor point neuron PTNRN10 showing tonic firing, phasic firing, and subthreshold areas in the *SC* vs. *c* plane. The neuron was simulated from time 0 to 500 ms, with current step input of amplitude *SC* from 0 to 500 ms. *SC* is from 0 to 30 in 0.1 steps and *C* is from 0 to 1 in 0.001 steps. *TTH* = 20 ms, *B*=5, *TGK* = 10 ms, THO = 10 mV, *TMEM* = 11 ms. Note that we chose the upper limit of *TGK* and *TMEM* that MacGregor provided in Appendix 1 of his book. Larger values of *TGK* and *TMEM* correspond to slower firing of the neurons. The three neurons shown have the same input *SC* value at 20, and different c values of 0.5, 0.55, and 0.9 corresponding to tonic firing, phasic firing, and subthreshold states, respectively.

From Eq. (3.2) and Fig. 3.3, we can derive that as *TTH* increases, this curve moves to the right with the same initial point (*TH*0, 0) and vice versa.

In cases where $SC < \dfrac{TH0}{1-c}$, which means below the curve, the neuron may not fire if *SC* is too low or *c* is too high. To obtain the boundary between phasic firing and subthreshold behavior analytically is a tedious job. Given the MacGregor model's fast-simulating property, we easily parse the parameter space shown in Fig. 3.4 by simulating the neuron while systematically varying parameters *SC* and *c*.

In these simulations, a step current *SC* was input during the simulation trial, which was 500 ms. With all the other parameters fixed, we varied *SC* from 0 to 30 in 0.1 steps and *c* from 0 to 1 in 0.01 steps. Thus, we obtained 30,000 different spike train outputs from the model neuron. If no firing occurred during the 500-ms trial, we define the neuron status as "subthreshold"; if no firing in the last 100 ms of the simulation, we define it as "phasic firing"; otherwise, we define it as "tonic firing." The contours of these three states are plotted in Fig. 3.4. Three spike trains, *PS*, are shown as examples of subthreshold, phasic, and tonic firing states. The three crossing points of the dotted lines in the figure denotes the specific *SC* and *c* values that we used to simulate these three spike train subplots. The parameters were chosen in a manner that is close to the boundary, which shows that the parameters SC and c are very sensitive near the boundary lines. Three spikes can be seen in the phasic firing plot, while no spikes at all are seen in the subthreshold plot. The potentials of the cell approach SC, here 20, in both phasic firing and subthreshold neurons.

The electrical circuit shown in Fig. 3.2c is based on the MacGregor single-point neuromime (MacGregor 1987). It is a parallel circuit containing membrane capacitance, leakage conductance, a potassium channel branch, and excitatory/inhibitory synaptic connection branches. Each excitatory or inhibitory input adds a branch to the point neuron, with the variable conductance controlled by the parameters conductance step, σ, and time constant, τ.

Event time is recorded when the membrane potential exceeds its threshold, θ and the potassium conductance is activated to induce a refractory period. Computation time is reduced as details of action potential generation are ignored. The following equations describe the neuromime model:

$$\tau_m \frac{dV_m}{dt} = -V_m - g_K(V_m - E_K) - g_{ex}(V_m - E_{ex}) - g_{\text{in}}(V_m - E_{\text{in}}), \tag{3.6}$$

$$\tau_K \frac{dg_K}{dt} = -g_K + b_K S, \tag{3.7}$$

$$S = \begin{Bmatrix} 0, V_m < \theta \\ 1, V_m \geq \theta \end{Bmatrix} \tag{3.8}$$

where $g_k = G_k/G$, $g_{ex} = G_{ex}/G$ and $g_{in} = G_{in}/G$. For target cell B, the variable conductance that represents synapses from source cells A is described by:

$$\tau_{A\to B} \frac{dg_{A\to B}}{dt} = -g_{A\to B} + \sigma_{A\to B} \times \sum_{i=1}^{N_{A\to B}} S_A \tag{3.9}$$

where S_A are the input spikes.

In this model, four intrinsic (τ_m, θ, b_K, and τ_K) and five connection parameters ($C_{A\to B}$, $BW_{A\to B}$, $N_{A\to B}$, $\sigma_{A\to B}$, and $\tau_{A\to B}$) are used. The intrinsic parameter set describes the membrane properties of the cell, which determines the steady firing rate and the threshold sound pressure level of a single cell. The connection parameter set describes the connections between two groups of cells; this is used to specify the connectivity within the neuronal circuitry of the model. The intrinsic values in the former work of Hancock and Voigt (1999) are used, as these have been found to provide satisfactory fits to many data sets.

MacGregor (1993) pointed out that the steady-state effect of one population on its target is proportional only to the product $N_{A\to B}\,\sigma_{A\to B}\,\tau_{A\to B}$, which allows one to select only one of these parameters to be a free parameter; here $\sigma_{A\to B}$, the synaptic strength from cell A to B was chosen. Thus, only three connection parameters, $\sigma_{A\to B}$, $BW_{A\to B}$ and $C_{A\to B}$ were considered in previous simulations (Zheng and Voigt 2006a).

3.2.2.4 The Auditory Nerve Model

AN fibers are based on the auditory nerve model described by Carney (1993). The input to an AN-fiber is a sound pressure signal and the output is a spike train. Each fiber consists of a gammatone filter that provides the frequency selectivity, a nonlinear process that generates an inner hair cell potential and a compartmental model of neurotransmitter release. This in turn creates an instantaneous firing probability. AN-fiber thresholds and SR were randomly assigned from physiological distributions of these parameters (Hancock et al. 1997). To save time, all stimuli are processed by the auditory nerve filter bank once and the AN-fibers' spike times saved for use in subsequent simulations of the DCN model.

3.2.2.5 Model Results

This DCN model has been successful in describing physiological data of various kinds, including RMs, cross-correlograms from simultaneously recorded pairs of units consisting of either a type II unit and a type IV unit or from pairs of type IV units, from both cat (Davis and Voigt 1994, 1996; Hancock et al. 1997; Hancock and Voigt 1999) and gerbil (Zheng and Voigt 2006a, b). The model can fit the notch noise data from both cat (Spirou and Young 1991; Hancock and Voigt 1999) and gerbil (Parsons et al. 2001; Zheng and Voigt 2006a). Figure 3.5a, from Zheng and Voigt (2006a), shows how the DCN model can fit the rate vs. cutoff frequency curve derived from a gerbil DCN type III unit's responses to notch noise stimuli. The model is very robust even though it lacks certain known components. The model does not include at this time cartwheel cells or granule cells with their parallel fibers. It will be important to include these components in the model because they are important for modeling sound localization in the median plane.

One important result using this model is a better understanding of the ways a neuron's RM is created. Simply by changing some synaptic strengths, a P-cell

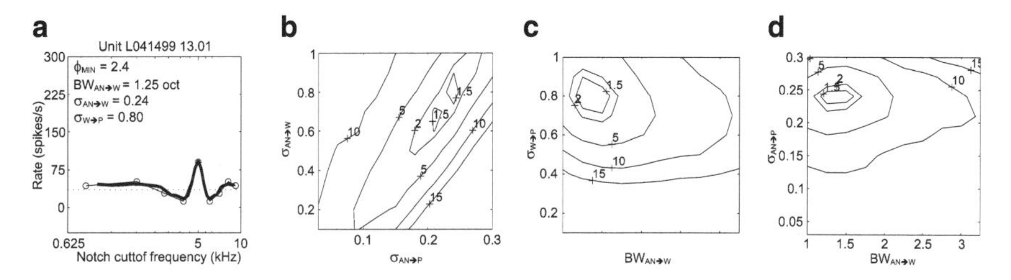

Fig. 3.5 Model fit to the notch noise response of unit L041499 13.01 and parameter sensitivity analysis. **(a)** Best fit of the model to physiological data. The circles represent the physiological data and the solid line represents the best fit of the model. The dashed line represents the average spontaneous rate of the physiological data. The minimum error Φ_{min} and the values of the three important parameters of the best fit are shown in the upper left. **(b–d)** Contours of equal Φ (1.5, 2, 5, 10, and 15, from inside to outside) showing the sensitivity of the fit to the parameters: $BW_{AN\rightarrow W}$, $\sigma_{AN\rightarrow P}$, and $\sigma_{W\rightarrow P}$. (Modified from Zheng and Voigt 2006a.)

could display type IV unit's RM or a type III unit's RM (Zheng and Voigt 2006b). In fact, P-cells can display type I/III, type III, type III-i, type IV, type IV-T, and type V unit RMs.

3.2.2.6 Simulation Protocol and Data Analysis for P-Cell Response Maps

All simulations were conducted on an IBM pSeries 655, which is a 48-processor system composed of six nodes. Each p655 node consists of 8 Power4 processors running at 1.1 GHz and sharing 16 GB of memory. There are three levels of cache on this machine. Each processor has a 32-kB L1 cache and then each pair of processors shares a 1.41-MB L2 cache and each p655 node shares a 128-MB L3 cache.

To create RMs, the model was stimulated by 200-ms tone bursts presented once per second with a 10-ms delay while sound pressure levels varied from 0 to 90 dB SPL in 2-dB steps and the frequency varied in 0.1-octave steps within a 1.5-octave band above and below 5 kHz. Thus there are 31 frequency slices in one RM simulation. The spikes of each trial's last 160 ms were averaged over time to compute the SR and the spikes of the last 160 ms of each tone burst were averaged to compute the driven rate.

Each RM is displayed as a two-dimensional (2D) matrix of frequency-level combinations with blue, red, and gray representing excitatory, inhibitory and SpAc responses respectively. The raw SR and driven rates were smoothed by a 3×3 low-pass spatial averaging filter with FIR coefficients 1/9. The SR of all 46 sound levels and 31 slices trials were used to calculate the average SR and its standard deviation for one RM. Since the P-cells generally have high SR (>2.5 spikes/s), a symmetric statistical criterion (two standard deviations above and below SR) was applied to the driven data to obtain excitatory and inhibitory thresholds. Excitatory regions were assigned the value 1 and inhibitory regions assigned the value –1; SpAc regions were assigned the value 0. A median spatial filter was then applied to the resulting RM to reduce "salt and pepper" noise (Davis et al. 1995).

In the broad band noise (BBN) simulations, the sound level was varied from 0 to 90 dB SPL in 2-dB SPL steps and the noise bursts were presented for 200 ms in 1000-ms trials with 10-ms delay. As for tonal simulations, the spikes of the last 160 ms of each trial were averaged over time to compute the SR and the spikes of last 160 ms of each BBN burst were averaged to compute the driven rate. In spike rate vs. sound level plots, a 3-point triangle filter with FIR coefficients [¼ ½ ¼] was applied to smooth the rate data.

3.2.2.7 RMs of P-Cells Simulated with the Nominal Parameter Set

The model is capable of simulating 100 P-cells responses at a time, that is, 100 RMs or 100 rate-level curve plots, by varying two chosen parameters. In Fig. 3.6, the two parameters $\sigma_{AN \to P}$ (varying from 0.1 to 1.0 across columns) and $\sigma_{\mathrm{I2} \to P}$

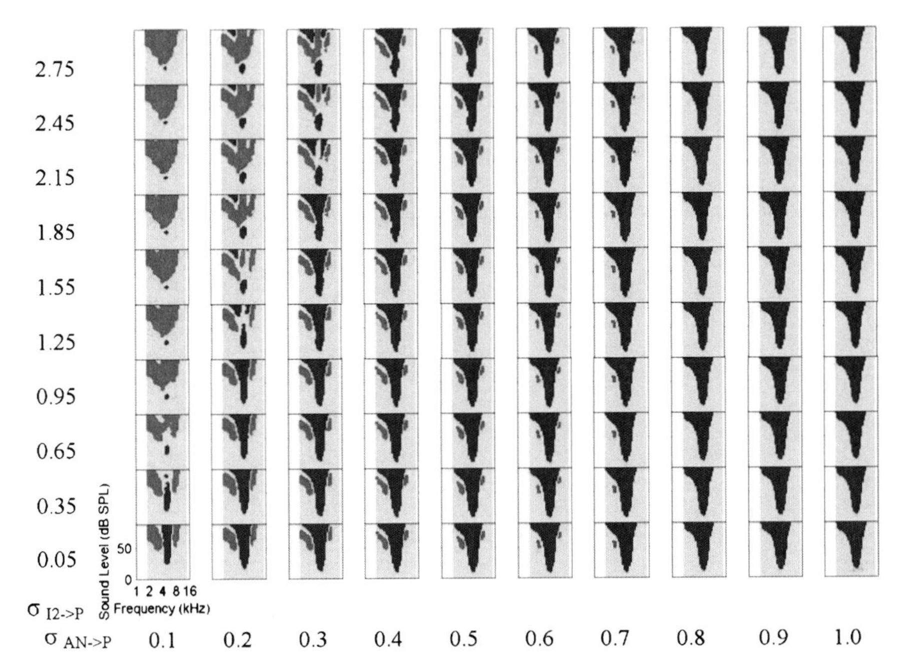

Fig. 3.6 RM matrix simulated using the nominal parameter value set: $BW_{AN\to W}$ = 2.0 oct., $BW_{AN\to I2}$ = 0.4 oct., $BW_{AN\to P}$=0.4 oct., $BW_{I2\to P}$=0.6 oct., $\sigma_{AN\to W}$=0.06, $\sigma_{AN\to I2}$ = 0.55, $\sigma_{W\to P}$=0.5, $\sigma_{NSA\to P}$=0.15. $\sigma_{AN\to P}$ increases systematically from 0.1 to 1.0 in 0.1 steps across the columns from left to right. $\sigma_{I2\to P}$ increases systematically from 0.05 to 2.75 in 0.3 steps across the rows from bottom to top. Within each RM, there are 31 frequency slices centered at 5 kHz with 1.5 octaves below and 1.5 octaves above. Each RM covers a range of 1.77 kHz to 14.14 kHz along the abscissa in 0.1 octave step, and the sound pressure levels range from 0 dB SPL to 90 dB SPL in 2 dB steps along the ordinate. The excitatory response regions are shown in blue, the inhibitory response regions are shown in red and the SpAc regions are shown in gray. In the matrix, type I, type III, type IV and type IV-T units are shown. (From Zheng and Voigt 2006b.)

(varying from 0.05 to 2.75 across rows) were chosen to vary systematically in 10 steps. Other parameters in this simulation were assigned as in Table 3.3. In Fig. 3.6, three columns of type I unit RMs are on the right side of the 10 by 10 matrix. Four columns in the middle show type III unit RMs with increasing sideband inhibition with decreasing $\sigma_{AN\to P}$. For the left-most three columns, type III, type IV, and type IV-T unit RMs are observed. As $\sigma_{AN\to P}$ increases, the excitatory region of each RM grows while the inhibitory region(s) decreases. In contrast, as $\sigma_{I2\to P}$ increases, the excitatory region of each RM decreases while the inhibitory region(s) grows. The mixed area of the first three columns shows a subtle change of RM types with changes in $\sigma_{AN\to P}$ and $\sigma_{I2\to P}$ parameter values. When $\sigma_{AN\to P}$ decreased to 0.1 and 0.2, the threshold to BF tones increased. Shown in Fig. 3.7 are four specific units with RM types type III-i, type III, type IV, and type IV-T respectively.

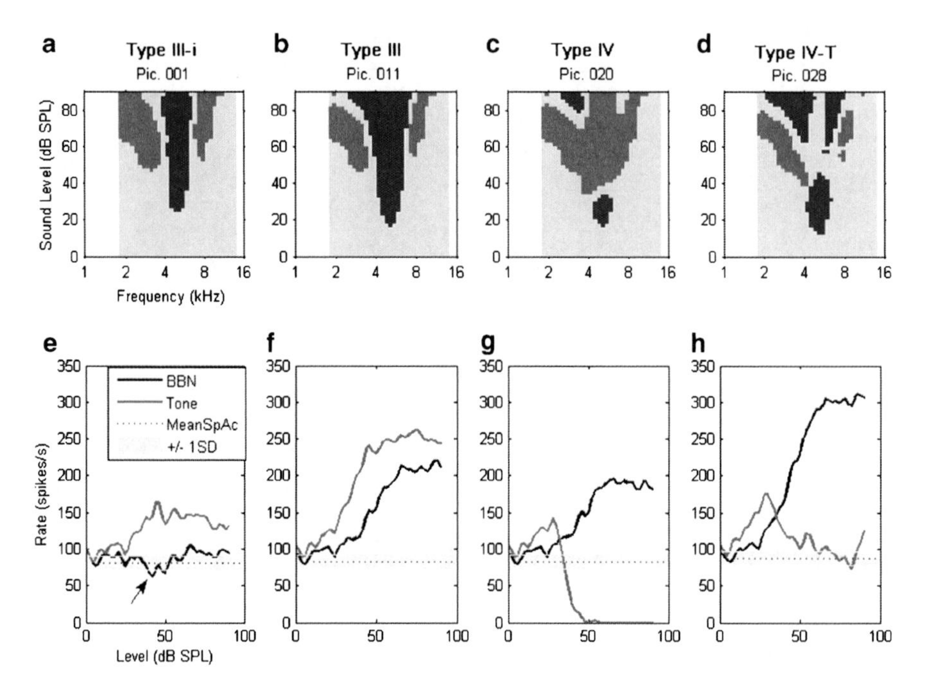

Fig. 3.7 Detailed RMs and discharge rate vs. sound level plots for four units from Fig. 3.6 are shown. **(a–d)** RMs of units with Pic. 001 ($\sigma_{AN \to P}=0.1$, $\sigma_{I2 \to P}=0.05$, as in Fig. 3.6), 011 ($\sigma_{AN \to P}=0.2$, $\sigma_{I2 \to P}=0.05$), 020 ($\sigma_{AN \to P}=0.2$, $\sigma_{I2 \to P}=2.75$) and 028 ($\sigma_{AN \to P}=0.3$, $\sigma_{I2 \to P}=2.15$) respectively. **(e–h)** Rate vs. sound level plots of units with Pic. nos. 001, 011, 020, and 028, respectively. **a** and **e** show a type III-i unit's responses. It has similar tonal responses as type III units but is inhibited by median level BBN (as indicated by arrow in **e**). **b** and **f** show a type III unit's responses. It has V-shaped excitatory RM with sideband inhibitory region and both tonal and BBN responses show excitation. **c** and **g** show a type IV unit's responses. It shows a RM with an inhibitory region above an excitatory island at *BF*. The responses to BBN are excitatory. **d** and **h** show type IV-T unit's responses. It has a similar nonmonotonic response to *BF* tones, like type IV units, but it does not show inhibition. (Modified from Zheng and Voigt 2006b.)

3.2.2.8 Implications and Speculations

DCN principal cells show a variety of RMs in a single species, and these RMs may appear in different species in different proportions (i.e., more type IV units in cat DCN and more type III units in gerbil DCN). The fundamental result of the modeling study is that a single DCN circuit can account for these variations both within and across species. An important implication of this is that a portion of the neural circuitry of the central auditory system is invariant across mammalian species.

By varying the connection parameters, different unit types that are seen in physiological experiments (e.g., type I, III-i, III, IV, IV-T, and V units) emerge. Type IV units are associated with fusiform (pyramidal) cells in the cat DCN (Young 1980), while type IV-i units have been recorded in identified giant cells in the gerbil DCN (Ding et al. 1999). Gerbil DCN fusiform cells have been recorded and marked with

HRP or neurobiotin and have primarily type III units response properties, although a few were found with type I/III, type III-i or type IV-T unit RMs (Ding et al. 1999). Davis et al. (1996) reported recording from 133 gerbil DCN cells. The dominant unit types in gerbil DCN are different than those observed in cat DCN, with more type III units (62.4%) than type IV units (11.3%) and two new unit subtypes (type IV-i, ~50% of the type IV population; type III-i, ~30% of the type III population). Type III-i units are similar to type III units except that type III-i units are inhibited by low levels of noise and excited by high levels of noise, whereas type III units have strictly excitatory responses to noise. Type IV-i units are similar to type IV units except that type IV-i units are excited by low levels of noise and become inhibited by high levels of noise, whereas type IV units have strictly excitatory responses to noise. The type IV-i unit has a nonmonotonic BBN response feature that needs type III-i units as inhibitory source rather than the I2-cell interneuron inhibitor used in our model. However, type III-i units emerge when the values of $\sigma_{AN\rightarrow P}$ and $\sigma_{I2\rightarrow P}$ are low.

With a single neural architecture, both cat and gerbil DCN P-cell responses are simulated by simply varying different connection parameters, specifically the synapse strength and the bandwidth of source cells to target cells. The anatomical neuron types in cat and gerbil DCN have similar morphologies: fusiform (pyramidal) cells, giant cells, cartwheel cells, granule cells, and so forth. Physiologically, however, many RMs are observed in fusiform (pyramidal) cells. Assuming that these different RMs are important to normal DCN function, it appears that rather than have a specific genetic code for each RM type, the developing nervous system could have a single set of genes that simply create fusiform cells. Random connections from source cells to their target fusiform cells are then responsible for the diversity of RM types observed.

Generally, a unit's RM is thought to be a relatively stable unit property. This chapter demonstrates that simple changes to some of the connection parameters, however, can cause large changes in a unit's RM. For example, if the synaptic strength of the AN to I2-unit increases, a unit's RM can change from a type III to a type IV very easily. Likewise, if W-cell synapses to P-cells are strengthened, the same transition is possible. There are several mechanisms known that results in increasing or decreasing synaptic efficacy in the central nervous system (see Zucker and Regehr 2002). Oertel and Young (2004) summarize the DCN evidence suggesting that synapses in the superficial DCN are adjustable. Parallel fiber synapses to both cartwheel cells and fusiform cells show both long-term potentiation (LTP) and long-term depression (LTD; Fujino and Oertel 2003). Tzounopoulos (2006) has demonstrated spike timing-dependent synaptic plasticity in these cells as well. As this part of the DCN circuit is not included in the current DCN computational model, the circuitry involving the parallel fiber system cannot be responsible for the dynamic RM type changes suggested in the preceding text. There is little evidence to date that any of the synapses in the DCN model are plastic. In fact, the auditory nerves to fusiform basal dendrite synapses do not show LTP or LTD (Fujino and Oertel 2003). Synaptic plasticity of the auditory nerve to tuberculoventral cell connections or wide-band inhibitor to fusiform cell connections is not known.

One possible mechanism for adjusting the synaptic strengths of DCN neurons may be the cannabinoid system and depolarization-induced suppression of inhibition (DSI; Straiker and Mackie 2006). Cannabinoid receptors are found in the DCN (Mailleux and Vanderhaeghen 1992). Cannabinoids are released by postsynaptic neurons that are sufficiently depolarized, diffuse to the presynaptic neurons where they inhibit neurotransmitter release (Straiker and Mackie 2006). If this mechanism were active in the DCN, it could result in a vastly more dynamic nucleus than is currently appreciated.

3.3 Summary

Computational modeling of neural systems, especially those that are anatomically inspired, are extremely useful tools for exploring the full implications of conceptual models. As shown above, the conceptual model of the DCN neural circuitry shows a very rich and expansive set of behaviors that were difficult to foresee prior to the simulation studies. In addition, the computational model, coupled with parameter estimation techniques and sensitivity analysis of those parameters, shows how robust such a model can be in accounting for the great variability observed in the physiological data. Finally, a computational model can help to explore alternative or competing neural architectures and suggest physiological experiments that can test and help distinguish among those alternatives.

Acknowledgments The authors would like to acknowledge the intellectual and programming contributions to the DCN computational model by Drs. K. Davis, K. Hancock and T. McMullen, and the financial support over the years by NIH and Boston University's Hearing Research Center and Biomedical Engineering department.

References

Arle JE, Kim DO (1991) Neural modeling of intrinsic and spike-discharge properties of cochlear nucleus neurons. Biol Cybern 64:273–283.

Bahmer A, Langner G (2006) Oscillating neurons in the cochlear nucleus: II. Simulation results. Biol Cybern 95:381–392.

Bahmer A, Langner G (2008) A simulation of chopper neurons in the cochlear nucleus with wide-band input from onset neurons. Biol Cybern doi: 10.1007/s00422-008-0276-3.

Banks MI, Sachs MB (1991) Regularity analysis in a compartmental model of chopper units in the anteroventral cochlear nucleus. J Neurophysiol 65:606–629.

Cai Y, Walsh EJ, McGee J (1997) Mechanisms of onset responses in octopus cells of the cochlear nucleus: implications of a model. J Neurophysiol 78:872–883.

Cai Y, McGee J, Walsh EJ (2000) Contributions of ion conductances to the onset responses of octopus cells in the ventral cochlear nucleus: simulation results. J Neurophysiol 83:301–314.

Carney LH (1993) A model for the responses of low-frequency auditory-nerve fibers in cat. J Acoust Soc Am 93:401–417.

Davis KA, Voigt HF (1994) Neural modeling of the dorsal cochlear nucleus: cross-correlation analysis of short-duration tone-burst responses. Biol Cybern 71:511–521.

Davis KA, Voigt HF (1996) Computer simulation of shared input among projection neurons in the dorsal cochlear nucleus. Biol Cybern 74:413–425.
Davis KA, Voigt HF (1997) Evidence of stimulus-dependent correlated activity in the dorsal cochlear nucleus of decerebrate gerbils. J Neurophysiol 78:229–247.
Davis KA, Gdowski GT, Voigt HF (1995) A statistically based method to generate response maps objectively. J Neurosci Methods 57:107–118.
Davis KA, Ding J, Benson TE, Voigt HF (1996) Response properties of units in the dorsal cochlear nucleus of unanesthetized decerebrate gerbil. J Neurophysiol 75:1411–1431.
Ding J, Benson TE, Voigt HF (1999) Acoustic and current-pulse responses of identified neurons in the dorsal cochlear nucleus of unanesthetized, decerebrate gerbils. J Neurophysiol 82:3434–3457.
Eager MA, Grayden DB, Burkitt A, Meffin H (2004) A neural circuit model of the ventral cochlear nucleus. In: Proceedings of the 10th Australian International Conference on Speech Science & Technology, pp. 539–544.
Ferragamo MJ, Oertel D (1998) Shaping of synaptic responses and action potentials in octopus cells. In: Proceedings of Assoc Res Otolaryngol Abstr 21:96.
Fex J (1962) Auditory activity in centrifugal and centripetal cochlear fibres in cat. A study of a feedback system. Acta Physiol Scand Suppl 189:1–68.
Fujino K, Oertel D (2003) Bidirectional synaptic plasticity in the cerebellum-like mammalian dorsal cochlear nucleus. Proc Natl Acad Sci USA 100:265–270.
Golding NL, Robertson D, Oertel D (1995) Recordings from slices indicate that octopus cells of the cochlear nucleus detect coincident firing of auditory nerve fibers with temporal precision. J Neurosci 15:3138–3153.
Golding NL, Ferragamo MJ, Oertel D (1999) Role of intrinsic conductances underlying responses to transients in octopus cells of the cochlear nucleus. J Neurosci 19:2897–2905.
Hancock KE, Voigt HF (1999) Wideband inhibition of dorsal cochlear nucleus type IV units in cat: a computational model. Ann Biomed Eng 27:73–87.
Hancock KE, Davis KA, Voigt HF (1997) Modeling inhibition of type II units in the dorsal cochlear nucleus. Biol Cybern 76:419–428.
Hawkins HL, McMullen TA, Popper AN, et al., eds (1996) Auditory Computation. New York: Springer Verlag.
Hemmert W, Holmberg M, Gerber M (2003) Coding of auditory information into nerve-action potentials. In: Fortschritte der Akustik, Oldenburg. Deutsche Gesellschaft für Akustik e.v, pp. 770–771.
Hewitt MJ, Meddis R (1991) An evaluation of eight computer models of mammalian inner hair-cell function. J Acoust Soc Am 90:904–917.
Hewitt MJ, Meddis R (1993) Regularity of cochlear nucleus stellate cells: a computational modeling study. J Acoust Soc Am 93:3390–3399.
Hewitt MJ, Meddis R (1995) A computer model of dorsal cochlear nucleus pyramidal cells: intrinsic membrane properties. J Acoust Soc Am 97:2405–2413.
Hewitt MJ, Meddis R, Shackleton TM (1992) A computer model of a cochlear-nucleus stellate cell: responses to amplitude-modulated and pure-tone stimuli. J Acoust Soc Am 91:2096–2109.
Hodgkin AL, Huxley AF (1952) A quantitative description of membrane current and its application to conduction and excitation in nerve. J Physiol 117:500–544.
Holmberg M, Hemmert W (2004) An auditory model for coding speech into nerve-action potentials. In: Proceedings of the Joint Congress CFA/DAGA, Strasbourg, France, March 18–20, 2003, pp. 773–774.
Kalluri S, Delgutte B (2003a) Mathematical models of cochlear nucleus onset neurons: II. model with dynamic spike-blocking state. J Comput Neurosci 14:91–110.
Kalluri S, Delgutte B (2003b) Mathematical models of cochlear nucleus onset neurons: I. Point neuron with many weak synaptic inputs. J Comput Neurosci 14:71–90.
Kane EC (1973) Octopus cells in the cochlear nucleus of the cat: heterotypic synapses upon homeotypic neurons. Int J Neurosci 5:251–279.

Kane EC (1974) Synaptic organization in the dorsal cochlear nucleus of the cat: a light and electron microscopic study. J Comp Neurol 155:301–329.

Kane ES (1977) Descending inputs to the octopus cell area of the cat cochlear nucleus: an electron microscopic study. J Comp Neurol 173:337–354.

Kanold PO, Manis PB (1999) Transient potassium currents regulate the discharge patterns of dorsal cochlear nucleus pyramidal cells. J Neurosci 19:2195–2208.

Kim DO, D'Angelo WR (2000) Computational model for the bushy cell of the cochlear nucleus. Neurocomputing 32–33:189–196.

Kipke DR, Levy KL (1997) Sensitivity of the cochlear nucleus octopus cell to synaptic and membrane properties: a modeling study. J Acoust Soc Am 102:403–412.

Levy KL, Kipke DR (1997) A computational model of the cochlear nucleus octopus cell. J Acoust Soc Am 102:391–402.

Liberman MC, Brown MC (1986) Physiology and anatomy of single olivocochlear neurons in the cat. Hear Res 24:17–36.

Lorente de Nó R (1981) The Primary Acoustic Nuclei. New York: Raven Press.

MacGregor RJ (1987) Neural and Brain Modeling. San Diego: Academic Press.

MacGregor RJ (1993) Theoretical Mechanics of Biological Neural Networks. San Diego: Academic Press.

Mailleux P, Vanderhaeghen JJ (1992) Distribution of neuronal cannabinoid receptor in the adult rat brain: a comparative receptor binding radioautography and in situ hybridization histochemistry. Neuroscience 48:655–668.

Manis PB (1990) Membrane properties and discharge characteristics of guinea pig dorsal cochlear nucleus neurons studied *in vitro*. J Neurosci 10:2338–2351.

Manis PB, Marx SO (1991) Outward currents in isolated ventral cochlear nucleus neurons. J Neurosci 11:2865–2880.

Nelken I, Young ED (1994) Two separate inhibitory mechanisms shape the responses of dorsal cochlear nucleus type IV units to narrowband and wideband stimuli. J Neurophysiol 71:2446–2462.

Oertel D, Young ED (2004) What's a cerebellar circuit doing in the auditory system? Trends Neurosci 27:104–110.

Oertel D, Wu SH, Garb MW, Dizack C (1990) Morphology and physiology of cells in slice preparations of the posteroventral cochlear nucleus of mice. J Comp Neurol 295:136–154.

Osen KK (1969) Cytoarchitecture of the cochlear nuclei in the cat. J Comp Neurol 136:453–484.

Osen KK (1970) Course and termination of the primary afferents in the cochlear nuclei of the cat. An experimental anatomical study. Arch Ital Biol 108:21–51.

Osen KK, Mugnaini E (1981) Neuronal circuits in the dorsal cochlear nucleus. In: Syka J, Aitkin L (ed), Neuronal Mechanisms in Hearing. New York: Plenum Press, pp. 119–125.

Ostapoff EM, Feng JJ, Morest DK (1994) A physiological and structural study of neuron types in the cochlear nucleus. II. Neuron types and their structural correlation with response properties. J Comp Neurol 346:19–42.

Parsons JE, Lim E, Voigt HF (2001) Type III units in the gerbil dorsal cochlear nucleus may be spectral notch detectors. Ann Biomed Eng 29:887–896.

Pathmanathan JS, Kim DO (2001) A computational model for the AVCN marginal shell with medial olivocochlear feedback: generation of a wide dynamic range. Neurocomputing 38–40:807–815.

Pont MJ, Damper RI (1991) A computational model of afferent neural activity from the cochlea to the dorsal acoustic stria. J Acoust Soc Am 89:1213–1228.

Popper AN, Fay RR (Eds.) (1992) The Mammalian Auditory Pathway: Neurophysiology. New York : Springer Verlag.

Reed MC, Blum JJ (1995) A computational model for signal processing by the dorsal cochlear nucleus. I. Responses to pure tones. J Acoust Soc Am 97:425–438.

Reed MC, Blum JJ (1997) Model calculations of the effects of wide-band inhibitors in the dorsal cochlear nucleus. J Acoust Soc Am 102:2238–2244.

Rhode WS, Oertel D, Smith PH (1983) Physiological response properties of cells labeled intracellularly with horseradish peroxidase in cat ventral cochlear nucleus. J Comp Neurol 213:448–463.

Rothman JS, Manis PB (2003) The roles potassium currents play in regulating the electrical activity of ventral cochlear nucleus neurons. J Neurophysiol 89:3097–3113.

Rothman JS, Young ED, Manis PB (1993) Convergence of auditory nerve fibers onto bushy cells in the ventral cochlear nucleus: implications of a computational model. J Neurophysiol 70:2562–2583.

Smith PH, Rhode WS (1989) Structural and functional properties distinguish two types of multipolar cells in the ventral cochlear nucleus. J Comp Neurol 282:595–616.

Spirou GA, Young ED (1991) Organization of dorsal cochlear nucleus type IV unit response maps and their relationship to activation by bandlimited noise. J Neurophysiol 66:1750–1768.

Straiker A, Mackie K (2006) Cannabinoids, electrophysiology, and retrograde messengers: challenges for the next 5 years. Aaps J 8:E272–276.

Tzounopoulos T (2006) Mechanisms underlying cell-specific synaptic plasticity in the dorsal cochlear nucleus. In: Proceedings of Assoc Res Otolaryngol Abstr 208.

van Schaik A, Fragnière E, Vittoz E (1996) An analogue electronic model of ventral cochlear nucleus neurons. In: Proceedings of the Fifth International Conference on Microelectronics for Neural Networks and Fuzzy Systems, Los Alamitos, CA, February 12–14, 1996, pp. 52–59.

Voigt HF, Young ED (1980) Evidence of inhibitory interactions between neurons in dorsal cochlear nucleus. J Neurophysiol 44:76–96.

Voigt HF, Young ED (1985) Stimulus dependent neural correlation: an example from the cochlear nucleus. Exp Brain Res 60:594–598.

Voigt HF, Young ED (1988) Neural correlations in the dorsal cochlear nucleus: pairs of units with similar response properties. J Neurophysiol 59:1014–1032.

Voigt HF, Davis KA (1996) Computation of neural correlations in dorsal cochlear nucleus. Adv Speech Hear Lang Process 3:351–375.

Webster DB, Popper AN, Fay RR, eds (1992) The Mammalian Auditory Pathway: Neuroanatomy. New York : Springer Verlag.

Wickesberg RE, Oertel D (1988) Tonotopic projection from the dorsal to the anteroventral cochlear nucleus of mice. J Comp Neurol 268:389–399.

Young ED (1980) Identification of response properties of ascending axons from dorsal cochlear nucleus. Brain Res 200:23–37.

Young ED, Brownell WE (1976) Responses to tones and noise of single cells in dorsal cochlear nucleus of unanesthetized cats. J Neurophysiol 39:282–300.

Young ED, Voigt HF (1982) Response properties of type II and type III units in dorsal cochlear nucleus. Hear Res 6:153–169.

Zheng X, Voigt HF (2006a) A modeling study of notch noise responses of type III units in the gerbil dorsal cochlear nucleus. Ann Biomed Eng 34:1935–1946.

Zheng X, Voigt HF (2006b) Computational model of response maps in the dorsal cochlear nucleus. Biol Cybern 95:233–242.

Zucker RS, Regehr WG (2002) Short-term synaptic plasticity. Annu Rev Physiol 64:355–405.

Chapter 4
Models of the Superior Olivary Complex

T.R. Jennings and H.S. Colburn

Abbreviations and Acronyms

AHP	Afterhyperpolarization
AVCN	Anteroventral cochlear nucleus
BD	Best delay
BP	Best phase
CD	Characteristic delay
CN	Cochlear nucleus
CP	Characteristic phase
EC	Equalization–cancellation
EE	Excitatory–excitatory
EI	Excitatory–inhibitory
EPSP	Excitatory postsynaptic potential
IC	Inferior colliculus
IE	Inhibitory–excitatory
ILD	Interaural level difference
IPD	Interaural phase difference
ISI	Interspike interval
ITD	Interaural time difference
JND	Just noticeable difference
LINF	Leaky integrate-and-fire
LNTB	Lateral nucleus of the trapezoid body
LSO	Lateral superior olive
MNTB	Medial nucleus of the trapezoid body
MSO	Medial superior olive
NA	Nucleus angularis
NL	Nucleus laminaris
NM	Nucleus magnocellularis

H.S. Colburn (✉)
Department of Biomedical Engineering, Boston University, Boston, MA 02215, USA
e-mail: colburn@bu.edu

R. Meddis et al. (eds.), *Computational Models of the Auditory System*,
Springer Handbook of Auditory Research 35, DOI 10.1007/978-1-4419-5934-8_4,

PSTH	Poststimulus time histogram
SOC	Superior olivary complex
SON	Superior olivary nucleus
VNLp	Ventral nucleus of the lateral lemniscus pars posterior

4.1 Introduction

Sounds in the real world originate from specific sources, either alone or in combination, so that a natural description of a sound includes its location and other spatial properties. The extraction of these spatial properties by the mammalian auditory system involves early extraction of interaural difference information in the superior olivary complex (SOC), and the modeling of this processing by the neurons in the SOC is the topic of this chapter. This chapter's focus on the SOC and on interaural difference information means that the monaural localization cues, notably the direction-dependent spectral filtering of the source waveform, are not addressed here, even though these cues also carry information about source location, especially its elevation. The spectral cues for location are discussed in Chapter 5, describing models of the inferior colliculus (IC) in relation to psychophysical abilities.

The physical attributes of a sound signal that provide azimuthal spatial cues are few in number, mathematically well-defined, and easily extracted using relatively simple operations that fall within the range of mathematical transformations that individual neurons can apply to a spike train. The primary attributes of the received stimulus waveform that carry azimuthal spatial information are the interaural time difference (ITD) and interaural level difference (ILD). The usefulness of these stimulus attributes for azimuthal sound localization was recognized more than a century ago (Strutt 1907) and supported by early observations and physical analysis. Simple mechanisms for the extraction of information about ITD and ILD were suggested by Jeffress (1948, 1958) and von Békésy (1930), namely coincidence detection and level cancellation. These mechanisms are closely related to two common mathematical operations, cross-correlation and subtraction, respectively. This fact has made the study of binaural analysis in these nuclei particularly well suited to mathematical and computational modeling, and as a result a large amount of effort has been placed in modeling these mechanisms. The purpose of this chapter is to give an overview of the principles behind the modeling of these nuclei and a review of some of the mathematical and computational models that were developed using these principles.

The early conceptual and mathematical models of ITD and ILD analysis provided fairly specific expectations regarding the properties of neurons that are sensitive to those cues. The rest of this paragraph discusses simple mechanisms of ITD sensitivity and the next paragraph discusses mechanisms for ILD sensitivity. According to the ITD analysis mechanism suggested by Jeffress (1948), coincidence-detecting neurons respond with an output when two excitatory inputs arrive close together in time. This simple mechanism, if present in a population of neurons in which the stimulus ITD that results in coincident inputs to each neuron is distributed across a range, allows the pattern of activity in the neural population to represent spatial information about the sound source. Specifically, when the stimulus location is varied, the stimulus ITD changes and as a result there is a corresponding change in the spatial distribution of activity over the population of neurons. Further, other spatial aspects of the source would also influence the distribution of activity. For example, a broader source would excite a broader distribution of stimulus ITDs and a broader distribution of activity in the neural population. This mechanism of Jeffress, which he called a "place theory of localization," implies that each ear provides an excitatory input to the coincidence neuron, and thus that this ITD-sensitive neural population would be composed of so-called excitatory–excitatory (EE) neurons. The medial superior olive (MSO) in mammals and the nucleus laminaris (NL) in birds have EE characteristics (Fig. 4.1) and thus were early candidates as likely sites of ITD analysis. Direct recordings

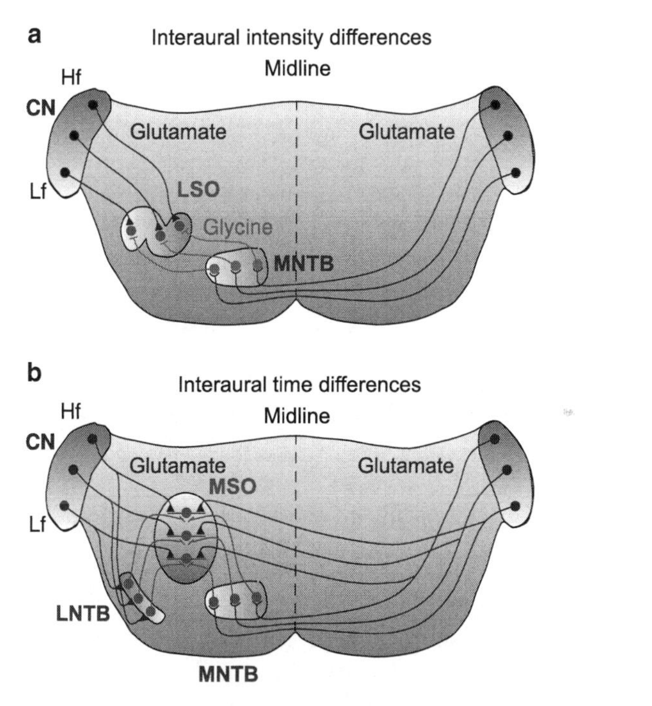

Fig. 4.1 Anatomy of the LSO and MSO showing excitatory and inhibitory pathways. The lightness gradients in the diagrams show the frequency map in the nuclei, with *darker areas* being higher frequency (Hf) and *lighter areas* being lower frequency (Lf). (**a**) Inputs to the LSO. The ipsilateral cochlear nucleus (CN) sends excitatory glutamatergic projections directly to the LSO, while the contralateral CN sends them to the ipsilateral MNTB, which in turn sends inhibitory glycinergic projections to the ipsilateral LSO. (**b**) Inputs to the MSO. Both the ipsilateral CN and contralateral CN send excitatory glutamatergic projections to the MSO and to the nuclei of the trapezoid body, with the ipsilateral CN sending projections to the lateral nucleus of the trapezoid body (LNTB) and the contralateral CN sending projections to the medial nucleus of the trapezoid body (MNTB). The nuclei of the trapezoid body in turn send inhibitory glycinergic projections to the MSO (From Kandler and Gillespie 2005.)

from these structures have shown patterns of activity very similar to the expected for left–right coincidence patterns, as discussed further later.

An early suggested mechanism for ILD analysis (von Békésy 1930; extended by van Bergeijk 1962) compares levels of excitation from right and left sides through a population of central neurons that are tuned right or left depending on the source of their excitation. One way to realize such a population is to have two groups of neurons, each excited by one side and inhibited by the other, so that the relative strength of left and right excitation determines the excitation level of each group. Mechanisms such as these lead to an association of ILD processing with neurons that have one excitatory input and one inhibitory input (EI or IE neurons). Such neurons have been observed in the lateral superior olive (LSO) in mammals and the ventral nucleus of the lateral lemniscus pars posterior (VNLp) in birds (Fig. 4.1). Direct recordings from these structures (Boudreau and Tsuchitani 1968; Moiseff and Konishi 1983) have similarly shown a pattern of activity very close to the pattern expected for neurons sensitive to ILD.

These two categories of neurons, separated into distinct nuclei, EE neurons in the MSO and EI neurons in the LSO, suggest that the MSO is critically important for ITD sensitivity and that the LSO is similarly important for ILD analysis. The apparent partitioning of the early auditory brain stem between ITD-focused and ILD-focused pathways matches an early conceptual model from Lord Rayleigh called the duplex model (Strutt 1907). According to this concept, ITD is handled by a low-frequency pathway, normally associated with the MSO, and ILD is handled by a second, high-frequency pathway, nominally the LSO; however, the reality is not this simple. The MSO and LSO, while biased toward lower and higher frequencies respectively, both include a broad range of frequencies (Guinan et al. 1972a, b). Also, modeling work and recent experimental work has indicated that the LSO is sensitive to ITD as well as to the ILD, as has been traditionally assumed (Tollin and Yin 2005).

Although the MSO and LSO make up the bulk of the SOC and comprise the focus of this chapter, the SOC also contains a handful of smaller nuclei often called the periolivary nuclei. These smaller nuclei have been largely ignored in computational modeling and so they are not described here.

Although the MSO and LSO have been of considerable interest, especially because they are the first nuclei in the ascending auditory pathway where left and right neurons converge onto individual neurons, neurophysiological data have been limited. It is recognized (e.g., Guinan et al. 1972a, b) that there are unusual difficulties associated with recording from single units in these regions, from the MSO in particular. These regions are not easily accessible, being deep in the brain, and the fact that many neurons fire with strong synchrony to the waveform results in large field potentials so that individual action potentials are difficult to isolate or even identify. Thus, it is difficult to characterize the activity of single neurons, particularly in the MSO, and considerably more data are available from the IC than from the SOC.

The agreement between the anatomy and physiology of the SOC and its conceptual models, along with the difficulty in recording from the structure, has led to a strong relationship between modeling and experimentation there. The predictions of models have often led to new goals for physiological and anatomical research, while new results from physiological and anatomical experiments have often been rapidly incorporated into new models. Further, because the structures and models are dealing specifically with important psychophysical cues, close agreement with psychophysical results is also expected, at least for simple cues.

The goal of this chapter is to summarize mathematical and computational models of the MSO and LSO in mammals and the corresponding nuclei in birds, as well as the conceptual underpinnings of these models. The remainder of this chapter is divided into three sections. The next section, Sect. 4.2, discusses models of neurons in the MSO. Section 4.3 addresses models of LSO neurons. Finally, Sect. 4.4 contains brief comments about models of perception that combine ITD and ILD analysis and that are not explicitly related to specific neural populations. Because the readers of this book are assumed to be mammals rather than birds, and because the corresponding structures in mammals and birds are similar in many ways, for simplicity only the MSO and LSO are referred to unless the models are dealing with properties or experimental results unique to birds.

4.2 Models of MSO Neurons

This section describes computational models of neurons in the MSO and NL. The discussion is organized around mechanisms for sensitivity to ITD and starts with Sect. 4.2.1 introducing the Jeffress (1948) model, which, although a conceptual model, forms the framework on which almost all modern physiological ITD models were built. The remaining three sections deal with specific implementations and variations on the ideas laid out by Jeffress. The second Sect. 4.2.2 deals with Jeffress-like EE models, as well as corresponding EI models. The third Sect. 4.2.3 deals with models that incorporate inhibitory inputs to EE neurons. The final Sect. 4.2.4 deals with issues regarding the distribution and layout of ITDs in the MSO and NL.

4.2.1 The Jeffress Model (Coincidence, Internal Delays, Cross-Correlation Functions)

As noted in the preceding text, Jeffress (1948) proposed a mechanism for ITD sensitivity in his landmark paper, when single-neuron recording was still in its infancy; nevertheless, it became and remains the conceptual foundation through which physiological and anatomical data are discussed. The basic ideas of the model are separated here into three components that are often discussed separately: the principle of coincidence detection, so that the degree of correspondence between two input spike trains can be measured; the concept of a distribution of internal delays compensating for external delays, so that external delay can be estimated from the internal delay that generates maximum coincidence; and the concept of a "space map" in which external location is mapped to a local geography, so that neighboring neurons tend to respond to external locations that are adjacent in space. This section also includes a description of the close relationship between a Jeffress-like coincidence mechanism and the cross-correlation function used in signal processing (to estimate relative delay, for example).

4.2.1.1 Coincidence Detector Neurons

In the context of neurons, a coincidence detector is a neuron that generates an action potential (often called a spike) when it receives two or more temporally "coincident" input spikes. In other words, a coincidence neuron responds only when it receives two or more spikes simultaneously, or almost simultaneously. More precisely, the probability of an action potential in the neuron is a function of the time between the current input and the most recent previous input. It is intuitively clear that a coincidence detector with two input neurons fires more frequently as the sequences of spike times of its inputs become more similar. This is measured mathematically by the correlation of the instantaneous rates of the input spike trains. This equivalence is demonstrated by the analysis of Colburn (1973) and more elegantly by Rieke et al. in Spikes (1997). If one assumes that a relative internal delay τ shifts one input spike train relative to the other, then the number of coincidences will estimate the cross-correlation function evaluated at the delay τ. Thus, a network of coincidence detector neurons with a distribution of internal delays can be thought of as providing the same information as the cross-correlation function.

By the time Jeffress published his paper there was already evidence that auditory nerve fibers respond to a small range of frequencies and that each fiber spikes preferentially at a consistent point on each cycle of the sound waveform for low-frequency sounds (Galambos and Davis 1943). This phenomenon is called phase-locking. Jeffress realized this spike pattern could be used with a series of coincidence detector neurons to determine the interaural phase difference (IPD) or, equivalently, the ongoing ITD. As outlined earlier, if a population of coincidence detector neurons has a distribution of internal delays then the average rate of firing of each coincidence detector neuron would estimate the cross-correlation function at its particular delay. The stimulus ITD that generates the maximum firing rate of a neuron, the ITD that cancels out the neuron's intrinsic delay, is called that neuron's best delay (BD). Similarly, the IPD that generates the maximum firing rate for a neuron at a specific frequency is called the best phase (BP) at that frequency. If the BD is determined by a fixed delay line, then when the stimulus frequency is varied, the BD should be consistent at every frequency as well as for bands of noise. Thus, if the BP is plotted as a function of frequency (the phase plot), it should form a straight line. The slope and zero-frequency intercept of this linear phase plot are called the characteristic delay (CD) and the characteristic phase (CP), respectively. For a coincidence-detector neuron, the CP is zero and the CD is equal to the BD, leading to a "peaker" response for which the rate-ITD curves all align at a peak. In contrast, when the rate-ITD share a common minimum or trough, the neuron is called a "trougher" and the phase plot would show a CP of π radians and a CD equal to ITD of the common minimum. Trougher responses are seen in the LSO and are dealt with in Sect. 4.3.

4.2.1.2 Internal Delay Distributions

Jeffress hypothesized the existence of an array of coincidence detectors in a brain stem nucleus with a branching pattern of input fibers such as in Fig. 4.2 so that the lengths of the input fibers determine the distribution of internal delays over the population of coincidence detectors. In the Jeffress illustration, each branch along the nucleus has a slightly longer length and thus imposes a slightly longer delay than the branch before it. A longer branch from one side of the head synapses on the same coincidence detector as a shorter branch from the other side, and the net delay at each coincidence detector varies systematically as you move from one end of the nucleus to the other. Then, as the stimulus ITD changes, the location along the array where there is the most activity also changes so that the location of the maximum response represents the ITD at the two ears.

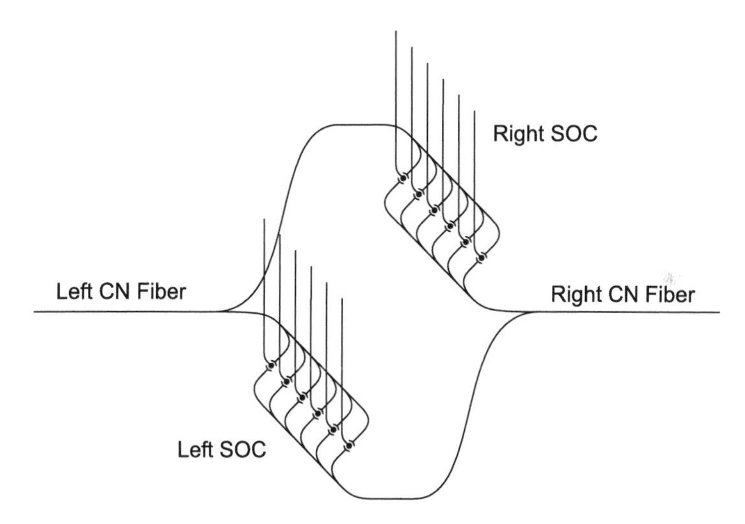

Fig. 4.2 Jeffress model showing the pattern of innervation. Ascending axons from the two sides split into two branches, one going to the ipsilateral coincidence detector nucleus (tertiary neurons) and the other to the corresponding contralateral nucleus. As the axons approach the nuclei, branches split off regularly to form paired synapses on coincidence detector neurons, with the axons from one side of the head forming one set of synapses on each neuron while the axons from the other side form another set of synapses on the same neurons. This results in a characteristic ladder-like pattern. Because the axons are approaching from opposite directions, an increase in the axon length from one side of the head generally matches with a corresponding decrease in the axon length from the other side of the head. This gives each coincidence detector a different relative axon length difference and therefore a different internal time delay. In the actual nucleus, the pattern would not necessarily be this regular, there might be multiple coincidence detectors with the same axon length difference, and there might be multiple AVCN neurons from the same side of the head forming synapses with a given coincidence detector (Based on a figure from Jeffress 1948.)

Three aspects of the distribution of delays are distinguished here and discussed separately: the spatial arrangement of the neurons with the varying BDs, the mechanism by which the delays are generated, and the shape of the delay distribution. The spatial arrangement of BDs as hypothesized by Jeffress and shown in Fig. 4.2 is consistent with the anatomy of the barn owl (*Tyto alba*) as shown in tracings of NL inputs in Fig. 4.3. These tracings match the hypothesized branching pattern and suggest that best-ITD varies systematically with location in the nucleus. Recordings from the NL (Carr and Konishi 1988, 1990) are also compatible with such a place map of ITD. However, although available evidence indicates that both MSO and NL neurons behave like coincidence detectors and that there is a distribution of BDs within both nuclei, there is little evidence supporting the existence of a Jeffress-style place map in mammals. Several mechanisms have been suggested to generate internal delays in addition to fiber length. Among other things, the diameter of the fiber, its membrane properties, and the presence and arrangement of its myelination (particularly the spacing of the nodes of Ranvier) can alter conduction velocities and thus induce delays. Synaptic delays and mechanical delays in the cochlea could also play a role (Armin, Edwin and David, in press). Brand et al. (2002) showed evidence for a delay mechanism that is influenced by inhibitory inputs and models have shown other possible mechanisms, as described in Sect. 4.2.3.

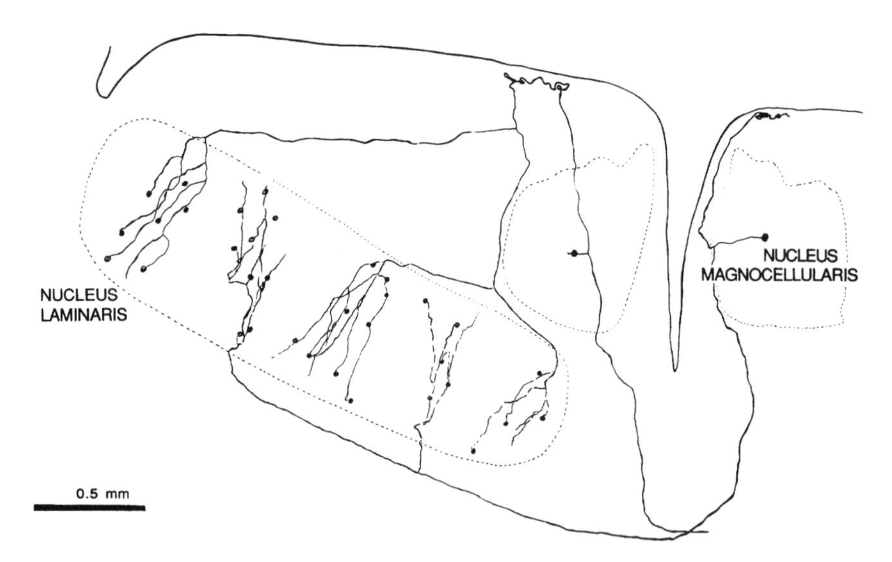

Fig. 4.3 A reconstruction of afferents projecting from bilateral nucleus magnocellularis (NM) to the nucleus laminaris (NL) in the barn owl (*Tyto alba*), based on stained slices. This illustration is based on separate reconstructions of a single ipsilateral NM neuron and a single contralateral NM neuron, both stained with horseradish peroxidase and combined in the figure for comparison. The axons from the contralateral NM (the ones approaching the NL from below) form a ladder-like branching pattern as predicted by Jeffress. The fibers approaching from the ipsilateral NM, in contrast, do not show any obvious pattern or have any specific relationship between position along the NL and axon length. Even though the delays are not imposed symmetrically, the overall pattern is consistent with a Jeffress-style coincidence detector (From Carr and Konishi 1988.)

The original Jeffress (1948) model proposed that the BDs of coincidence detector neurons are distributed across the ITD range to which an animal is sensitive. Recordings from the barn owl match this pattern, with the BDs measured in the NL being spread across ITD the range the owl would encounter in nature (Carr and Konishi 1988, 1990). Recordings from small mammals, however, specifically guinea pigs and gerbils, do not match this pattern (McAlpine et al. 2001; Shackleton et al. 2003; Pecka et al. 2008). Instead, the BDs recorded from the MSOs of these animals cluster around 1/8 cycle of the frequency each MSO neuron is most sensitive to (a phase of $\pi/4$).

One possible reason for the discrepancy between small mammals and barn owls is that they simply evolved different strategies to accomplish the same task. Mammals and birds diverged around 300 million years ago (Donoghue and Benton 2007), tens of millions of years before they (and several other groups of animals) independently evolved the tympanic auditory system they use today (Clack 1997). Another possible explanation for this discrepancy lies in the information content of the rate-ITD curves. The clustering around 1/8 cycle does not make sense if one assumes that the important point in a neuron's rate-ITD curve is the peak of that curve. From an information standpoint, however, the peak is not the important part of the curve. The part of the curve that contains the most information, that is, the part where a change in neuronal firing rates says the most about a change in ITD, is the part of the curve with the steepest slope relative to the variability. So if the distribution of rate-ITD curves in different species evolved not to spread out the peak across the physiological range, but instead to spread out the slopes, that may account for the patterns seen in barn owls and small mammals.

To test this directly, the optimal distribution of BDs was computed for different animals, including the gerbil, barn owl, human, and cat (Harper and McAlpine 2004). The optimal curves matched the recordings seen in the real animals. In particular, for small animals, such as the gerbil, a distribution with BDs at approximately ±1/8 cycle was generated for most of their frequency range. For medium-sized animals, like the barn owl, a distribution with ITDs spread across the physiological range was generated for most of their frequency range. However, due to their head size, chickens should be more similar to gerbils based on optimal coding strategies. However, in recordings from the chicken NL they matched the pattern seen in barn owls instead (Koppl and Carr 2008). These questions, especially differences across species, are not yet resolved.

4.2.1.3 Cross-Correlation Function

The responses of coincidence detector neurons with two inputs can be related to the cross-correlation function of the unconditional rates of firings of the inputs, at least in the case that the coincidence window is relatively narrow and the responses are averaged over some time. In this case, the expected number of coincidences is equal to $\int\int r_1(t)r_2(t-\tau)f(\tau)\mathrm{d}\tau\,\mathrm{d}t$ where $f(\tau)$ is a function that specifies the coincidence window (the probability of a response to inputs separated by τ). If $f(\tau)$ is a narrow function, then $f(\tau)$ behaves like a shifted impulse and the double integral becomes a single integral over t. The result is that the expected number of coincidences is equal to the area of $f(\tau)$ times the cross-correlation (the integral of the product of the rates over duration). If one of the rates is delayed relative to the other for different CD neurons, then the set of neurons would provide the cross-correlation function.

4.2.2 *Pure EE Models of MSO and NL Neurons*

Models in this section comprise the simplest class of ITD models, those based on two sets of excitatory inputs, which have been very successful in reproducing the responses of individual MSO and NL neurons. These models include one or more excitatory inputs from each side of the head; models with both excitatory and inhibitory inputs are reviewed in Sect. 4.2.3.

Our discussion of purely excitatory models is divided according to the complexity of the description of the neuron membrane. The models in the first set (Sect. 4.2.2.1) do not explicitly model membrane potential; they are based on simple mathematical or statistical descriptions. The models in the second set (Sect. 4.2.2.2) are based on membrane potential, and generate an output action potential when the membrane potential crosses a threshold; however, the consequences of inputs are excitatory postsynaptic potentials (EPSPs) in the form of additive voltage pulses and refractory behavior is also oversimplified. The models in the third set (Sect. 4.2.2.3) describe the nonlinear, voltage-dependent characteristics of the membrane ion channels explicitly with models in the style of Hodgkin and Huxley (1952) models. The models in the fourth set (Sect. 4.2.2.4) further incorporate spatial properties of the neuron including the location of neuron components such as the axon, soma, and/or dendrites and the distribution of ion channels.

4.2.2.1 EE Models Without Explicit Membrane Potential

The simplest and most abstract method of modeling coincidence detector neurons is to focus on the mathematics of coincidence detection. This results in a form of black box model, wherein each neuron is treated as a black box that generates an output when input pulses occur within a coincidence window.

One of the earlier descriptions of the patterns resulting from coincidence detection models is found in Colburn (1973). Unlike many later models that cannot be represented in closed form, this model was derived in the form of an explicit formula for output rate of firing and its dependence on the rates of firing of the input neurons. Although this was a model for binaural temporal information content and was not explicitly a model of the MSO, it gives a mathematical representation of the Jeffress model.

A more recent black-box model includes both EE and EI ITD information (Marsálek and Lansky 2005). The properties of their EI model are discussed in Sect. 4.3.4. This model is unusual, in part, because it treats the internal time delay as a random distribution of delays between zero and a maximum value. The combination of the EE and EI models was also able to account for a dip in the psychophysical just noticeable difference (JND) for tones at approximately 2 kHz (Mills 1960) by combining the output of the EE and EI models, but otherwise was not compared to experimental data.

4.2.2.2 EE Leaky Integrate-and-Fire Models

The models described in this section, leaky-integrate-and-fire (LINF) models, have been used to describe neural processing in a number of systems (Dayan and Abbott 2001). These models are formulated in terms of membrane potential, with excitatory inputs generating "membrane depolarizations." In this type of model, action potentials are generated when the membrane potential crosses a threshold value, usually followed by some refractory mechanism such as a reset of potential to rest and/or a period where no inputs are accepted. This allows these models to ignore many complexities that surround an action potential, such as changes in voltage-sensitive conductances.

A simple example of an LINF model is a shot-noise model. Shot noise is generated by filtering Poisson impulses, usually with a first-order filter. If inputs are described by Poisson processes and each input creates an exponential response, then the membrane potential can be described as shot noise, although the effects of threshold crossing make the overall behavior more complicated than a simple shot noise. Nevertheless, techniques from the analysis of shot noise can be used to characterize the time to threshold (time between events in the LINF model) and other statistical measures of the firing patterns. An EE version of a shot-noise model for MSO responses was suggested by Colburn et al. (1990), who used a 1 ms refractory period and sinusoidally varying input firing rates. The models showed very good agreement with the published spike trains from physiological recordings. If the assumption is made that all inputs have the same strength, this class of model has the advantage of having only two parameters, the amplitude of the exponential relative to the threshold and its time constant. This model had the interesting result that, despite having no inhibitory inputs, at the worst delays firing rates were suppressed below the monaural rate, a feature seen in real neurons. The model had this property because, even with monaural input, the temporal synchronization makes the inputs drop to zero in one half of the cycle, allowing no left–right coincidences at bad phase. Of course, the neuron was still being driven by spontaneous input from the nonstimulated ear in the monaural case, leading to random coincidences that were not present when driven binaurally.

4.2.2.3 Two-Input Point-Neuron Models

The LINF model can be extended to be more physiologically realistic by including voltage-sensitive conductances in the membrane, notably active sodium and potassium channels. These extended, point-neuron models include some physiological membrane properties of the neuron while neglecting any role of the neuron's size, shape, arrangement, the distribution of ion channels over the surface of the neuron, or any other spatial properties. The most basic point-neuron models are based on the traditional Hodgkin and Huxley (1952) model of action potential generation, which takes into account leak currents, inactivating voltage-gated sodium channels, and non-inactivating voltage-gated potassium channels, as well as excitatory synaptic currents. Each of these currents has several parameters, leading to a much larger number of adjustable parameters compared to simple LINF models. Even restricting the parameters to the known ranges still leaves a wide variety of combinations. With so many parameters, predicted responses can be matched to empirical responses by manipulating parameters and it is difficult to evaluate the significance of even good fits.

A Hodgkin–Huxley point-neuron model was used to describe MSO responses (Han and Colburn 1993) with generally compatible results. Performance was similar to the LINF model in Colburn et al. (1990), but required that the time constant of the excitatory synaptic conductances be smaller than is physiologically likely. If the time constant was set to a realistic value, the neuron was sensitive to spikes that arrive too far apart temporally.

The Hodgkin–Huxley model of Han and Colburn (1993) does not include the full collection of ion channels found in SOC and NL neurons, for example, the low-threshold potassium channel was not included. Similar to the Hodgkin–Huxley potassium channels, which serve to shorten the depolarization time constant of an action potential and bring it back to its resting state, the low-threshold potassium channels serve to shorten the time constant of EPSPs and narrow the window over which temporal summation of spikes can occur. A point-neuron model incorporating the traditional Hodgkin–Huxley ion channels along with a low-threshold potassium channel (Brughera et al. 1996) was able to attain tight temporal tuning while still maintaining realistically parameters for all of the ion channels. This allowed the model to generate correct responses to tone stimuli while still maintaining physiological relevance.

The model of Brughera et al. (1996) was based on data regarding the membrane characteristics of anteroventral cochlear nucleus (AVCN) bushy cells (Rothman et al. 1993), a class of neurons that have tight temporal tuning similar to MSO neurons. The model was able to reproduce closely both the rate-ITD curves and synchronization-index-ITD curves from physiological recordings when stimulated by model AVCN neurons as inputs. (The model was also used to test the response to clicks, but in this case the model was targeted at reproducing IC data and thus is not relevant to this chapter.)

4.2.2.4 Multicompartment Models

The neuron models described so far assume a single membrane potential, as if the inside of the neuron were an equipotential region. In this section, this assumption is relaxed so that model neurons have an extended spatial dimension with variation in the intracellular potential across the neuron. This more accurate model increases the number of parameters, but avoids limitations of simpler point-neuron models.

There are three primary limitations of point-neuron models that multi-compartment models can overcome. First, point-neuron models cannot reproduce the interaction between different parts of a neuron with different electrical properties, such as dendrites and axons. Second, point-neuron models cannot reproduce the impact of having different components spatially separated from each other, for example, the consequences of having different synaptic inputs on different compartments. Third, point-neuron models cannot reproduce the impact the shape and volume of a neuron has on the neuron's properties.

A multicompartment model for the chick NL was developed (Simon et al. 1999) and included a soma, an axon, and two multiple-compartment dendrites. This model included the standard complement of Hodgkin–Huxley channels, excitatory synaptic ion channels, and additional high- and low-threshold potassium channels and was able to produce results similar to recordings of NL neurons, provided that the high-threshold potassium current was varied with frequency and dendritic length. Performance was sensitive to the locations of the synapses along the dendrites. Spacing the synapses equally along the dendrites led to the best performance. Concentrating the synapses toward the middle of each dendrite caused small decreases in performance, particularly for large dendrites and low frequencies. The performance was even worse for large dendrites and low frequencies if the synapses were concentrated at the base of the dendrites, and the performance was extremely poor when the dendrites were removed entirely and the synapses were placed directly on the soma.

A similar model of the NL was constructed by Dasika et al. (2007). This model assumed a single-compartment soma and two dendrites. The soma was either passive, using only leak currents, or active, based on the Rothman et al. (1993) model with Hodgkin–Huxley channels and additional potassium channels that narrowed the temporal tuning of the model as described above. The dendrites were passive and either single-compartment, multicompartment, or cable-type (a continuum of compartments). The potential at the soma was compared against the conductance of the two dendrites, with maximum potential occurring when both dendrites had equal conductance and falling off as the conductance became unbalanced. Further, models with longer dendrites were more sensitive to binaural coincidences relative to monaural coincidences. Longer dendrites also had the effect of increasing the coincidence detection window and the refractory period, reducing the temporal tuning.

An even simpler model was used to study the interaction between the soma and axon (Ashida et al. 2007). This model included only two Hodgkin–Huxley compartments, a soma and a node of Ranvier. Unlike most of the models described in this chapter, inputs, which were on the soma, were not modeled as spikes; rather, they were modeled as sinusoidal (AC) changes in conductance with a DC offset. The model rate-ITD curve had fairly sharp cutoffs, but more realistic rate-ITD curves were generated by including noise in the model. In the model neuron, having a passive soma that had a large volume compared to the active node of Ranvier led to higher ITD sensitivity, higher tolerance to noise, and improved sensitivity to high-frequency stimuli compared to cases in which the soma was active and/or was closer in size to the node of Ranvier.

4.2.3 EE Models of MSO Including Inhibitory Inputs

The models of MSO and NL described in this section consider EE coincidence detector models with additional inhibitory inputs that serve to change the properties of the neuron in various ways, depending on the model.

Although up until recently models of the MSO generally have included only excitatory inputs, it has been known for a while that the MSO also receives inhibitory inputs. Both neuropharmacological (Adams and Mugnaini 1990; Cant and Hyson 1992; Schwartz 1992) and physiological slice preparations (Grothe and Sanes 1993) showed evidence of glycinergic inhibition in the MSO. The MSO receives inhibitory inputs from the two nuclei of the trapezoid body, the medial nucleus of the trapezoid body (MNTB), and the lateral nucleus of the trapezoid body (LNTB; see Fig. 4.1). The NL also receives GABAergic inhibitory inputs from superior olivary nucleus (SON). However, models that included only excitatory inputs had been able to reproduce the data being recorded from the MSO and NL. This made the conclusion that the role of inhibition in the MSO is limited at best the most parsimonious explanation, pending more direct experimental evidence on the subject.

In a variation of the Brughera et al. (1996) model described in Sect. 4.2.2.1, inhibitory inputs from three different types of AVCN bushy cells with different temporal response patterns were used in addition to highly phase-locked EE inputs to test whether invoking inhibition was necessary to explain physiological recordings from the MSO (Goldberg and Brown 1969). Specifically, the plausible sources of inhibition to the MSO were three types of AVCN neurons: highly phased-locked, onset-type, and pri-notch type (which have an onset and a lower sustained response separated by a gap). Each of these was assumed to pass through an MNTB neuron that converted them from excitatory to inhibitory, although this neuron was not explicitly modeled. Although the model's response amplitude was significantly affected by presence of inhibition, the shape of the rate-ITD curve was not changed in a way that made it a better or worse fit for the physiological recordings. This reinforced the idea that, at least for tones, inhibition was not required for modeling the responses seen in MSO neurons.

A major stimulus to the development of models with inhibition came from measurements (Yang et al. 1999; Brand et al. 2002) showing that, at least in some EE MSO and NL neurons, inhibitory inputs have large effects on the responses of the neurons. For the MSO neurons inhibition changes the BD of the neuron (Brand et al. 2002), while for the NL neurons it changes the sharpness of the temporal tuning (Yang et al. 1999).

In the study of Brand et al. (2002), blocking inhibition led to an increase in the neurons' firing rates, as expected; unexpectedly, however, blocking the activity of glycine in MSO neurons also caused the BD of the neuron to shift to zero, at least in the small number of neurons tested. This contradicts the fixed internal delay hypothesis of the Jeffress model, which hypothesizes that delays are determined by fixed physical properties of the input fibers to the MSO. This result implies that not only is the delay at least partially determined by inhibition at least in some cases, but also that modulating inhibition might be able to dynamically modulate the temporal tuning of such an MSO neuron. A qualification on the Brand study is that the amounts of the shifts observed were relatively small so that there could be multiple contributions to the delays for some neurons.

In the same paper, Brand et al. (2002) proposed a model to account for these results. In this model, which was based on the Brughera et al. (1996) model described in Sect. 4.2.2.3, the excitatory inputs were immediately preceded by phase-locked contralateral glycinergic inhibition with a very short time constant ($\tau = 0.1$ ms). This inhibition blocked the early portion of the contralateral EPSP, resulting in a net delay in the peak of the MSO neuron's EPSP. By changing the level of inhibition, the EPSP shape was modified and the model was able to modulate the best ITD of the neuron.

Further analysis of the Brand et al. (2002) model shed more light on the specific roles that inhibition and low-threshold potassium currents can play (Svirskis et al. 2003). This analysis showed that increasing the low-threshold potassium current narrowed the temporal tuning of the model MSO neuron while at the same time lowering its overall response level. This led to a trade-off between sharp tuning and the ability to respond to coincident spikes. It also reproduced the results seen by Brand et al. (2002). Although the effects of inhibition and the effects of low-threshold potassium currents seem to be largely independent, they both lower response levels; thus, their combined effect could hamper an MSO neuron's ability to respond to appropriate stimuli.

More detailed anatomical data on MSO neurons has led to an alternative approach to modeling the physiological results of Brand et al. (2002) described in the preceding text. Many of the principal cells in the MSO appear to have two symmetrical dendrites with a single axon originating not on the soma but on one of the two dendrites (Smith 1995). This lends an asymmetry to the neuron and makes the site of action potential generation, the axon hillock, closer to one set of dendrites than the other. Further, the inhibitory time constant used by Brand et al. (2002) is an order of magnitude smaller than the time constant seen in physiological data (Smith et al. 2000; Chirila et al. 2007).

A multicompartment model of MSO neurons was developed to explore the possible role this asymmetric structure may play (cf., Brew 1998) in the context of inhibitory synapses on the soma of a neuron that may affect the conduction of EPSPs (Zhou et al. 2005). This model includes an explicit description of the passive and active ion channels but also includes the anatomical arrangement of the neuron and the locations of different types of ion channels and synapses on the neuron.

The model of Zhou et al. (2005) used an active soma compartment, symmetrical passive dendrite compartments (one receiving ipsilateral input, the other contralateral), and an active axon originating from one of the two dendrites. The dendrites receive excitatory inputs spaced out along their length, while the soma receives inhibitory input across its surface. The soma includes only voltage-gated noninactivating sodium channels, while the axon includes all of the channels seen in Rothman et al. (1993) and further has an inward-rectifying hyperpolarization-triggered current. The inhibitory time constant was set at 2 ms, which is reasonable based on physiological data on the membrane properties of MSO neurons.

The mechanism by which the inhibition alters the ITD tuning of the neuron in this model is completely different than the mechanism described in Brand et al. (2002). Because of the offset of the axon, the EPSPs from the dendrites on the opposite side of the soma have to cross over the soma to reach the axon while the EPSPs from the dendrites on the same side do not. In the absence of inhibition and with the active component of the soma turned off, the now passive soma shunts current from the EPSPs. This slows the rise time of the membrane voltage at the axon hillock, creating a delay in the peak of the EPSP. The active soma counters this, bringing in current to regenerate the EPSP as it travels along the soma. This speeds up the rise time of the EPSP, reducing or even eliminating the effective delay. The inhibition has the opposite effect, countering the active ion channels in the soma and increasing the effect of the shunting. Figure 4.4 illustrates these principles. By modulating the inhibition, the delay can be moved between the two extremes. This model was able to reproduce the same results seen in Brand et al. (2002), but was able to reproduce them using an inhibitory time constant an order of magnitude larger and much larger variability in the arrival time of the inhibitory spikes.

The NL of birds also receives inhibitory inputs, but these inputs are very different than those in the MSO of mammals. First, the inhibition is GABAergic instead of glycinergic (Yang et al. 1999). More importantly, however, is that the inhibition forms a feedback circuit, where excitatory fibers from the NL to the SON synapse on inhibitory fibers leading back from the SON to the NL. Both the NL and SON receive secondary fibers, the NL receiving binaural fibers from the nucleus magnocellularis (NM) and the SON receiving monaural fibers from the nucleus angularis (NA). Besides the NL, the SON also sends inhibitory fibers to the NM, the NA, and the contralateral SON. The inhibition does not need to be phase-locked, and unlike the Brand et al. (2002) model, appears to increase shunting in the NM. Finally, these inputs have the interesting property of being depolarizing while still being inhibitory.

Measurements from neurons in the NM indicate that the inhibition may serve as a gain control mechanism to modulate temporal tuning (Yang et al. 1999). A model including all of these nuclei was built to test this hypothesis (Dasika et al. 2005). The model used a shot-noise LINF model that includes adaptation. To model the adaptation, an inhibitory input resulted in a decrease in the time constant of the exponential decay and an increase in the threshold for firing. The model matched the responses from physiological recordings well, although the onset response decayed to the steady-state response much more slowly than in physiological recordings. In the model, inhibitory feedback was important for keeping the response of the NL neurons independent of overall sound level. Further, it was important for maintaining tight temporal tuning. The inhibitory connections between the two SON kept the inhibition level from the SON from getting too high and suppressing the NL too much.

A version of the model of Simon et al. (1999) that adds inhibitory synapses has also been used for the NL (Grau-Serrat et al. 2003). Unlike the multicompartment MSO model, this model was not developed to explain possible fundamental limitations of a single-compartment approach or a purely excitatory multicompartment approach. Instead, the model was developed with the goal of making a model as physiologically accurate as possible based on currently available information. Besides the

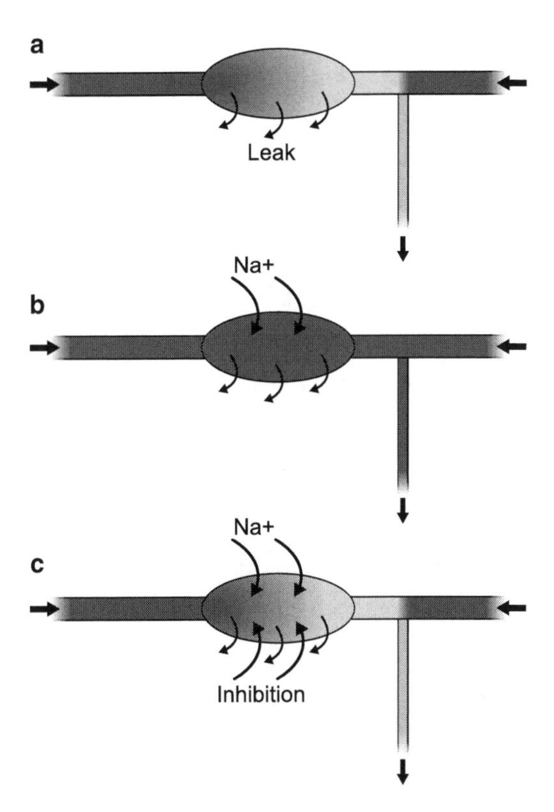

Fig. 4.4 Multicompartment, asymmetric, inhibition-dependent MSO model. In this diagram the *ellipse* is the soma, the *thin vertical rectangle* is the axon, and the medium *horizontal rectangles* are the dendrites. Note that the axon is connected to the right dendrite, not to the soma. The *shading* represents the degree of depolarization at that location, with *darker areas* being more depolarized and *lighter areas* being less depolarized. In (**a**), the dendrites and soma are both passive. Due to somatic shunting by the leak current, the EPSP from the left dendrites decays as it moves across the soma toward the axon. As a result, at the axon the left EPSP is reduced, leading to a lower total firing rate in the axon. In (**b**), the soma has an active sodium current that counteracts the leak current, compensating for the shunting effect. In this case, the EPSPs from both sides arrive at full strength. In (**c**), the soma is active but also has inhibitory synapses. The inhibitory synapses open potassium channels that have the opposite effect of the active sodium channels, effectively canceling them out and leaving the net shunting effect as it was in (**a**). By changing the relative strength of inhibition, sodium current, and leak current situations between these extremes can be produced (Based on a figure from Zhou et al. 2005.)

inhibition, this model was essentially that of Simon et al. (1999), except that it had excitatory synapses only on the dendrites, not on the soma. Instead, the soma received the inhibitory synapses, which were driven by model SON neurons.

The model of Grau-Serrat et al. (2003) was able to produce realistic rate-ITD curves for low frequencies, with appropriate drops in performance for high frequencies. However, the model output improved phase locking of the output compared to the input even at the worst ITD in some cases, which may be a flaw in the model.

The model was also able to explain several nonlinearities in NL neurons. As the dendrites length got longer, ITD discrimination got better. However, this improvement saturated, and the length at which it saturates decreased as frequency increased. Another nonlinearity is related to the reduction in firing rate at worst ITDs as compared to monaural inputs, which the model of Colburn et al. (1990) was also able to explain by spontaneous activity. The Grau-Serrat et al. (2003) model, however, was able to explain this result even without any spontaneous activity from the unstimulated ear. The low-threshold potassium channel acted as a current sink to suppress depolarization further at worst delays.

4.2.3.1 Inhibition and Human Anatomy

These inhibition-based models all assume inhibitory inputs are available to the MSO. Dale's law, a rule that appears to be nearly (although not completely) uniform across all neurons that have been looked at so far, states that no neuron is able to produce multiple difference collections of synaptic neurotransmitters (Strata and Harvey 1999). This means that if a neuron releases excitatory neurotransmitters at one of its synapses it cannot release inhibitory neurotransmitters at another. The AVCN, being a collection of various types of excitatory neurons, does not appear to produce any inhibitory inputs of its own. Instead, it forms excitatory synapses on neurons in another nucleus, and these neurons then proceed to form inhibitory synapses elsewhere. A specialized type of neuron, found in the MNTB among other areas, is characterized by having a very specialized synapse called a calyx of Held. The axon from a contralateral AVCN neuron completely envelopes the soma of the MNTB neuron, providing a nearly one-to-one correspondence between spikes in the excitatory AVCN neuron and those in the inhibitory MNTB neuron.

Almost all of the aforementioned inhibitory models assume that this nearly perfect conversion between excitation and inhibition takes place, and the place this is known to happen is the MNTB. That this occurs is not in serious question in most mammals. However, in humans the data on the presence or absence of the MNTB is sketchy and contradictory. Some studies have not found an MNTB at all; others have found a small and loose collection of neurons that may have calyces of Held in the rough area of the MNTB (Richter et al. 1983), and yet others have found a clear MNTB that lacks any calyces of Held at all (Kulesza 2008). Determining the presence or absence of the MNTB in humans, as well as the potential role of a similar structure (the LNTB), is important to relating these models to human physiology and ultimately human psychophysics.

4.3 LSO Models

In contrast to the EE neurons of the MSO, LSO neurons are consistently reported as EI type. EI neurons are particularly sensitive to changes in ILD, so the focus of this section is on ILD models of the LSO. However, the LSO also has sensitivity to ITD, and models of this behavior are discussed later in this section.

This section starts with a discussion of the conceptual principles behind ILD modeling, followed by a section on models of steady-state ILD processing. Many of these models have structures similar to the EE models (except that one input is inhibitory and hyperpolarizes membrane potential), and are organized here in a similar manner, in order of increasing detail in the modeling of the membrane. The temporal structure of the LSO responses, considered as a random point process, is discussed in a separate section. Finally, models of ITD in the LSO are discussed at the end of this section.

4.3.1 Level Difference Models

Like the Jeffress model for ITD responses, there is a canonical model for ILD responses. Compared to the Jeffress model, the model of ILD responses is relatively simple. If the intensity level of two sounds is known, the simplest method to determine the difference between those two levels to take a simple subtraction of one level from the other ($L_1 - L_2$). An alternative, mathematically identical method is to add one level to the opposite of the other level ($L_1 + -1 * L_2$). Although the difference between these equations is trivial from a mathematical perspective, this difference becomes important when dealing with neurons. There is no simple, direct way to do a subtraction in a single neuron. However, a combination of excitatory and inhibitory inputs will cancel in a way that is similar to subtraction. By using a neuron that converts an excitatory input to an inhibitory input, a sign inverter, the difference in the firing rate between two excitatory neurons can be calculated in a fourth, EI neuron. This is the principle on which the LSO detector models are based.

There are some complications dealing with the nonlinear properties of neurons that are worth addressing. First, neurons that encode information using firing rate, which is the assumption for the ILD model, cannot encode anything below a firing rate of zero. Further, due to their refractory behavior, the average firing rates of such neurons saturate. This suggests that each ILD-sensitive neuron has a limited range of ILDs it can encode, giving the neuron's rate-ILD curve a characteristic sigmoidal shape.

The LSO is the earliest structure in the auditory pathway that has EI inputs. It receives excitatory input from the ipsilateral AVCN and inhibitory input from the contralateral AVCN by way of the MNTB (Glendenning et al. 1985). Based on these inputs it would be logical for the LSO to be the site of ILD discrimination. Direct recordings from the LSO in mammals have confirmed this hypothesis (Boudreau and Tsuchitani 1968), while recordings from the VNLp of birds have shown a similar ILD sensitivity in that structure (Moiseff and Konishi 1983).

4.3.2 Steady-State Models

Although the LSO has a characteristic temporal response to the onset of a sustained stimulus called a chopper response, after a brief period of time this response decays into a steady-state response that remains approximately constant as long as there is no change in the stimulus, so that the average rate of firing and its dependence on the stimulus characteristics like the ILD is the focus of most LSO models. Models in this section ignore the temporal modulation of the firing rate at the onset of a response; the chopper response is discussed in the following section. In the rest of this section, LSO models are described and discussed in order of increasing membrane complexity and specificity.

A number of models have provided mathematical descriptions of the firing rate and the statistical pattern of responses in LSO neurons. Early attention was given to the shape of the interspike interval (ISI) histogram by Guinan et al. (1972a, b). They distinguished Poisson-like (exponentially shaped) and Gaussian-like (shifted mean) distributions of ISIs, noting the regularity of the Gaussian-like and the irregularity of the Poisson-like distributions. They noted that the relative irregularity varied depending on the nucleus in the SOC, and specified statistics for the various nuclei. Their statistics were used explicitly by several models, including the simple shot-noise model of Colburn and Moss (1981) that is described later.

This attention to ISI distributions was prominent in a series of point-process models of the LSO (Yue and Johnson 1997). The first model represents the LSO as a simple Poisson spike generator whose firing rate is the ratio (in decibels) of sound level between the two ears. The second model uses Gaussian ISIs. The Poisson model was used with linear, logarithmic, and square root transformations between the ratio and the firing rate while for the Gaussian ISI case only the linear transformation was used. Statistical properties of the Gaussian ISI model, specifically the hazard function and correlation between adjacent ISIs, were compared against physiological data and they matched fairly well. Yue and Johnson also addressed the information in the patterns of LSO responses related to estimation of sound direction, but this is not discussed here.

A more mechanistic model of the LSO generates firing rates for fibers coming from each of the two AVCNs and subtracts them. This is the approach in the rate-combination models for ILD sensitivity in the LSO (Reed and Blum 1990; Blum and Reed 1991). These models assume that neurons in the LSO are laid out tonotopically along one axis and according to the intensity threshold of the input fibers along another, so that in a single isofrequency band neurons are arranged so there is a steady transition from low-threshold input fibers to high-threshold input fibers. Further, the direction of the change in the threshold for fibers coming from one side of the head is opposite the direction for fibers coming from the other, so that low-threshold ipsilateral inputs are paired with high-threshold contralateral inputs and vice versa. This is analogous to how short-delay fibers from one side of the head are paired with long-delay fibers from the other side in the Jeffress (1948) model. Assuming that the LSO neurons have the standard sigmoidal response with a physiologically reasonable range, and assuming that due to the proposed arrangement of inputs there is a steady transition of ILDs to which the neurons will respond, then the ILD level can be determined by looking at how many neurons are firing above their median firing rate and how many are firing below their median firing rate.

The model was used to generate both physiological (Blum and Reed 1991) and psychophysical (Reed and Blum 1990) predictions. The properties of the model's responses were generally consistent with the properties seen in physiological recordings of the LSO, including the physiological response to overall sound level in addition to the response to ILD. Direct comparisons to physiological data were not attempted, but the model did produce rate-ILD curves with similar shapes and properties to physiological curves.

This model was modified to use a continuous distribution of input thresholds instead of a finite number of neurons with different inputs thresholds (Solodovnikov and Reed 2001). This does not have a neural correlate but allows for more direct mathematical analysis of the model's properties. This analysis explained various properties of the model, such as why it was mostly independent of overall sound level. It also showed robustness to changes in the firing thresholds of the model neurons.

As with all neurons, the simplest membrane-based model of LSO neurons is an LINF model with excitatory and inhibitory inputs, so that the model membrane potential is the difference of two shot-noise processes. Such a model was used by Colburn and Moss (1981) and by Diranieh (1992). This model, like the similar MSO model described in Sect. 4.2.2.3 (Colburn et al. 1990), treated input spike trains as impulses filtered with an exponential decay. To take into account inhibitory inputs, the jumps in the inhibitory spike train had negative amplitude while the jumps in the excitatory spike train were positive. Because the potential could not get above threshold in the positive direction but could get arbitrarily large in the negative direction, a minimum potential equal to the opposite of the threshold was imposed to prevent negative spike from pushing the potential too far in the negative direction (more or less playing the role of the Nernst potential for the implied inhibitory channels). The input firing times were computed using a Poisson spike generator with firing rates matched to the firing rates of auditory nerve fibers. If it is assumed that the excitatory and inhibitory spikes have the same magnitude and the same decay time constant, then the only two parameters of the model are the amplitude of a spike relative to the threshold and the decay time constant.

One version of this model used by Diranieh (1992) is really somewhere in-between an LINF and point-neuron model, in that it neglects the voltage-gated sodium current, relying solely on a voltage-gated potassium channel. During an action potential, the depolarization is treated as a fixed increment in the potential after it crosses threshold, but repolarization is handled by the potassium current. The responses of this model were very similar to those of the simpler passive membrane model of Diranieh (1992).

4.3.3 Models Focused on the Temporal Structure of the Onset Response

As discussed in the preceding text, LSO neurons have a characteristic relationship between their steady-state firing rates and the ILD. However, LSO neurons also have a complex and unusual post-stimulus time histogram (PSTH) in response to tone-burst stimuli (Tsuchitani 1977, 1988). The PSTH estimates the short-term average rate of response after the onset of a stimulus (a sustained tone in this case). This response pattern for LSO neurons, which is called a chopper response, is characterized by brief periods of high firing rate followed by short periods of low firing rate, as seen in Fig. 4.5. This pattern of response gradually decays into a more steady sustained response. Chopper firing patterns in the LSO also have a feature called negative serial dependence, where a short ISI tends to be followed by a long one and vice versa.

Initial attempts to model this behavior included only monaural excitatory inputs (Johnson et al. 1986). This model was a black-box descriptive model, based on the statistical properties of LSO spike trains. By treating a spike train as a point process whose properties are dependent on the level of the input at the two ears, an output spike train can be generated (either as a series of spike times or as a series of ISIs).

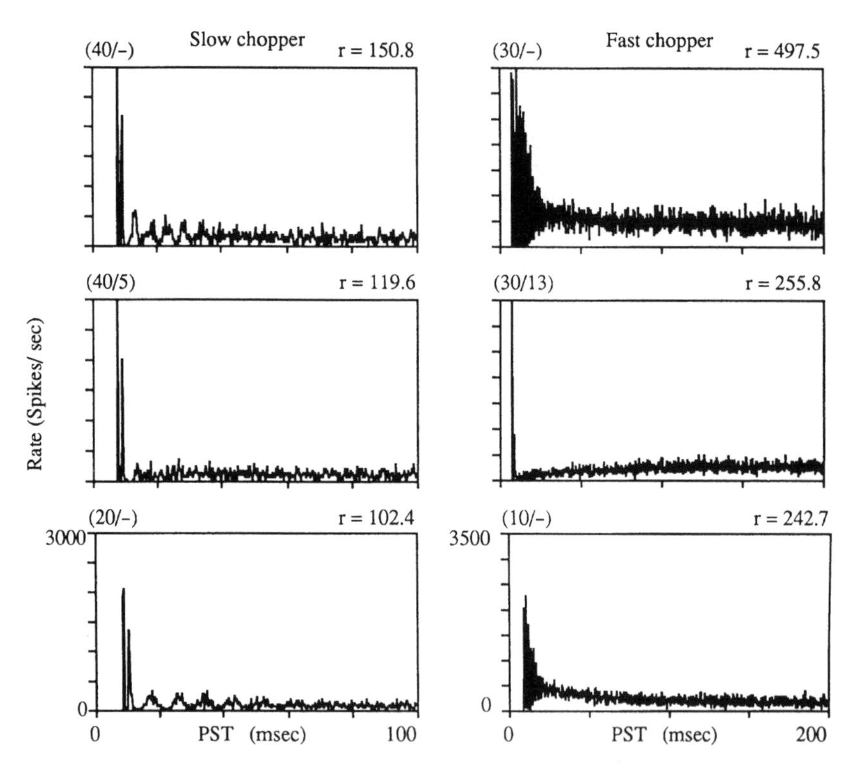

Fig. 4.5 Chopper response illustrations for LSO neurons. The *left* three plots are PSTHs for an LSO neuron with a slow chopper response, and the *right* three plots are PSTHs for an LSO neuron with a fast chopper response. The horizontal axis is the time since the stimulus onset while the vertical axis is spikes per second in a given bin (250 μs bin width for the *left* and 500 μs bin width for the *right*). For each plot, the stimulus intensities are indicated in the *upper left parentheses*: the number to the *left* of the slash is the ipsilateral level in dB above threshold, and the *right* number or *dash* indicates the contralateral level, a *dash* indicating no contralateral stimulation. The *r* number in the *upper right* of each plot is the steady-state firing rate, which can be seen to be larger for the fast chopper neurons than the slow chopper neurons (From Zacksenhouse et al. 1992.)

The spike train was modeled as a point process in which, unlike a Poisson process, the probability of firing in a given time window was dependent not only on the length of the time windows but also on the history of the spike train. To account for negative serial dependence, a delay is imposed on the recovery function after a short ISI but not after a long ISI. The model worked for low firing rates and responses with unimodal ISI histograms, but had trouble tracking high firing rates and responses with bimodal ISI histograms.

An extension of this model incorporated both excitatory and inhibitory inputs to allow for ILD dependence (Zacksenhouse et al. 1992, 1993). In this version of the model, the high firing rates and bimodal ISIs could be accurately modeled by changing the level of the inhibitory input. More specifically, the model indicated that the strength of the recovery function was scaled by the inhibitory input level while the excitatory level affected both the strength of the recovery function and the amplitude of the shifting function. However, this model is purely descriptive and does not postulate physiological mechanisms.

In addition to analyzing the rate-ILD curves of LSO neuron, the Diranieh (1992) point-neuron models described earlier were also used in an attempt to reproduce the chopper response. They had similar results to the Zacksenhouse et al. (1992, 1993) model in that the inhibition appears to scale the recovery function. However, neither of the Diranieh (1992) models showed negative serial dependence.

To make a physiologically accurate model that can account for chopper response properties like negative serial dependence, a multicompartment Hodgkin–Huxley type model was developed (Zacksenhouse et al. 1998). The model had passive dendrites with excitatory synapses, an active axon using only Hodgkin–Huxley sodium and potassium channels, and an active soma with inhibitory synapses and Hodgkin–Huxley sodium and potassium channels. In some cases the soma also included voltage-gated calcium and calcium-gated potassium channels that worked together to generate a cumulative afterhyperpolarization (AHP) effect. This had been previously hypothesized as the source of negative serial dependence (Zacksenhouse et al. 1992).

Like the Diranieh (1992) models, the Zacksenhouse et al. (1998) model lacking the AHP channels also lacked negative serial dependence, although it was able to generate chopper responses that were similar to, but not identical to, LSO responses. By adding the AHP channels, however, the responses were brought in line with LSO responses and the response featured negative serial dependence. However, the Zacksenhouse et al. (1998) model was not able to match the ratio of the amplitudes between the onset and sustained responses.

The model of Zhou and Colburn (2009) described LSO neurons with a relatively simple model designed to explore the effects of AHP channels and their characteristics. This model, in addition to excitatory and inhibitory inputs from each side, included an AHP channel that can be thought of as a calcium-dependent potassium channel. In this study, they explored contributions of membrane AHP channels to the generation of discharge patterns, including serial dependence in the spike trains, as well as the effects on the encoding of ILD information in the LSO. The AHP effect was varied from neuron to neuron by variation in the increment (G_{AHP}) and decay time constant (τ_{AHP}) of the adaptation channel. Model neurons with different values of G_{AHP} and τ_{AHP} simulated the multiple distinct chopper response patterns and level-dependent ISI statistics as observed in vivo (Tsuchitani and Johnson 1985). In the ILD simulations, the gain, regularity, and serial correlations of model discharges also show different ILD dependencies. This model also explains observed differences in binaural and monaural firing patterns that yield the same firing rate but different degrees of regularity (regularity decreases when the stimulation is binaural). They hypothesized that differences in the model AHP time course (τ_{AHP}) explained the differential expression of Kv-potassium channels observed in slices (Barnes-Davies et al. 2004), and differences in the model AHP amplitude (G_{AHP}) reflected different potassium channel densities in the LSO. This study suggests a possible neuron classification based on heterogeneous membrane properties in the LSO. The spatial distribution of certain membrane properties may also link to neural representations of ILD information in distributed responses of LSO neurons.

4.3.4 Models of ITD in the LSO

Although EI neurons in general and LSO neurons in particular have been traditionally associated with ILD cues, they are also sensitive to the ITD of a stimulus. For a stimulus with an ITD, there will be a brief period at the onset of the stimulus where one ear is receiving sound and the other is not. This means that the onset of a stimulus containing an ITD will also have a brief ILD. Further, ongoing ITD determination can be done with EI coincidence detectors in place of EE coincidence detectors. In EE coincidence detection, the neuron requires spikes from two inputs simultaneously in order to fire. In an EI coincidence detector, only excitatory spikes from one input are needed for the neuron to fire. The other input, the inhibitory input, will suppress the excitatory spike only when the two spikes arrive close together in time. In this sort of neuron, instead of firing most vigorously to two spike trains that line up in time, it fires least vigorously. So an EI ITD detector would be similar to an EE one except it would respond minimally at its "best" delay. This has proven important as recordings from the LSO have provided evidence that this nucleus, which receives EI inputs, is sensitive to ITD (Tollin and Yin 2005).

When the phase plots (described in Sect. 4.2.1.1) are analyzed, or equivalently the rate-ITD curves are viewed for different tone frequencies, there is a common minimum. In other words, LSO neurons are consistently "trough-type" or "trougher" neurons. This trougher description applies to responses to amplitude modulation as well, which is especially important because the LSO neurons have predominantly high best frequencies. Since much of the early recordings from LSO neurons were made using high-frequency tonal stimuli, there was no synchronization to the stimulus fine structure and the models were concerned primarily with the dependence of average firing rate on the levels at the two ears and particularly with the ILD.

The Diranieh (1992) models, a LINF model and a point-neuron model which have been discussed in the preceding sections in the context of ILD determination, were also used to model responses to ITD in the LSO. In these models the rate-ITD curves for different sound levels and the time course of the response at different ITDs were looked at. The models showed good agreement with the general shape of rate-ITD curves from physiological recordings, as well as looking at the interaction of ITD and ILD in the model.

Another study looking at ITD analysis in EI neurons was the Marsálek and Lansky (2005) model discussed in Sect. 4.2.2.1. In this work, both EE and EI models were built. In the EE model the order of the spikes is not important, while the EI model was set to produce an output spike only if the inhibitory spike arrives before the excitatory spike. The specific physiological mechanism for the latter approach was not discussed.

4.4 Models of Perceptual Phenomena

In this section, the information encoded by the MSO and LSO is considered in terms of auditory perception and the modeling of perceptual phenomena. The discussion is divided into two categories, azimuthal localization, especially the phenomenon of time-intensity trading, and general binaural processing models, several of which have been discussed in terms of coincidence detectors or difference processing. This discussion is relatively brief; the purpose is to make the reader aware of this work rather than review it in detail.

4.4.1 Time-Intensity Trading

Location is a prominent aspect of sound sensation and it is often perceived strongly for headphone sounds whether or not there is an effort to make the stimuli contain realistic combinations of spatial cues as in virtual displays. Because ITD and ILD are both spatial cues that, at least in mammals, code for azimuthal (horizontal) angle, they can be manipulated separately to give reinforcing or conflicting cues. In psychophysical studies with many stimuli, including tones, the perceptual effects of changing one cue can be compensated for by changing the other cue. This phenomenon is called time-intensity trading (Kuroki and Kaga 2006). A number of models were developed that attempt to explain this phenomenon by predicting a single binaural statistic that is sensitive to both ITD and ILD.

In his article "A place theory of localization," Jeffress (1948) hypothesized that increases in level at the ear would result in a reduced latency of neural response, and thereby hypothesized the level differences would be encoded as time differences (the "latency hypothesis" for time-intensity trading). This hypothesis has not gotten much general support, from either physiological or psychophysical experiments, but it may be a factor in some cases.

Another hypothesis for time-intensity trading might be called the count-comparison hypothesis. This is based on the concept that there are two populations of neurons, for example, on the left and right sides of the brain, and the location of a sound is determined by a comparison of the counts of active neurons on the two sides, with larger differences leading to more lateralized images. This is consistent with the model of van Bergeijk (1962) noted earlier, which he ascribed to von Békésy (1930) and which is illustrated in Fig. 4.6. This basic model was pursued empirically by Hall (1965), who recorded responses to click stimuli in "the accessory nucleus" of the SOC and interpreted his responses in terms of the van Bergeijk model.

Other researchers have specified count-comparison models with different basic mechanisms. For example, Sayers and Cherry (1957) suggested a count-comparison model based on the interaural cross-correlation function. Specifically, the positive-delays are integrated together, the negative-delays are integrated together, and these integrations are weighted by the relative levels at the two ears to form the left and right counts. An explicit physiological realization of this model was used successfully by Colburn and Latimer (1978) to form decision variables for discrimination of ITD.

Two other models that are based on weighted cross-correlation functions were suggested by Stern and Colburn (1978) and by Lindemann (1986a, b). The Stern model estimates lateral position from the centroid of a cross-correlation function that is weighted by a function determined by the ILD. The Lindemann model has a complex structure, also based on the cross-correlation mechanism that includes attenuation along the delay axis. The attenuation is determined by the strength of the signal from the other input. This structure is shown in Fig. 4.7. The inhibitory weighting is seen in the multipliers along the delay line in Fig. 4.7a, and the computation of the inhibitory weighting factor, which depends on the propagating signal and on the local correlation, is shown in Fig. 4.7b.

4.4.2 General Binaural Models

Many models of binaural hearing can be related to the information available in the patterns of MSO and LSO neurons. For example, the Colburn (1977) model is based on a set of coincidence counts and performance in psychophysical experiments can be predicted assuming optimum use of these counts. The most applied general model of binaural processing is probably the equalization-cancellation (EC) model of Durlach (1963, 1972). Although this model is usually specified in terms of the processing of signal waveforms, it has been described in terms of an EI processor by Breebaart et al. (2001a, b, c). They discuss an implementation based on internal delays and attenuations that might be implemented by neural circuitry related to the MSO and LSO. A more detailed description of older general models of binaural processing can be found in previous chapters by Colburn and Durlach (1978) and by Colburn (1996).

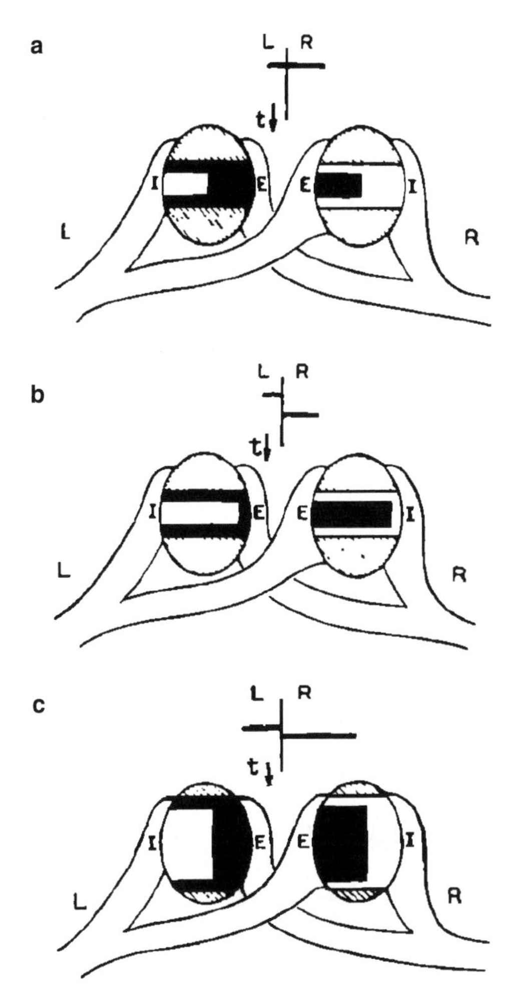

Fig. 4.6 Diagram showing how a count-comparison code for ITD and ILD in the SOC could function. *Black areas* represent neurons that are receiving excitatory inputs while *white areas* represent neurons that are receiving inhibitory inputs. In this diagram the signals from the input fibers move horizontally across the nucleus, exciting or inhibiting the neurons as they go. When the inhibition and excitation meet they cancel and the signals stop. Neurons closer to the midline in the vertical axis have lower thresholds for activation while the threshold increases for neurons higher or lower than the midline. The position of the sound is determined by comparing the number of active neurons in each SOC. More activity in the left SOC indicates that the sound is located to the right while more activity in the right SOC indicates the opposite. In (**a**), the sound has zero ITD but the level in the right ear is greater. In this case, for neurons close to the vertical midline, those that are sensitive to stimulus levels present in the left ear, the excitation and inhibition meet in the middle. However, for neurons slightly higher and lower only the intensity in the right ear is strong enough to trigger a response and so the signal from the right ear travels all the way across the nucleus without encountering the oppose wave from the left ear. This leads to more neurons responding in the left SOC. In (**b**), the ITD is shifted toward the left while the ILD is the same as in the previous case. Because of the ITD the binaurally stimulated neurons have excitation and inhibition meeting at a location offset to the right, but this is canceled by the neurons that are stimulated only by the right, leading to an equal number of neurons excited in both SOCs and the perception of the sound being from the midline. (**c**) also has excitation and inhibition balanced despite there being a significant ITD and ILD. However, in this case the overall sound level is higher leading to more total excitation in both nuclei (From van Bergeijk 1962.)

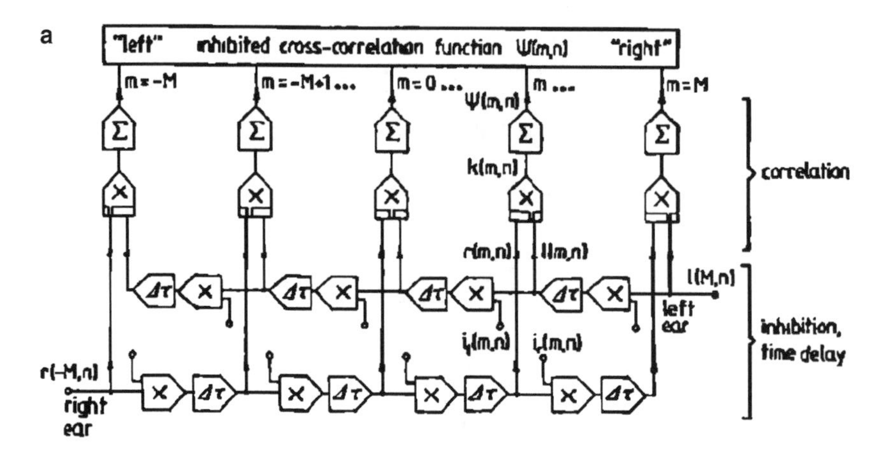

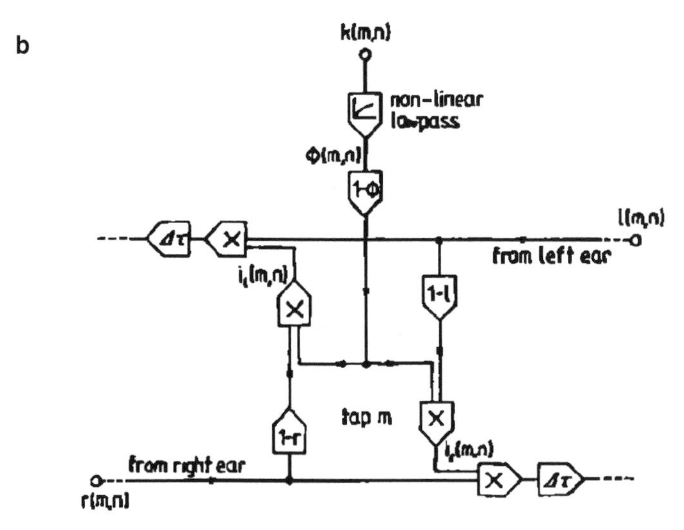

Fig. 4.7 Lindemann's inhibited cross-correlation model, combining ITD and ILD analysis. In this diagram, the variable *m* represents a discrete ITD between the constants –M and M, the variable *n* represents a discrete time point, $\Delta\tau$ is the time spacing between ITD values, X is multiplication, Σ is summation, r and l are the signals from the left and right CN, respectively, i is the inhibitory feed forward signal, k is the inhibitory feedback signal, Φ is the output of a nonlinear low-pass filter, and ψ is the resulting overall function. (**a**) Overall model structure with the cross-correlation prominent. This is largely the same as the conventional Jeffress delay line, with a ladder-like pattern of axons and with branches separated by time delays. The difference lies in the multiplication that is inserted before each time delay. Here ILD-sensitive signals are combined with the excitatory signals to form the inhibitory multiplier, allowing for ILD discrimination. This ILD influence accumulates, so that every branch of the neuron is influenced by both its own inhibitory input as well as the inhibitory inputs from every previous branch. (**b**) The inhibitory component in more detail. There are two parts to the inhibitory component. The first, k, is the combination of the outputs of neighboring cross-correlation units, creating a lateral inhibition system that sharpens the response curve by suppressing neurons that are slightly offset from the most strongly firing neuron. The other component combines contralateral inhibitory components with ipsilateral excitatory components to create an ILD-sensitive IE component. The horizontal multiplication and delay components in this figure are the same as the components in (**a**) (From Lindemann 1986a.)

4.5 Summary and Comments

This review is focused on models of the medial and lateral nuclei of the superior olivary complex, that is, the MSO and LSO, in mammals and the analogous nuclei in birds, that is, the nucleus laminaris (NL) and the ventral nucleus of the lateral lemniscus pars posterior (VNLp). Primary attention was directed toward the sensitivity of neurons in these regions to interaural time and level differences (ITDs and ILDs).

Considering the MSO and NL, almost all available models of MSO and NL neurons are based on coincidence detection, which is accepted as the basic model for these neurons. Simple implementations such as EE neurons with simple EPSPs from each side show the basic properties observed, although it is clear that this model does not capture all aspects of the empirical responses. With our current state of knowledge, it appears that there may be important differences from species to species in the importance, role, and properties of inhibition as well as the mechanism for and the distribution of effective internal delays that determine the BDs of neurons.

Partly because of uncertainty about interspecies differences and partly due to a paucity of empirical results, the role of inhibition in the generation of response patterns in the MSO remains to be determined. There is clear evidence for a role of inhibition in several species at least in some neurons, there are clearly differences among species, and there are multiple theoretical interpretations for many observations.

Considering the LSO and VNLp, models universally reflect basic EI characteristics so that ipsilateral inputs are primarily excitatory and contralateral inputs are inhibitory. There is some uncertainty about the presence of this basic network in some primates, because of the lack of supportive neuroanatomy, but this is also still an unresolved issue. In the temporal domain, LSO neurons appear to have chopper characteristics that reflect AHP effects, which also lead to sequential correlations in the firing patterns of these neurons. Finally, it is also clear that LSO (and presumably VNLp) neurons are sensitive to the stimulus ITD, both onset/offset and ongoing delays change the rate of firing of these neurons, as one predicts for simple EI neurons with time-synchronized fluctuations in the input rates. However, what role, if any, this sensitivity plays in perception is unknown.

References

Adams JC, Mugnaini E (1990) Immunocytochemical evidence for inhibitory and disinhibitory circuits in the superior olive. Hear Res 49:281–298.

Armin HS, Edwin WR, David MH Mechanisms for adjusting interaural time differences to achieve binaural coincidence detection (in press).

Ashida G, Abe K, Funabiki K, Konishi M (2007) Passive soma facilitates submillisecond coincidence detection in the owl's auditory system. J Neurophysiol 97:2267–2282.

Barnes-Davies M, Barker MC, Osmani Forsythe ID (2004) Kv1 currents mediate a gradient of principal neuron excitability across the tonotopic axis in the rat lateral superior olive. Eur J Neurosci 19:325–333.
Blum JJ, Reed MC (1991) Further studies of a model for azimuthal encoding: lateral superior olive neuron response curves and developmental processes. J Acoust Soc Am 90:1968–1978.
Boudreau JC, Tsuchitani C (1968) Binaural interaction in the cat superior olive s segment. J Neurophysiol 31:442–454.
Brand A, Behrend O, Marquardt T, McAlpine D, Grothe B (2002) Precise inhibition is essential for microsecond interaural time difference coding. Nature 417:543–547.
Breebaart J, van de Par S, Kohlrausch A (2001a) Binaural processing model based on contralateral inhibition. I. Model structure. J Acoust Soc Am 110:1074–1088.
Breebaart J, van de Par S, Kohlrausch A (2001b) Binaural processing model based on contralateral inhibition. II. Dependence on spectral parameters. J Acoust Soc Am 110:1089–1104.
Breebaart J, van de Par S, Kohlrausch A (2001c) Binaural processing model based on contralateral inhibition. III. Dependence on temporal parameters. J Acoust Soc Am 110:1105–1117.
Brew H (1998) Modeling of interaural time difference detection by neurons of mammalian superior olivary nucleus. Assoc Res Otolaryngol 25:680.
Brughera AR, Stutman ER, Carney LH, Colburn HS (1996) A model with excitation and inhibition for cells in the medial superior olive. Audit Neurosci 2:219–233.
Cant NB, Hyson RL (1992) Projections from the lateral nucleus of the trapezoid body to the medial superior olivary nucleus in the gerbil. Hear Res 58:26–34.
Carr CE, Konishi M (1988) Axonal delay lines for time measurement in the owl's brainstem. Proc Natl Acad Sci U S A 85:8311–8315.
Carr CE, Konishi M (1990) A circuit for detection of interaural time differences in the brain stem of the barn owl. J Neurosci 10:3227–3246.
Chirila FV, Rowland KC, Thompson JM, Spirou GA (2007) Development of gerbil medial superior olive: integration of temporally delayed excitation and inhibition at physiological temperature. J Physiol 584:167–190.
Clack JA (1997) The evolution of tetrapod ears and the fossil record. Brain Behav Evol 50:198–212.
Colburn HS (1973) Theory of binaural interaction based on auditory-nerve data. I. General strategy and preliminary results on interaural discrimination. J Acoust Soc Am 6:1458–1470.
Colburn HS (1977) Theory of binaural interaction based on auditory-nerve data. II. Detection of tones in noise. J Acoust Soc Am 54:525–533.
Colburn HS (1996) Computational models of binaural processing. In: Hawkins HL, McMullen TA, Popper AN, Fay RR (eds), Auditory Computation. New York: Springer, pp. 332–400.
Colburn HS, Durlach NI (1978) Models of binaural interaction. In: Carterette EC, Freidman M (eds), Handbook of Perception, Vol. IV. New York: Academic, pp. 467–518.
Colburn HS, Latimer JS (1978) Theory of binaural interaction based on auditory-nerve data. III. Joint dependence on interaural time and amplitude differences in discrimination and detections. J Acoust Soc Am 64:95–106.
Colburn HS, Moss PJ (1981) Binaural interaction models and mechanisms. In: Syka J, Aitkin L (eds), Neuronal Mechanisms of Hearing. New York: Plenum, pp. 283–288.
Colburn HS, Han YA, Culotta CP (1990) Coincidence model of MSO responses. Hear Res 49:335–346.
Dasika VK, White JA, Carney LH, Colburn HS (2005) Effects of inhibitory feedback in a network model of avian brain stem. J Neurophysiol 94:400–414.
Dasika VK, White JA, Colburn HS (2007) Simple models show the general advantages of dendrites in coincidence detection. J Neurophysiol 97:3449.
Dayan P, Abbott LF (2001) Theoretical Neuroscience. Cambridge, MA: MIT Press, pp. 162–166.
Diranieh YM (1992) Computer-based neural models of single lateral superior olivary neurons. M.S. Thesis, Boston University, Boston, MA.
Donoghue PC, Benton MJ (2007) Rocks and clocks: calibrating the tree of life using fossils and molecules. Trends Ecol Evol 22:424–431.
Durlach NI (1963) Equalization and cancellation theory of binaural masking-level differences. J Acoust Soc Am 35:1206–1218.

Durlach NI (1972) Binaural signal detection: equalization and cancellation theory. In: Tobias J (ed), Foundations of Modern Auditory Theory. New York: Academic, pp. 371–462.
Galambos R, Davis H (1943) The response of single auditory-nerve fibers to acoustic stimulation. J Neurophysiol 6:39–57.
Glendenning KK, Hutson KA, Nudo RJ, Masterton RB (1985) Acoustic chiasm II: Anatomical basis of binaurality in lateral superior olive of cat. J Comp Neurol 232:261–285.
Goldberg JM, Brown PB (1969) Response of binaural neurons of dog superior olivary complex to dichotic tonal stimuli: some physiological mechanisms of sound localization. J Neurophysiol 32:613–636.
Grau-Serrat V, Carr CE, Simon JZ (2003) Modeling coincidence detection in nucleus laminaris. Biol Cybern 89:388–396.
Grothe B, Sanes DH (1993) Bilateral inhibition by glycinergic afferents in the medial superior olive. J Neurophysiol 69:1192–1196.
Guinan JJ, Guinan SS, Norriss BE (1972a) Single auditory units in the superior olivary complex I: responses to sounds and classifications based on physiological properties. Int J Neurosci 4:101–120.
Guinan JJ, Norriss BE, Guinan SS (1972b) Single auditory units in the superior olivary complex II: locations of unit categories and tonotopic organization. Int J Neurosci 4:147–166.
Hall JL (1965) Binaural interaction in the accessory superior-olivary nucleus of the cat. J Acoust Soc Am 37:814–823.
Han Y, Colburn HS (1993) Point-neuron model for binaural interaction in MSO. Hear Res 68:115–130.
Harper NS, McAlpine D (2004) Optimal neural population coding of an auditory spatial cue. Nature 430:682–686.
Hodgkin AL, Huxley AF (1952) Propagation of electrical signals along giant nerve fibers. Proc R Soc Lond B Biol Sci 140:177–183.
Jeffress LA (1948) A place theory of sound localization. J Comp Physiol Psychol 41:35–39.
Jeffress LA (1958) Medial geniculate body – a disavowal. J Acoust Soc Am 30:802–803.
Johnson DH, Tsuchitani C, Linebarger DA, Johnson MJ (1986) Application of a point process model to responses of cat lateral superior olive units to ipsilateral tones. Hear Res 21:135–159.
Kandler K, Gillespie D (2005) Developmental refinement of inhibitory sound-localization circuits. Trends Neurosci 28:290–296.
Koppl C, Carr CE (2008) Maps of interaural time difference in the chicken's brainstem nucleus laminaris. Biol Cybern 98:541–559.
Kulesza RJ (2008) Cytoarchitecture of the human superior olivary complex: nuclei of the trapezoid body and posterior tier. Hear Res 241:52–63.
Kuroki S, Kaga K (2006) Better time-intensity trade revealed by bilateral giant magnetostrictive bone conduction. Neuroreport 17:27.
Lindemann W (1986a) Extension of a binaural cross-correlation model by contralateral inhibition. I. Simulation of lateralization for stationary signals. J Acoust Soc Am 80:1608–1622.
Lindemann W (1986b) Extension of a binaural cross-correlation model by contralateral inhibition. II. The law of the first wave front. J Acoust Soc Am 80:1623–1630.
Marsálek P, Lansky P (2005) Proposed mechanisms for coincidence detection in the auditory brainstem. Biol Cybern 92:445–451.
McAlpine D, Jiang D, Palmer AR (2001) A neural code for low-frequency sound localization in mammals. Nat Neurosci 4:396–401.
Mills AW (1960) Lateralization of high-frequency tones. J Acoust Soc Am 32:132–134.
Moiseff A, Konishi M (1983) Binaural characteristics of units in the owl's brainstem auditory pathway: precursors of restricted spatial receptive fields. J Neurosci 3:2553–2562.
Pecka M, Brand A, Behrend O, Grothe B (2008) Interaural time difference processing in the mammalian medial superior olive: the role of glycinergic inhibition. J Neurosci 28:6914–6925.
Reed MC, Blum JJ (1990) A model for the computation and encoding of azimuthal information by the lateral superior olive. J Acoust Soc Am 88:1442–1453.
Richter EA, Norris BE, Fullerton BC, Levine RA, Kiang NY (1983) Is there a medial nucleus of the trapezoid body in humans? Am J Anat 168:157–166.

Rieke F, Warland D, van Steveninck R, Bialek W (1997) Spikes: Exploring the Neural Code. Cambridge, MA: MIT Press.

Rothman JS, Young ED, Manis PB (1993) Convergence of auditory nerve fibers onto bushy cells in the ventral cochlear nucleus: implications of a computational model. J Neurophysiol 70:2562–2583.

Sayers BM, Cherry EC (1957) Mechanism of binaural fusion in the hearing of speech. J Acoust Soc Am 29:973–987.

Schwartz IR (1992) The superior olivary complex and lateral lemniscal nuclei. In: Webster DB, Popper AN, Fay RR (eds), The Mammalian Auditory Pathway: Neuroanatomy. New York: Springer, pp. 117–167.

Shackleton TM, Skottun BC, Arnott RH, Palmer AR (2003) Interaural time difference discrimination thresholds for single neurons in the inferior colliculus of guinea pigs. J Neurosci 23:716–724.

Simon JZ, Carr CE, Shamma SA (1999) A dendritic model of coincidence detection in the avian brainstem. Neurocomputing 26–27:263–269.

Smith AJ, Owens S, Forsythe ID (2000) Characterisation of inhibitory and excitatory postsynaptic currents of the rat medial superior olive. J Physiol 529:681–698.

Smith PH (1995) Structural and functional differences distinguish principal from nonprincipal cells in the guinea pig MSO slice. J Neurophysiol 73:1653–1667.

Solodovnikov A, Reed MC (2001) Robustness of a neural network model for differencing. J Comput Neurosci 11:165–173.

Stern R, Colburn H (1978) Theory of binaural interaction based on auditory-nerve data. IV. A model for subjective lateral position. J Acoust Soc Am 64:127–140.

Strata P, Harvey R (1999) Dale's principle. Brain Res Bull 50:349–350.

Strutt JW (1907) On our perception of sound direction. Philos Mag 13:214–232.

Svirskis G, Dodla R, Rinzel J (2003) Subthreshold outward currents enhance temporal integration in auditory neurons. Biol Cybern 89:333–340.

Tollin DJ, Yin TCT (2005) Interaural phase and level difference sensitivity in low-frequency neurons in the lateral superior olive. J Neurosci 25:10648–10657.

Tsuchitani C (1977) Functional organization of lateral cell groups of cat superior olivary complex. J Neurophysiol 40:296–318.

Tsuchitani C (1988) The inhibition of cat lateral superior olive unit excitatory responses to binaural tone bursts. I. The transient chopper response. J Neurophysiol 59:164–183.

Tsuchitani C, Johnson DH (1985) The effects of ipsilateral tone burst stimulus level on the discharge patterns of cat lateral superior olivary units. J Acoust Soc Am 77:1484–1496.

van Bergeijk WA (1962) Variation on a theme of Békésy: a model of binaural interaction. J Acoust Soc Am 34:1431–1437.

von Békésy G (1930) Zur theorie des hörens; über das richtungshören bei einer zeitdifferenz oder lautstärkenungleichheit der beiderseitigen schalleinwirkungen. Physik Zeits 31:824–835.

Yang L, Monsivais P, Rubel EW (1999) The superior olivary nucleus and its influence on nucleus laminaris: a source of inhibitory feedback for coincidence detection in the avian auditory brainstem. J Neurosci 19:2313–2325.

Yue L, Johnson DH (1997) Optimal binaural processing based on point process models of preprocessed cues. J Acoust Soc Am 101:982–992.

Zacksenhouse M, Johnson DH, Tsuchitani C (1992) Excitatory/inhibitory interaction in the LSO revealed by point process modeling. Hear Res 62:105–123.

Zacksenhouse M, Johnson DH, Tsuchitani C (1993) Excitation effects on LSO unit sustained responses: point process characterization. Hear Res 68:202–216.

Zacksenhouse M, Johnson DH, Williams J, Tsuchitani C (1998) Single-neuron modeling of LSO unit responses. J Neurophysiol 79:3098–3110.

Zhou Y, Colburn HS (2009) Effects of membrane afterhyperpolarization on interval statistics and interaural level difference coding in the lateral superior olive. J Neurophysiol (in review).

Zhou Y, Carney LH, Colburn HS (2005) A model for interaural time difference sensitivity in the medial superior olive: Interaction of excitatory and inhibitory synaptic inputs, channel dynamics, and cellular morphology. J Neurosci 25:3046–3058.

Chapter 5
The Auditory Cortex: The Final Frontier

Jos J. Eggermont

5.1 Introduction

The auditory cortex consists of 10–15 interconnected areas or fields whose neurons receive a modest input from the thalamus and about 10–100 times more input from other auditory cortical areas and nonauditory cortical fields from the same and contralateral hemisphere. Modeling this conglomerate as a black box functional network model is potentially doable (Stephan et al. 2000), but that does not give us much insight into how individual cortical areas compute and the nature of the output from those areas to cognitive and motor systems. At the other end of the scale, there is the challenge of realistic modeling of the canonical cortical neural network that is typically based on primary visual cortex (Martin 2002). When implemented for primary auditory cortical columns this needs detailed modeling of 10–15 different cell types (Watts and Thomson 2005) with different ion channels and neural transmitter and modulatory systems; even such minimal circuits present daunting complexities. The main problem for the neuroscientist is of course to identify the computational problem that the auditory cortex has to solve. This chapter reviews the basic structural and functional elements for such models on the basis of what is currently known about auditory cortical function and processing. The emphasis here is on vocalizations, speech, and music. Some promising analytic and modeling approaches that have been proposed recently are discussed in light of two views of cortical function: as an information processing system and as a representational system.

J.J. Eggermont (✉)
Department of Psychology, University of Calgary, Calgary, AB, Canada T2N 1N4
e-mail: eggermon@ucalgary.ca

R. Meddis et al. (eds.), *Computational Models of the Auditory System*,
Springer Handbook of Auditory Research 35, DOI 10.1007/978-1-4419-5934-8_5,

5.2 The Primate Auditory Cortex

The auditory cortex comprises the cortical areas that are the preferential targets of neurons in either the ventral or dorsal divisions of the medial geniculate body (MGB) in the thalamus. By this definition (de la Mothe et al. 2006), three regions of the superior temporal cortex comprise the auditory cortex in primates: core, belt, and parabelt. The core, belt, and parabelt regions represent hierarchical information processing stages in cortex. Each of these three major auditory cortical regions consists of two or more areas, or subdivisions, in which thalamic and cortical inputs are processed in parallel.

Evidence from lesion studies in animals and strokes in humans suggests that the discrimination of temporally structured sounds such as animal vocalizations and human speech requires auditory cortex. Specifically, the local cooling experiments of Lomber and Malhotra (2008) found that in cats the primary auditory cortex (A1) and the anterior auditory field (AAF) are crucial for behavioral sound discrimination, whereas posterior auditory field (PAF) is required for sound localization.

5.3 Representation of Vocalizations and Complex Sounds in the Auditory Cortex

Natural sounds, including animal vocalizations and speech, have modulation spectra confined to low temporal and spectral frequencies. One could argue that the auditory system is adapted to process behaviorally relevant and thus natural sounds. Therefore, neural representations and computations in the auditory system will reflect the statistics of behaviorally relevant sounds. Animal vocalizations and other environmental sounds are both well represented by the filtering process in the cochlea and the hair cells, suggesting that the initial stage of auditory processing could have evolved to optimally represent the different statistics of these two important groups of natural sounds (Lewicki 2002).

Natural as well as spectrotemporally morphed cat vocalizations are represented in cat auditory cortex (Gehr et al. 2000; Gourévitch and Eggermont 2007b). In the study of Gehr et al., about 40% of the neurons recorded in A1 showed time-locked responses to major peaks in the vocalization envelope, 60% responded only at the onset. Simultaneously recorded multiunit (MU) activity of these peak-tracking neurons on separate electrodes was significantly more synchronous during stimulation than under silence. Thus, the representation of the vocalizations is likely synchronously distributed across the cortex. The sum of the responses to the low- (<2.5 kHz) and high-frequency (≥2.5 kHz) part of a natural meow, was larger than the neuronal response to the natural meow itself, suggesting that strong lateral inhibition is shaping the response to the natural meow. In this sense, the neurons are combination sensitive. Analysis of the mutual information in the firing rate suggested that the activity of at least 95 recording sites in A1 would be needed to reliably distinguish between nine different vocalizations. This suggests that a distributed representation based on temporal stimulus aspects may be more efficient than one based on firing rate.

In a follow-up of this study, Gourévitch and Eggermont (2007b) showed the neural representation of the same natural and morphed cat vocalizations in four different areas of the auditory cortex. Again, MU activity recorded in A1 of anesthetized cats occurred mainly at onsets (<200 ms latency) and at subsequent major peaks of the vocalization envelope, and was significantly inhibited during the stationary course of the stimuli. The first 200 ms of processing appeared crucial for discrimination of a vocalization by neurons in A1. The dorsal part of AI showed inhibition for carrier-frequency altered and temporally expanded vocalizations. The ventral and posterior parts of AI were both more strongly activated by the time-reversed meow. Thus A1 as a whole does not respond uniformly to natural and morphed cat vocalizations.

Sustained firing neurons in the posterior ectosylvian gyrus (EP) discriminated temporal envelope alterations of the meow and time reversion thereof on the basis of neural synchrony. These findings suggest a potentially important role of EP in the detection of information conveyed by the alterations of vocalizations. Discrimination of the neural responses to different alterations of vocalizations, could be based on either firing rate, type of temporal response, or neural synchrony, suggesting that all these are likely simultaneously used in processing of natural and altered conspecific vocalizations.

The spectrotemporal discharge pattern of spatially distributed neuron populations in A1 of marmosets was correlated with the spectrotemporal pattern of a complex natural vocalization (Wang et al. 1995). In the time domain, responses of individual A1 units were locked to the envelope of a portion of a complex vocalization. Neuronal responses to a complex vocalization were distributed very widely across A1, but the spatially distributed neural responses were strongly synchronized to both the stimuli and to each other. As a result, a coherent representation of the integrated spectrotemporal characteristics of a particular vocalization was present in the population response. Marmoset A1 neurons often responded more vigorously to natural than to time-reversed twitter calls, although the spectral energy distribution in the natural and time-reversed signals is the same. In contrast, neurons recorded in cat A1 showed no such selectivity for natural marmoset calls (Kadia and Wang 2003). This was interpreted as call selectivity being the result of natural exposure to such sounds.

To investigate whether call selectivity in A1 can arise purely as a result of auditory experience, Schnupp et al. (2006) recorded responses to marmoset calls in A1 of naive ferrets, as well as in ferrets that had been trained to recognize these natural marmoset calls. Training did not induce call selectivity for the trained vocalizations in A1. However, trained animals recognized marmoset twitters with a high degree of accuracy, and represented the vocalizations similarly to those of naïve animals. These results cast doubt on the functional significance of call-selective neurons in the processing of animal communication sounds at the level of A1. As Schnupp et al. (2006) wrote, "the role of A1 in the processing of vocalization stimuli would be predominantly that of representing acoustic features of stimuli through temporal pattern codes in a nonselective manner, whereas a more auditory object-selective representation of the acoustic environment may well emerge in higher-order areas of auditory cortex."

These studies all suggest that the representation of species-specific vocalizations in auditory cortex is temporally integrated and spectrally distributed by spatially dispersed and synchronized cortical neural populations that potentially respond to individual vocalizations in a specific way. Some parts of auditory cortex are better suited for that task than others.

5.4 Cortical Spatial Maps

5.4.1 Tonotopic Maps

Several auditory cortical areas show an orderly mapping of characteristic frequency (CF) on a spatial dimension: the tonotopic map, shown for cat A1 in Fig. 5.1a. In cat cortex there are at least five tonotopically organized areas, and likely also at least five in core and belt areas, but not in parabelt areas of primates.

The tonotopic organization within cortical areas is not static but is dynamically maintained and stability results from a balance of excitation and inhibition. Receptors activate cortical space to an extent that is influenced by competition between inputs and relative synaptic use, so that increasing use probably increases cortical space and decreasing use probably decreases this. Such a mechanism could account for the improvements in perceptual and motor skills that occur with practice (Kaas 1987). Peripheral lesions, such as sensorineural hearing loss, dramatically change the cortical organization (Robertson and Irvine 1989; Rajan et al. 1993; Eggermont and Komiya 2000) in such a way that cortical neurons with CFs in the frequency range where the hearing loss is greater than 30 dB acquire new CFs that are all similar to those of the normal edge frequencies of the audiogram (Fig. 5.1b). This process requires a few weeks, and is again caused by the imbalance of excitation levels across the array of auditory nerve fibers with those in the hearing loss range having typically lower spontaneous and driven firing rates. This reorganization process can be prevented by applying an acoustic environment with frequency content limited to the hearing loss range immediately after the trauma (Noreña and Eggermont 2005). In contrast to the suggestion (Kaas 1987) that "increasing use probably increases cortical space and decreasing use probably decreases cortical space," Noreña et al. (2006) have recently found that presenting a 4- to 20-kHz random multitone sound for an extended period of time, and at a level that does not produce hearing loss, in fact creates a functional lesion. Here, only a few of the cortical neurons that would normally be tuned to the 4- to 20-kHz range still do, the other ones have become tuned to CFs at the higher and lower edge of the exposed frequency range with increased sensitivity. The current plasticity dogma states that cortical map expansion or retraction occurs only when the sounds are attended to or are in any other way relevant to the animal (Keuroghlian and

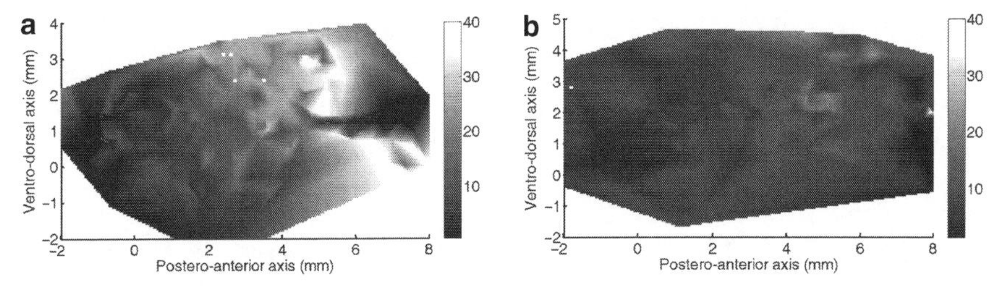

Fig. 5.1 Interpolated tonotopic maps showing CF as a function of distance from the tip of the posterior ectosylvian sulcus (coordinate 0, 0) for all recordings from control (**a**) and noise-exposed (**b**) cats. Basically, the CF of each recording site was plotted and these values were then linearly interpolated. The *gray-scale bar* indicates the CF of the recording sites. The primary auditory cortex is roughly localized between −1 and +5 mm along the posterior–anterior axis and between −1 and +4 mm along the ventrodorsal axis. The map in control cats (**a**) shows a CF gradient from left to right and then reverses (at around the 5- to 6-mm mark along the posterior–anterior axis) where the A1 borders the AAF. The tonotopic map following noise trauma is shown in (**b**). One notes that the map is clearly reorganized and that nearly one-half of A1 is now dedicated to CFs of about 10–15 kHz, which represented the cutoff frequency of the noise-induced hearing loss (Recalculated from data of Noreña and Eggermont 2005.)

Knudsen 2007), and this appears to be not generally valid (Kral and Eggermont 2007). It is hard to see why cats would pay attention to the frequency ranges outside the stimulated range: the 20- to 40-kHz range is likely used to detect moving prey but is behaviorally not very meaningful in a laboratory environment. In contrast, the below 4 kHz range is meaningful as it covers the dominant harmonics of vocalizations; because there is no downward spread of masking, this would be no different from the absence of the enhanced acoustic environment. Bottom-up explanations for this phenomenon are based on a Hebbian type of competition between strongly activated thalamocortical inputs resulting in synaptic gain decrease (Rothman et al. 2009) and horizontal fiber inputs from the nonaffected frequency regions that are upregulated because of decreasing lateral inhibition (Noreña et al. 2006). As mentioned above, the global convergence of input from thalamic and cortical cells outside the targeted frequency domain in A1 and AAF (Lee et al. 2004a) could form the basis for this context-dependent processing and for intracortical plasticity and reorganization.

5.4.2 *Pitch Neurons*

The ability to process pitch is of major importance to music and speech perception. Although the acoustic cues necessary for the perception of pitch are mainly of a temporal nature, human subjects describe pitch as having a spatial dimension (i.e., musical scale). The processing of pitch, a central aspect of music perception, uses different neural representations from other perceptual functions. Studies using behavior combined with lesion techniques, as well as brain imaging methods, demonstrate that tonal processing recruits mechanisms in areas of the right auditory cortex. Specifically, the core auditory area in the right hemisphere appears to be crucial for fine-grained representation of pitch information. Processing of pitch patterns, such as occurs in melodies, requires higher-order cortical areas, and interactions with the frontal cortex (Zatorre 2001).

Absolute pitch (AP) is the ability to identify the pitch of any tone in the absence of a musical context or a reference tone (Takeuchi and Hulse 1993). The mechanism and neural basis underlying this ability remain unclear. There is evidence that AP correlates highly with early exposure to music and it has been suggested that AP manifests itself in a critical period of brain development. One region that is of interest in regard to AP is the planum temporale (PT), a cortical parabelt region historically associated with language and auditory processing (Itoh et al. 2005). The PT is found to be asymmetric in normal right-handed individuals, being larger in the left hemisphere (Keenan et al. 2001). In addition, studies of melody and rhythm perception have elucidated mechanisms of hemispheric specialization. As one might expect, the neural processing of basic musical pitch information takes place within auditory processing areas particularly the right superior temporal gyrus (Limb 2006).

Patterson et al. (2002) used functional magnetic resonance imaging (fMRI) to identify the main stages of melody processing in the auditory pathway. Spectrally matched sounds, based on iterated-ripple noise that either produced no pitch, a fixed pitch, or a melody were all found to activate core and belt areas on Heschl's gyrus (HG) and the PT. Only in the lateral half of HG did sounds with pitch produce more activation than those without pitch. When the pitch was varied to produce a melody, there was activation in regions beyond HG and PT, specifically in the superior temporal gyrus (STG) and planum polare (PP). The results suggest a hierarchy of pitch processing in which the center of activity moves anterolaterally away from primary auditory cortex as the processing of melodic sounds proceeds. The road to pitch extraction suggested by Patterson et al. was as follows. (1) The extraction of time-interval information from the neural firing pattern in the auditory nerve probably occurs in the brain stem, midbrain, and thalamus. (2) Determining the specific value of a pitch and its salience from the time-interval information probably occurs in lateral HG. (3) Determining that the pitch changes in discrete steps and tracking the changes in a melody probably occurs beyond auditory cortex in STG and/or lateral PP. These latter processes associated with melody processing rather than pitch extraction per se likely give rise to the hemispheric asymmetries observed in neuropsychological and functional neuroimaging studies.

The perception of pitch is not unique to humans and has been experimentally demonstrated in several animal species. Bendor and Wang (2005) showed the existence of neurons in the auditory cortex of marmoset monkeys that respond to both pure tones and missing fundamental harmonic complex sounds with the same f_0, providing a neural correlate for pitch as well as pitch constancy. These pitch-selective neurons were located in a restricted low-frequency cortical region near the anterolateral border of the primary auditory cortex in both the left and right hemisphere. This location is consistent with the findings of Patterson et al. (2002) in humans.

5.5 Speech, Language, and Music

Music and speech are likely the most cognitively complex sounds used by humans. Music and speech processing domains share a dependence on modulations of acoustic information bearing parameters (Suga 1989). As we have seen, the auditory cortices in the two hemispheres are relatively specialized, such that temporal resolution is better in left auditory cortical areas and spectral resolution is better in right auditory cortical areas. The assumption of an intimate connection between music and speech is corroborated by the findings of overlapping and shared neural resources for music and language processing in both adults and children. In this respect it appears that the human brain, at least at an early age, does not treat language and music as strictly separate domains, but rather treats language as a special case of music (Koelsch and Siebel 2005). Arguments that music is a special case of speech have also been made (Ross et al. 2007).

Functional neuroimaging techniques have demonstrated a close relationship between the neural processing of music and language, both syntactically and semantically. Greater neural activity and increased volume of gray matter in Heschl's gyrus has been associated with musical aptitude. Activation of Broca's area, a region traditionally considered a language area, is also important in interpreting whether a note is on or off key. Perception of phonetic categories often depends on the relative timing of acoustic events within this range (such as the time between a noise burst and the onset of phonation in a syllable such as /pa/ (Eggermont 1995)). Shannon et al. (1995) degraded speech by removing most spectral information but retaining temporal variations and showed that after some training relatively good speech comprehension could be obtained with as few as two spectral channels. This indicates that the temporal changes contained within these two noise bands were sufficient to allow the speech decoding mechanism to function adequately under optimal listening conditions.

Pitch variations are the basis for melodies. Pitch changes in music tend to occur on a time scale that is an order of magnitude slower than those of consonants in speech: melodies with note durations shorter than about 160 ms are very difficult to identify. Liégeois-Chauvel et al. (1999) recorded electrical potentials from the human auditory cortex using implanted electrodes. They observed that responses from left, but not right, Heschl's gyrus distinguished brief temporal differences both in speech and non-speech sounds. More recently, these authors presented data that confirms the sharper tuning to frequency in the right auditory cortex than in the left (Liégeois-Chauvel et al. 2001). These data are therefore consistent with the idea that left auditory cortex responses are optimal for a range of temporal responses relevant for speech distinctions, but are not highly selective for frequency, whereas right auditory responses have a greater spectral resolution but are less sensitive to brief temporal events.

Cerebral blood flow in the anterior auditory region on the right side showed a greater response to increasing spectral compared to temporal variation, whereas the corresponding area on the left showed the reverse pattern. A third area, located in the right superior temporal sulcus, also showed a significant response to the spectral parameter, but showed no change to the temporal parameter. Finally, a direct comparison of the temporal and spectral conditions revealed greater left cortical activity for the combined temporal tasks, but greater right activity for the combined spectral tasks (Zatorre et al. 2002). A vast body of data indicates that certain aspects of speech decoding depend critically on left auditory cortical regions. Current evidence suggests that there are functional hierarchies, such that early stages of processing depend on core auditory areas bilaterally, with "higher" word-recognition mechanisms being associated with more anterior and ventral auditory regions in the left hemisphere (Hickok and Poeppel 2000).

Hemispheric differences may have emerged as a consequence of differences in processing these cues. In particular, Zatorre et al. (2002) suggest that low-level specializations evolved as a general solution to the computational problem posed by the need for precision in both spectral and temporal processing. Subsequent stages of processing might initially have developed from these low-level specializations, as small differences in processing are likely to be amplified when one goes to higher levels in the functional architecture. In this context, it is noteworthy that many of the above examples of domain-specific processing involve regions outside the auditory cortex, especially frontal areas, which could thus be independent of low-level auditory specializations.

5.6 The Spectrotemporal Receptive Field

The spectrotemporal receptive field (STRF) represents those spectrotemporal features in sound that, on average, cause the neuron to fire. The neuron and its peripheral connections filter out those aspects of the sound to which it is tuned. The following is partly based on two sets of seminal methodological papers (Klein et al. 2000, 2006; Theunissen et al. 2000, 2004) that among others reviewed the history of the STRF, to which the current author has contributed from the very beginning. Rather than rehearse my nearly three decades old papers this history and underlying theory is abstracted here in the words of these two modern prophets.

5.6.1 The White-Noise-Based STRF

It has been more than 25 years since the STRF was first used to describe the joint sensitivity to the spectral and temporal dimensions of a Gaussian white-noise (GWN) in the auditory midbrain (Aertsen and Johannesma 1981; Eggermont et al. 1981; Hermes et al. 1981). The STRF was specifically associated with stimuli characterized by randomly varying spectrotemporal features, and was calculated using a generalization of reverse correlation (or spike-triggered averaging; de Boer and de Jongh 1978). The STRF of a neuron can be interpreted as the best linear model that transforms any time-varying stimulus into a prediction of the firing rate of a neuron. Only if the auditory spectrotemporal dimensions are sampled uniformly and randomly, then one can estimate the STRF simply by averaging the stimuli before each spike (reverse correlation). Otherwise a deconvolution with the spectrotemporal autocorrelation function of the stimulus is needed. In reverse correlation, the spectrogram of the sound that precedes the spike in a given time window (typically up to 100 ms before the spike) is averaged. Parts of the sound that consistently activate the neuron, that is, frequency components to which the neuron is tuned and occur at a more or less fixed time before the spike (determined by the spike latency), will add up more than occurrences of frequency components that do not activate the neuron but put the background activity into the analysis window. After the neuron has fired it will be temporary less sensitive (because of short-term synaptic depression, feedforward inhibition, and/or after-hyperpolarization) and the average spectrotemporal energy in the excitatory frequency band will be less than the background activity. Because we are looking at the average spectrogram preceding the spikes, this postactivation suppression in the spectrogram will occur at larger times preceding the spikes than the excitatory part (cf. Fig. 5.2b). Lateral inhibitory areas, if present, typically flank the excitatory ones at approximately equal latencies (Valentine and Eggermont 2004).

For random multitone stimuli (see Sect. 5.3) the STRF can, in addition to reverse correlation, also be obtained as the frequency-dependent poststimulus time histogram

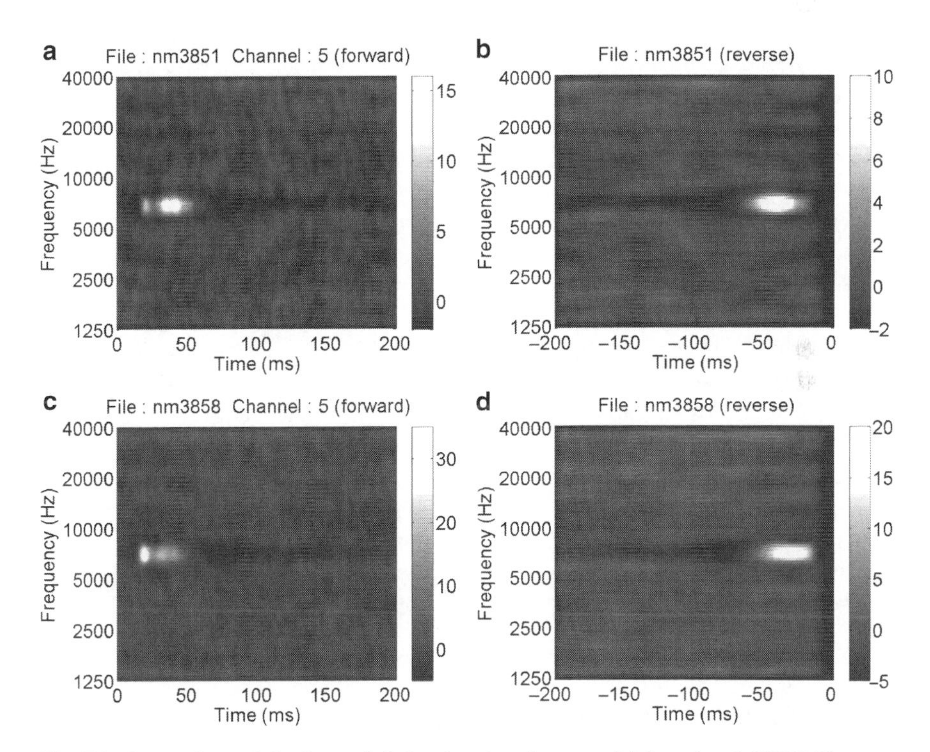

Fig. 5.2 Comparison of the forward (*left column*) and reverse (*right column*) STRF. The zero point in the *left column* represents the onset of individual tone pips, whereas in the *right-hand column* it indicates the time of the spikes. Shown are two examples illustrating the smoothing effect of the reverse correlation calculation of the STRF. This smoothing has the advantage that inhibitory effects (*dark areas*) become clearer but also the disadvantage that fine-temporal structure in the excitatory part of the STRF (*light areas*) is lost (Recalculated from data used in Valentine and Eggermont 2004.)

(forward correlation), which represents the spike times following the onsets of every tone pip over a 100-ms time window. The STRFs obtained with these two methods, reverse and forward correlation, differ owing to the smoothing provided by the tone pip envelope. Figure 5.2 presents two examples; as can be seen, the reverse and forward STRFs differ in their temporal resolution. The forward correlation shows a burst type of response with a more precise firing of the first spike in the burst, whereas the reverse correlation is unable to separate early from late activity. The effect of smoothing in the reverse correlation is also evident from the smaller peak response values (see gray scale bar).

The STRF has its theoretical underpinnings in time-frequency analysis (Eggermont and Smith 1990) and nonlinear systems theory (Eggermont 1993). The general methodology, through which the functionality of a neuron can be explored, is known as the white-noise approach. The white-noise approach is closely related to the cross-correlation method (Palm and Pöpel 1985) for developing a Wiener-series model of a nonlinear system. Estimating the first- and higher-order Wiener kernels of a system can be done by applying a GWN input and computing a series of first- and higher-order input–output cross-correlations. It was very early on recognized that the reverse-correlation function between spikes and the GWN waveform is basically identical to the first-order Wiener kernel of the system (Eggermont et al. 1983b). Therefore, the reverse-correlation function does not just reflect linear processing but instead should be considered the best linear fit to the observed input–output transformation (Palm and Pöpel 1985; Eggermont 1993). However, if the response of a neuron is not phase locked to fine details of the stimulus waveform, a first-order stimulus-response cross-correlation is not productive: the first-order Wiener kernel cannot be recovered from the noise.

Given the relative importance of the second-order Wiener kernel in central parts of the brain where phase-locking is present only at low frequencies, one could treat the dynamic spectrum rather than the waveform as the effective stimulus. The parallel auditory processing along the tonotopically organized afferents, separates the dynamic spectrum into energy fluctuations within discrete frequency bands, e.g., one-third octave bands (Eggermont et al. 1983a). The STRF can then be interpreted as the first-order Wiener kernel of a multiple-input system, composed of the system's impulse responses to these different frequency inputs. It is important to remember that spectrotemporal reverse correlation (with an appropriate time-frequency representation) is equal to the single Fourier transform of the second-order Wiener kernel (Eggermont et al. 1983b; Eggermont 1993).

5.6.2 STRFs for Modified Noise Stimuli

Shamma (1996) reviewed the use of a variant of the Gaussian white-noise approach for STRFs based on stimulation with broadband sounds having dynamic spectra. These sounds were named moving ripples. The ripple spectrum has a sinusoidal profile on a logarithmic frequency axis. The carrier of the profile can be a broadband noise or a series of tones that are equally spaced in frequency (f). The ripple has a ripple-modulation amplitude (ΔA), a ripple frequency (Ω) in units of cycles/octave, and a ripple phase (ψ) in radians. The moving ripple spectrum (e.g., $\Omega = 0.5$ cycles per octave) travels at a constant velocity ω in units of cycles/second. These parameters define the profile of the dynamic ripple spectrum as:

$$p(f,t) = 1 + \Delta A * \sin(2\pi(\omega t + \Omega f) + \psi). \tag{5.1}$$

White noise has power over all possible ripple components. Auditory nerve fibers are able to respond to an extended set of ripple components; however, at the level of A1, the range of densities and modulation rates to which neurons are responsive is quite narrow. Thus, the power of the dynamic spectrum should be focused over this relevant range of ripple components. The drawback is that the various ripple stimuli are no longer uncorrelated. A partial solution is found in a stimulus with temporally orthogonal ripple combinations (TORCs). This is a broadband sound with the same carrier as a single ripple, except that its modulation spectrum consists of a linear combination of several temporally orthogonal ripple envelopes (Elhilali et al. 2004). Although uncorrelated, these orthogonal ripples are not independent (Christianson et al. 2008), which can cause nonlinear interactions between them to show up in the STRF.

Klein et al. (2006) demonstrated that STRFs measured using a diverse array of broadband stimuli – such as dynamic ripples, spectrotemporally white noise (STWN), and temporally orthogonal ripple combinations (TORCs) – are similar, confirming earlier findings that the STRF is a robust linear descriptor of the cell. These broadband stimuli likely bias cortical cells in a manner different from that of narrowband or transient stimuli such as tones and clicks. Consequently, predicting details of tone and click responses from the broadband-based STRF has proven to be problematic.

5.6.3 Random Multifrequency Stimuli

deCharms and Merzenich (1996) used random chord tone pips that varied in their short-term spectrotemporal density to measure STRFs in the alert monkey A1. In a better realization of such a stimulus (Blake and Merzenich 2002; Tomita and Eggermont 2005), the temporal sequence of tone pips of a given frequency is drawn from a Poisson distribution, and different frequencies have temporal sequences independent of each other. As a result, the overall temporal sequence of the stimulus is also a realization of a Poisson process. The density is basically the average overlap in frequency and time of the tone pips. As density increased, A1 excitatory receptive fields systematically changed, that is, they became narrower in the frequency domain and had slightly longer latencies as shown here for anesthetized cat A1 (Fig. 5.3). Receptive field sensitivity, expressed as the expected change in firing rate after a tone pip onset, decreased by an order of magnitude with increasing density. However, spectral selectivity more than doubled. Inhibitory subfields,

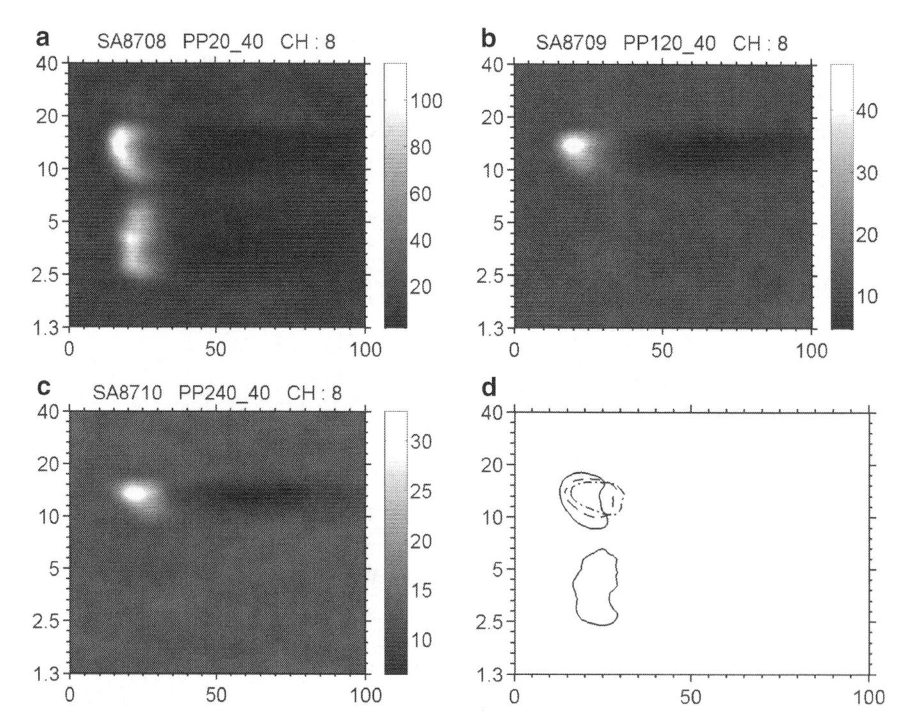

Fig. 5.3 STRFs obtained from the same unit for multifrequency stimuli with different densities: 20/s (**a**), 120/s (**b**) and 240/s (**c**). Each STRF is scaled onto its own peak response, the *gray-scale bars* besides each STRF indicate the range between 0 and maximum (in spikes/s/stimulus). The peak response decreases with increasing density. In (**d**) the contour lines calculated for 30% between maximum and mean value are shown. One observes that the STRF shapes are dependent on the stimulus density; here a double peaked STRF is present for low densities and a single peaked one for higher densities. The postactivation inhibition (the *darkest region* in the STRF) is most clearly seen for the higher densities. Minimum latencies for the STRF increased with increasing stimulus density (**d**) (Eggermont unpublished observations.)

which were rarely recorded at low sound densities, emerged at higher sound densities (Valentine and Eggermont 2004; Tomita and Eggermont 2005; Eggermont 2006; Noreña et al. 2008).

5.6.4 Vocalizations

Theunissen et al. (2000) found that in many cases the STRFs derived using natural sounds were strikingly different from the STRFs that were obtained using an ensemble of random tone pips. The results showed that the STRF model is an incomplete description of response properties of nonlinear auditory neurons, but that linear receptive fields are still useful models for understanding higher level sensory processing, as long as the STRFs are estimated from the responses to relevant complex stimuli. The STRF is nevertheless an informative and efficient tool (Theunissen et al. 2004) because the STRF describes the basic tuning of the neuron during the processing of complex sounds such as birdsong. This emphasizes that the STRF is context dependent and not an invariant characterization of a neuron (Gourévitch et al. 2009). The convolution between the spectrotemporal receptive field (STRF) and the spectrogram of the sound gives an estimate of the time-varying firing rate of the neuron.

Singh and Theunissen (2003) hypothesized that the STRFs of auditory neurons would be matched to the modulation spectrum of behaviorally relevant sounds. In mammals and birds, spectral modulation tuning is low-pass and temporal modulation tuning is band-pass and may therefore be a general property of vertebrate auditory systems. The STRF and its corresponding modulation transfer function (MTF), the 2-dimensional Fourier transform of the STRF, show the spectral and temporal modulation frequencies that best drive the neuron (see also Chi et al. 1999; Miller et al. 2002). One needs to keep in mind that the spectrotemporal density of the stimuli will largely determine the STRF obtained. A neuron could thus respond differently to similar vocalizations embedded in a different sound background. In other words, STRFs are context dependent, and so is cortical processing. Thus modeling cortical processing will require adaptive filters.

5.7 Cortical Circuitry

Most of the input (up to 99.9%) to a given cortical area originates from other cortical areas. This leaves only a tiny fraction to specific afferent input from the sensory systems. Nevertheless, the numerically small thalamic input has a profound role in the way the cortex works. The major afferent input to primary sensory regions of the cortex is from the thalamus. Ascending lemniscal thalamocortical input arrives mainly in lower layer III and layer IV (Fig. 5.4). Hundreds of thalamic axons ramify within a cortical column; yet each layer III or IV neuron receives input from only a fraction of them. On average, 20–30 thalamic cells converge onto a simple stellate cell in primary visual cortex (Alonso et al. 2001) and onto a small pyramidal cell in primary auditory cortex (Miller et al. 2001).

Connection diagrams (such as in Fig. 5.4) and estimates of the underlying microcircuits are the result of either anatomical tracings (indicating an anatomical and potentially also a functional connection) or of simultaneous recording neural activity from connected neurons (indicating functional connectivity). Whereas there are two general types of excitatory cells in cortex, pyramidal cells and spiny-stellate cells (but the latter likely absent in auditory cortex and replaced by small pyramidal cells; Smith and Populin 2001), there are up to 16 types of interneurons (Watts and Thomson 2005). Pyramid–pyramid connections and pyramidal inputs to interneurons that synapse on the cell soma and axon hillock, and represent mostly feedback inhibition, generally display paired-pulse and frequency-dependent

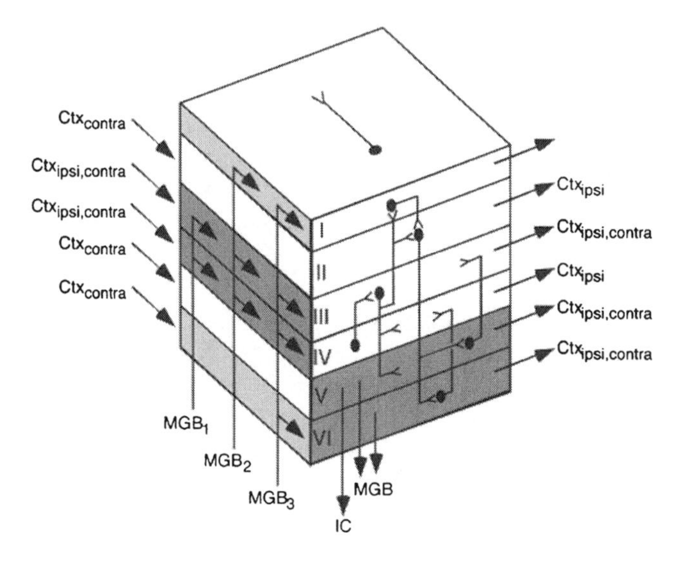

Fig. 5.4 Connections within a putative primary auditory cortex column. The *left face of the cube* shows thalamic and corticocortical inputs; the *right face* displays interlaminar connections as well as thalamic, collicular, and corticocortical outputs. *Shading* indicates layers that receive thalamocortical input (*left*) or produce corticothalamic outputs (*right*). Lemniscal thalamic inputs (MGB1) are in *dark shading*, while nonlemniscal inputs also activate layers I and VI, indicated in *light shading*. MGB2 represents the deep and caudal dorsal MGB; and MGB3 is the medial division. Inputs from the ipsilateral cortex (Ctxipsi) terminate in the middle layers, but those from the contralateral cortex (Ctxcontra) are distributed among layers II to VI. Within the column, small pyramidal cells in layer IV and lower layer III receive the major lemniscal thalamic input, initiating a flow of information into the supragranular layers and then down to the infragranular layers. Pyramidal cells in layers II and III also extend long-range horizontal connections with other cortical columns (symbolized by neuronal projection on top of cube). Feedback to the auditory thalamus (MGB) originates primarily in layer VI but also in layer V, and projections to the inferior colliculus (IC) emerge from layer V. Major corticocortical projections to both the ipsilateral and contralateral hemispheres emerge from layers II and III, but layers IV to VI also provide some corticocortical output (From Linden and Schreiner 2003, by permission from Oxford University Press.)

synaptic depression. Pyramidal inputs to interneurons that synapse on dendrites and that mostly represent feedforward inhibition show strong synaptic facilitation. Thus depending on the relative input- and output-firing rates of a pyramidal cell other balancing mechanisms come into play.

The concept of cortical microcircuits is still heavily inspired by the more than four decades old investigations in cat primary visual cortex (Hubel and Wiesel 1962). This framework offers a feedforward hierarchy of neural connections to explain the formation of more and more specialized receptive fields. This focus on the thalamocortical synapse obscures that, even in layer IV of the cat, 95% of the synapses on a simple cell in primary visual cortex come from other cortical neurons, that is, reflect some form of feedback activity. It is this collective computation in the network of cortical cells that needs to be incorporated in modeling studies (Martin 2002), not just the pattern of feedforward thalamic connections that contribute only a few percent of cortical synapses and a small fraction of the total synaptic input.

Connections over large spatial divisions of auditory cortex are provided by the thalamic cell axonal divergence and convergence, often estimated to cover between 2 and 5 mm at the cortical level (Lee et al. 2004a, b), and intracortically through horizontal fibers that connect distant pyramidal cells up to 8 mm apart (Wallace et al. 1991; Clarke et al. 1993). In the visual cortex, the spatially periodic effects of the patchy connections that the horizontal fibers make also have been illustrated by cross-correlation (Ts'o et al. 1986). These horizontal fiber connections in auditory cortex run dominantly within isofrequency lamina but are for a sizable part heterotopic, that is, do connect cell groups with characteristic frequencies (CFs) differing by more than one octave. Similarly connections between different auditory cortical areas are frequently heterotopic as well (Lee et al. 2004b). These connections between neurons with vastly different frequency selectivity form the basis for the use-dependent plasticity in auditory cortex.

5.8 Cortical Computations

5.8.1 Feature Extraction by Edge Detection

Is the primary auditory cortex a sophisticated, spectrotemporal filter bank working on the preprocessing done by subcortical structures (Chi et al. 2005)? In other words, is the STRF a feature-extracting filter? That would require that the STRF is stimulus invariant and that is all but clear (Klein et al. 2006). There is convincing evidence that a simple change in stimulus density results in different STRFs (Blake and Merzenich 2002; Valentine and Eggermont 2004), suggesting that the acoustic context wherein a particular stimulus is present will require a different STRF to extract its relevant features. The spectrotemporal cortical filter is therefore context dependent and likely adaptive (cf. Fig. 5.3). This has been elaborated recently by Gourévitch et al. (2009), who showed that "STRFs evoked by narrowband sounds revealed context dependency through at least two distinct phenomena: spectral-edge sensitivity of neurons and appearance of residual peaks far from the BF."

Spectral edge effects rely on the balance between inhibition and excitation and have led to two simplified models for the synaptic mechanisms shaping the receptive fields (Oswald et al. 2006): (1) lateral inhibition and (2) co-tuning. Recent in vivo intracellular recordings in auditory cortical areas using voltage-clamp techniques found co-tuned inhibitory and excitatory inputs (Wehr and Zador 2005). On this basis Gourévitch and Eggermont (2008) modeled effects of changing the bandwidth of the stimulus ensemble on the STRF. The results had at least two implications: (1) A preference for spectral edges and the unmasking of residual peaks were both well reproduced by the spike-generation model in response to narrowband sounds. This suggested that both phenomena were relative simple consequences of the weights of inhibitory and excitatory inputs as well as forward suppression mechanisms considered in the model. In particular, these latter neural properties were sufficient to explain the level of release from inhibition observed without the need for additional mechanisms such as synaptic depression. (2) Spectral edge preference and unmasking of residual peaks were both compatible with lateral inhibition and co-tuning models for synaptic excitatory and inhibitory inputs. This also suggested that both phenomena occur whenever the inhibitory synaptic input is equal or larger than the excitatory one. In particular, the spectral edge preference might be a general mechanism occurring as soon as the stimulus spectrum includes well-marked energy bands.

By assuming that primary auditory cortex can detect edges in time and frequency using a putative delay mechanism (Fishbach et al. 2001, 2003), some important properties of the putative spectrotemporal cortical filter bank have been elucidated. It is known that amplitude gradients affect auditory perception in general and auditory source segregation in particular. Examination of the responses of many cortical and subcortical neurons to amplitude transients suggested that the neural response is sensitive to the derivative of the stimulus intensity over time and therefore their neural responses may be interpreted as reflecting a temporal edge detection computation (Fig. 5.5).

The basic operation of this model is the calculation of the smoothed time derivative of the log-compressed envelope of the stimulus (Fishbach et al. 2001). In this model, a standard neural representation of the auditory input is being progressively delayed by a sequence of neurons that form a "delay layer." The time-derivative computation is implemented as a weighted sum of the activity along the delay layer, with weights that form an ON–OFF receptive field. The edge-detection model suggests that the sensitivity of the auditory system to amplitude transients is a consequence of auditory temporal edge calculation, and that this computation has a primary role in neural auditory processing. In an extension of the model, a set of neurons, tuned in the spectrotemporal domain, is created by means of neural delays and frequency filtering. The sensitivity of the model to frequency and amplitude transitions is achieved by applying a 2-dimensional frequency-delay operator to the set of spectrotemporally tuned neurons (Fishbach et al. 2003). The operator, which in its basic form is a separable function of a Gaussian in the spectral dimension and a first-order derivative of a Gaussian in the time-delay dimension, can be rotated in the spectrotemporal plane. The STRF operators used constitute a restricted subset of all possible STRFs, and can be characterized by only two free parameters: their spectral and temporal widths. These differences between the model's STRF operator and the linearly estimated STRFs from cortical neurons are important. Typically, the STRFs estimated for AI neurons show much more sluggish characteristics than what is needed to produce sensitivity to high-velocity transitions. This may be partly the result of the smoothing process resulting from reverse correlation (see earlier) or from the smoothing imposed even on forward correlation STRFs;

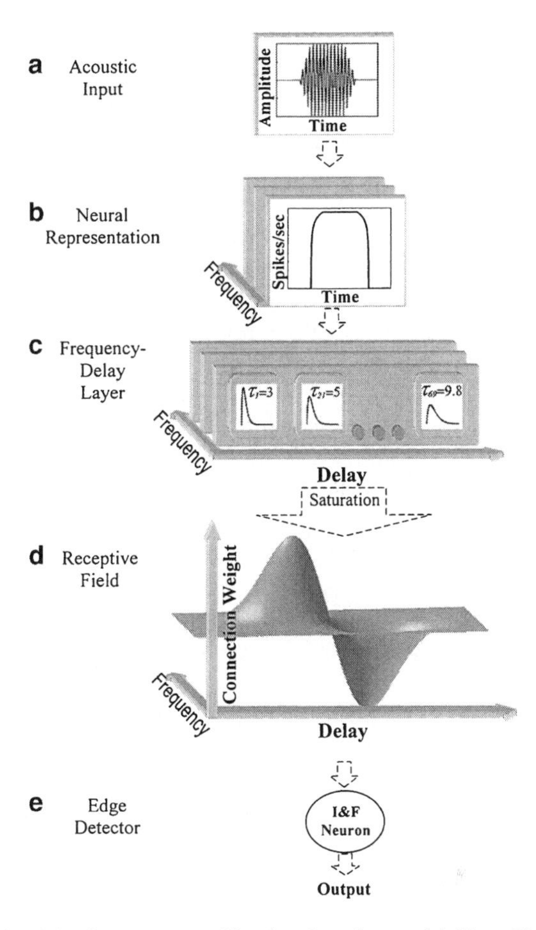

Fig. 5.5 Schematic of the frequency-specific edge-detection model (From Fishbach et al. 2003. Reproduced with permission from the *Journal of Neurophysiology*.)

typically using a 5×5 uniform filter. The time bins are typically 1 or 2 ms, so one can appreciate that fine temporal aspects are lost in these representations.

5.8.2 Perceptual Grouping by Synchronization

The basic models discussed in the preceding text do not address how the extracted spectral and temporal information is combined beyond the assumption that the spectral and temporal (delay) filters are independent. Because sensory systems likely extract distinct features of the environment, in the auditory system putatively by using the STRF filters of spatially distinct sets of neurons, perception requires a regrouping of those features that belong together (Singer and Gray 1995). In that respect, "components of an auditory scene appear to be perceptually grouped if they are harmonically related, start and end at the same time, share a common rate of AM, or if they are proximate in time and frequency" (Cooke 1993). Acoustic transients, such as plosive consonants, can be viewed as analogous to visual contours and thus may represent the gross shape of a sound. Large parts of primary auditory cortex, regardless of their frequency tuning, respond vigorously to such transients (Phillips 1993). It is likely that a combination of temporal aspects of sound and a common harmonic structure may provide for "feature binding" on which perceptual grouping can be based. Feature binding may be based on enhanced neural firing synchrony between groups of neurons that extract different stimulus features (Singer and Gray 1995; von der Malsburg 1995). The best candidates for binding features extracted in different cortical areas by neural synchrony are the features that produce the largest changes in the correlation of activity between neurons in those areas compared with a nonstimulus condition. Among those features, stimulus onsets are the prime candidate (Eggermont 1994; Brosch and Schreiner 1999). Synchronization between areas or between distant (heterotopic) points in the same cortical area may be more important than local synchronization to provide feature binding. Local activity appears to be synchronized regardless the level of stimulation and forms distinct correlation clusters (Eggermont 2006), whereas specifically the very low between-area synchronization under spontaneous conditions is considerably elevated during stimulation. Specifically this applies to broadband stimuli, such as amplitude-modulated noise or Poisson-distributed clicks that are able to synchronize across a wide range of CFs. Communication sounds often contain noise, for example, plosive consonants in speech, and these sections are imminently suited as grouping features on the basis of enhanced synchrony (Eggermont 2000).

It is useful to reemphasize that auditory features can be categorized into contour and texture components of sound (Eggermont 2001). This division in contours and texture corresponds roughly to Rosen's (1992) three-way partition of the temporal structure of speech into envelope (2–50 Hz), periodicity (50–500 Hz), and fine structure (600Hz–10 kHz). Contour components are those temporal aspects of sound that covary across frequency, and are likely exclusively coded in the temporal domain. Onsets, noise bursts, and common rates of slow (<50 Hz) AM and FM, that is, the region where rhythm and flutter are dominant (Bendor and Wang 2007), are clearly contours that delineate for instance sound duration and separation between noise bursts and formant transitions. Common higher rates of AM (50–500 Hz), in the roughness and pitch range, common changes in formant frequencies, and harmonicity, are aspects of sound texture. These texture aspects of sound thus relate to constant or slowly changing spectral representations in a cortical rate-place code. Pitch is likely coded in a place representation (Bendor and Wang 2005), whereas timbre and roughness are potentially reflections of the sound representation across different cortical areas. One expects texture and contour components thus to have largely independent neural representations in auditory cortex or potentially also in subcortical areas. Sound texture cannot be represented in a temporal code at the level of AI, but instead requires a rate-place code (Lu et al. 2001; Bendor and Wang 2006).

5.8.3 Auditory Cortex Acts Both as a Coincidence Detector and Integrator

The experimentally observed correlation of spiking activity in auditory cortex is represented in a histogram of the firing times of one neuron relative to those of another: the cross-correlogram. The most commonly obtained cross-correlogram for the firings of two neurons in auditory cortex is that with a peak centered around zero time lag and with a width at half-peak amplitude of at least 10 ms for dual electrode pairs. This is observed both in silence (Eggermont 1992) and during stimulation (Eggermont 1994, 2000; Brosch and Schreiner 1999) and is thought to reflect the convergent–divergent connectivity between thalamic and cortical neurons and potentially also the more abundant inputs of cortical origin. The latter inputs (based on their much lower conduction velocities) are likely arriving not as synchronously as the thalamocortical ones and may have, therefore, in particular at the onset of a stimulus, a smaller direct effect on the firing of the target neurons compared to the thalamocortical ones (Abeles 1991). In case neurons fire in a more sustained way, for example, in awake animals (Wang et al. 2005) or for certain stimuli in anesthetized animals (Gourévitch and Eggermont 2007b), the intracortical horizontal fiber activity may induce strong correlations (Noreña et al. 2006). So the time window in which the inputs are co-occurring and can be integrated by the recipient neuron is the important measure here.

Whether neocortical pyramidal cells are acting as coincidence detectors or temporal integrators depends on the degree of synchrony among the synaptic inputs, including those arriving from intracortical fibers (Aertsen et al. 1994; Rudolph and Destexhe 2003; Grande et al. 2004). High input synchrony leads to the more efficient coincidence detection, whereas low input synchrony leads to temporal integration. As discussed in the preceding text, about 20–30 thalamic inputs have to be simultaneously present to fire an A1 neuron (Miller et al. 2001). Correlation and connectivity may ensure efficient propagation and preservation of the temporal precision of neural firing downstream along the neural pathway (Kistler and Gerstner 2002; Kimpo et al. 2003; Reyes 2003). Correlation also increases the signal to noise ratio in the population STRF (Eggermont 2006). Temporally correlated activity is important in spike-time dependent plasticity (Holmgren and Zilberter 2001; Karmarkar and Buonomano 2002; Sjöström et al. 2003) and may cause either LTP or LTD depending on the timing differences between pre- and postsynaptic activity.

5.8.4 Neural Synchrony Clusters

Eggermont (2006) recorded spiking activity from cat auditory cortex using multi-electrode arrays. Cross-correlograms were calculated for spikes recorded on separate microelectrodes. The pairwise cross-correlation matrix was constructed for the peak values of the cross-correlograms. Hierarchical clustering was performed on the cross-correlation matrices obtained for six stimulus conditions. These were silence, three multitone stimulus ensembles with different spectral densities, low-pass amplitude-modulated noise, and Poisson-distributed click trains; each stimulus lasted 15 min. The resulting neuron clusters represented patches in the cortex of up to several square millimeters in size that expanded and contracted in response to different stimuli. Cluster positions and size were very similar for spontaneous activity and multitone stimulus-evoked activity but differed between those conditions and the amplitude-modulated noise and Poisson-distributed click stimuli. Cluster size was significantly larger in PAF compared with A1, whereas the fraction of common spikes (within a 10-ms window) across all electrode activity participating in a cluster was significantly higher in A1 compared with PAF. Clusters crossed area boundaries only in approximately 5% of simultaneous recordings in A1 and PAF. Clusters are therefore similar to but not synonymous with the traditional view of neural assemblies that encompass various cortical areas. Common-spike STRFs based on coincident activity showed generally spectral and temporal sharpening. The coincident and noncoincident output of the clusters could potentially act in parallel and may serve different modes of stimulus coding.

5.8.5 Transfer Entropy

Over the past decades, most investigations using spike trains have focused on one of the main issues in systems neuroscience, the understanding of the neural code. Basically, the two main approaches depict neurons either as sensitive to input firing rates and act as leaky integrators, or as coincident detectors that are sensitive to the temporal patterns of the input (Gerstein and Kirkland 2001). Given that the synchronization between multiple single-unit recordings has a modest dependence on the sensory stimulus used (Eggermont 1994), coincident firing has been investigated and used mainly in the search for neural assemblies or neuron clusters able to show a stimulus-induced modulation of their synchronized firing (Eggermont 2006). Studies based on firing rates have instead mainly focused on the processing abilities of single neurons as a consequence of their unique physiological specialization (type of cell, tuning properties), how they integrate their input activity, and eventually how a population of neurons possibly encode features of the stimulus by a rate code (Pouget et al. 2000). Consequently, there has been a steady rise in interest in the estimation of "information" carried by a single neuron, multiple single units, or neural populations to specific stimuli. The concept of information in relation to neuronal activity, although stemming from a well-defined concept in Shannon's theory, has no standardized meaning in neuroscience and has been used in several different ways (Borst and Theunissen 1999). For instance, information rates have been estimated at a microscopic physiological scale within a single synapse (London et al. 2002). The mutual information between the spike-train responses and a set of stimuli can be estimated to investigate the discrimination abilities of neurons (Werner and Mountcastle 1965; Borst and Theunissen 1999; Gehr et al. 2000; Chechik et al. 2006).

Gourévitch and Eggermont (2007a) presented "transfer entropy" as a new exploratory tool that provides a bridge between the study of neural assemblies and the information about stimuli carried by individual neurons. More precisely, the transfer entropy estimates that part of the activity of a neuron that is not dependent on its own past but dependent on the past activity of another neuron. In a nutshell, it estimates the information transferred in a certain time window between two neurons in both directions. To a certain extent, this tool is able to distinguish information resulting from common history and exclude it by appropriate conditioning of the entropy. Transfer entropy also detects asymmetry in neural relations, allowing studies of possible feedback in neural circuits. Last but not least, transfer entropy takes into account linear and nonlinear flows and thus may represent a very general way to define the causality strength between two spikes trains. In particular, the window size for which maximum information is transferred may be useful to study neural integrative properties. The information transfer through a network of 16 simultaneous multiunit recordings in cat's auditory cortex was examined for a large number of acoustic stimulus types. Specific flow directions within AI could be distinguished; among those transfer from neurons with monotonic rate-level functions to neurons with nonmonotonic rate-level functions (Fig. 5.6). Application of the transfer entropy to a large database of recorded multiple single-unit activity from cat's primary auditory cortex revealed that most windows of temporal integration found during spontaneous activity range between 2 and 15 ms. This time window is consistent with the rise time and duration of cortical EPSPs and also with the width of the peaks in cross-correlograms. The normalized transfer entropy shows similarities and differences with the strength of cross-correlation and can form the basis for revisiting the neural assembly concept.

5.9 Cortical Output

5.9.1 Feedforward Projections: What and Where Pathways

Some areas in auditory cortex are specifically involved in sound localization; in the cat these are AI (including the dorsal zone), PAF, and AES (Malhotra et al. 2004). Other areas (such as AAF) are potentially involved in the "what" of sound and likely involved in processing of environmental sounds including vocalizations (Lomber et al. 2007; Lomber and Malhotra 2008). Both pathways are activated within 70–150 ms from stimulus onset, but the "where" pathway is activated approximately 30 ms earlier than the "what" pathway, possibly enabling the brain to use top-down spatial information in auditory object perception (Ahveninen et al. 2006; Barrett and Hall 2006; Altmann et al. 2007). The separation of both pathways thus occurs already in primary auditory cortex, and follows a posterior route for the "where" and an anterior route for the "what" pathway. The "where" and "what" pathways converge in the prefrontal cortex, where they are integrated with other sensory modalities.

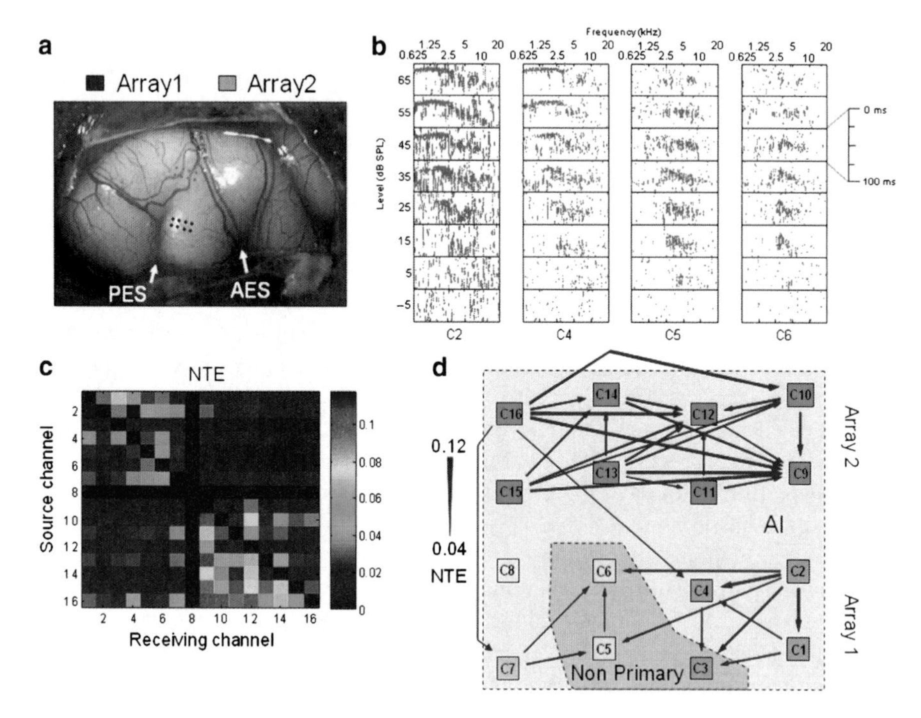

Fig. 5.6 Normalized transfer entropy (NTE) applied to a set of 16 recording sites. (**a**) Electrode positions on the surface of auditory cortex. Posterior and anterior ectosylvian sulci (PES, AES) are indicated. Frontal locations are to the right, dorsal locations to the top. (**b**) Frequency-tuning properties for channels C2, C4, C5, and C6 in the form of frequency-intensity dot displays of multiple single-unit activity. Dot displays are obtained for six intensities and 27 frequencies from 0.625 to 20 kHz. Each *subpanel* shows, for a fixed intensity level, responses as a function of frequency (horizontal axis) and time after tone pip onset (vertical axis). (**c**) Matrix of NTE value between all pairs of the 16 electrodes. No activity was recorded on electrode C8, whose NTE values were thus fixed at zero, along with NTE values between the same electrodes (diagonal). (**d**) Network of information transfer within the set of 16 recording sites. *Thickness of the arrows* is proportional to the strength of the transfer entropy. Recordings from channels C3, C5, and C6 are distinguished as showing nonprimary-like behavior. Clusters of electrode sites, based on cross-correlation values (see Sect. 5.8.5), are indicated by *background shading of small squares* representing each channel (From Gourévitch and Eggermont 2007a. Reproduced with permission from the *Journal of Neurophysiology*.)

5.9.2 Feedback Projections

The auditory system is not just a projection system from periphery to cortex but a reentrant system characterized by multiple, loosely interconnected, regional feedback loops (Spangler and Warr 1991). At the lowest level, a loop between cochlea and the superior olivary complex in the brain stem by way of the olivocochlear bundle is found. The olivocochlear bundle projects via its medial branch to the outer hair cells (OHCs), thus regulating the slow motility of the outer hair cells and thereby the stiffness of the basilar membrane. Via its lateral branch it especially affects the low spontaneously active auditory nerve fibers (ANFs) synapsing with the IHCs. Activation of the olivocochlear bundle appears to improve the discrimination of complex sounds such as speech, in the presence of broad-band noise by lowering the noise-induced background firing rate and increasing the dynamic range (Liberman 1988). Chronic cochlear de-efferentation in adult chinchillas resulted in reduced spontaneous ANF firing rates, increased driven discharge rates, decreased dynamic range, increased onset to steady-state discharge rate, and hypersensitive tails of the frequency tuning curves (Zheng et al. 1999).

A second loop exists between the lower brainstem nuclei and the inferior colliculus (IC). A third loop is formed between the IC and the thalamocortical system, which, in itself, consists of a feedback loop between the MGB and cortex. More specifically, the auditory cortex projects back to the MGB with at least ten times as many fibers than the number of afferents from the MGB to auditory cortex. The effect of cortical activity on the auditory thalamus was first demonstrated by cooling the surface of A1 (Villa et al. 1991) resulting in a decrease of spontaneous activity and changes in the CF of the frequency-tuning curve by up to 0.5 octaves. Recently, the role of the corticofugal systems has received new attention especially through the work of Suga, Yan, and collaborators. This modulatory action of the cortex is potentially involved in the adjustment to long-term changes in the overall functional organization of the IC, MGB, and auditory cortex.

The auditory cortex also connects with the IC, but with exclusion of its central nucleus (Winer and Lee 2007). The central and external IC subnuclei both project back to the dorsal cochlear nucleus (DCN). The DCN in turn feeds back to the ventral cochlear nucleus (VCN). This corticofugal activity was shown to be involved in subcortical frequency tuning in bats as well as in mice and cats. This was likely caused by positive feedback, which, in combination with widespread lateral inhibition, sharpens and adjusts the tuning of neurons in the IC and MGB (Zhang et al. 1997; Yan and Suga 1998; Yan and Ehret 2001; Yan et al. 2005) and even in the cochlear nucleus (Luo et al. 2008).

5.10 The Cortex as a Nonlinear Dynamic System

The role of this nested set of reentrant systems can at the moment only be speculated upon. One possible role is an involvement in expectancy about the probability of various environmental occurrences stored in its internal representation and incoming sensory information is continuously checked whether it corresponds, within a certain error, to this internal representation. This requires reentrant systems (Johannesma et al. 1986) with a definite role for thalamocortical and corticocortical loops. The presence of the feedback loops thus creates a nonlinear dynamic system. A crucial property of nonlinear dynamic systems is the exponentially increasing amount of uncertainty associated with predicting the future time evolution of the system and its related information content. This is determined by the sensitive dependence of the system on the initial conditions so that it takes only a very small perturbation to switch the system from one unstable state into another. Any stimulation could act as such a disturbance. We have to accept that brain activity reflects such a nonlinear dynamic system. If we assume that behavior is the only relevant correlate of brain activity, then the brain may be considered to operate at the edge of chaos. This is a state where very small changes in the controlling parameters (e.g., sensory stimuli) may cause large macroscopic state changes such as from one quasi-periodic state into another (Schiff et al. 1994). The only relevance of neural firing in this point of view is to be found in its capacity to induce transitions between the many unstable macrostates found in the brain, and the relevant questions are related to what determines the various transition probabilities. State switching is likely reflected in synchrony changes between individual neuron activities (Vaadia et al. 1995). Time-dependent changes in neural synchrony (Aertsen et al. 1989; Grün et al. 2002) may signal the building up or breaking down of assemblies. When control parameters (e.g., concentration of neurotransmitters, drugs, etc.) change beyond a critical value, the system suddenly forms a new macroscopic state that is quite different from the previous one.

Thus, the neural correlates of transitions between brain states may reveal more of neural coding than an exhaustive description of single-unit properties in relation to these brain states. Relational order, feature binding, and the emergence of wholes ("Gestalts") as revealed by synchronous population neural activity, may supersede the single- or few-unit activity description of brain function. Relational order between individual neurons is equivalent to the existence and strength of correlation maps (Koenderink 1984) and views assemblies as more than a statistical result of the outcome of a large number of weak interactions among discrete neurons (Schneidman et al. 2006). Through the reciprocity of connections between the neurons in the map, which may provide a positive feedback, very small changes in just a few pair correlations may give rise to a sudden state change characterized by the formation of a new assembly. That only very small changes are required to induce the transitions between states is suggested by the sudden increase in intercortical area synchronization between silence and low-level acoustic stimulation (Eggermont 2000). Are such state transitions operative when perceiving speech and music, or are they only operative when switching attention to sounds signifying danger or potential food? More likely it will depend on the emotional content of the perceived sound, which incorporates both communication and survival aspects.

5.11 The Cortex: Information Processor or Representational System?

The interpretation of the activity in the cortex depends on the view of what the cortex is supposed to do. In the information processing point of view the fidelity of encoding stimuli and decoding spike trains from a single neuron or neuronal population is at the heart of the problem. Information seems to be processed more efficiently and faithfully when the statistics of the stimuli match those found in the animals natural environment (Lewicki 2002). The alternative, representational point of view is best expressed by the statement that "organisms can only learn when events violate their expectations" (Rescorla and Wagner 1972). Here again the expectations may be based on prolonged exposure to the animal's natural environment, but the emphasis is completely different. Taking into account that the neurons in auditory cortex receive more than 99% of their input from other cortical neurons in the same or different area or hemisphere, the representational point of view may carry the day. This concurs with the prescient statement that "the cortex to a large extent is a thinking machine working on its own output" (Braitenberg 1974).

The research relying on evoked potential or magnetic field recordings in humans puts great emphasis on the use of infrequent (odd-ball or deviant) stimuli to probe the capacities of the cortex to discriminate small differences in stimulus parameters. Typically, one stimulus (the odd-ball) is presented with low probability (10–15%) and the frequent stimulus the remaining times. Calculated responses to such stimuli (by subtracting the frequent response from the odd-ball one) include the mismatch negativity (MMN, latencies around 150 ms) that does not require attention, and the P300 that is evoked for the same stimuli when attention is paid or a response is required (Picton et al. 2000). Here the discrimination of the odd-ball from the frequent stimuli is probed; there is only a different response when the event (odd-ball) violates the expectation (the frequent stimulus). Thus the paradigm reflects clearly that of cortex as a representational system. This is further highlighted in a proposed neurocognitive model of sentence comprehension based on evoked potentials/fields (Friederici 2005). Within the early time window of 100–300 ms, an early left anterior negativity (ELAN) correlates with rapidly detectable word category errors. In the 300- to 500-ms poststimulus time window, semantic process violations are represented in a negative wave (N400) that peaks about 400 ms after word onset and occurs in response to words that cannot be integrated semantically into the preceding context. In the 500- to 1,000-ms poststimulus time window the semantic and syntactic types of information are integrated and reflected in a positive wave with 600-ms latency (P600). The P600 correlates with outright syntactic violations, with sentences that require syntactic revision, and with processing of syntactically complex sentences.

Modeling the auditory cortex will be very different depending on the way the system is viewed: as a stimulus driven information processor or as an experience-driven representational system. The first one is easier to handle, but the second is probably the correct one.

5.12 Summary

The auditory cortex is a complex interconnected set of fields with different functional properties but rather similar structural features. Modeling such a system or even small subsets thereof requires as much simplification as possible, but not more, so as to capture the essence of what this subset of cortex does. That is the first problem, because we do not know this essence. We know what it represents of stimuli provided, but we usually do not know enough to use the activity patterns to predict the responses to other different stimuli. Most characterizations of stimuli, such as the STRF, are context dependent and predictions based thereupon are valid only within the same context, suggesting that the cortical neurons work with adaptive spectrotemporal filters. Thus modeling should incorporate a set of STRFs obtained for a wide range of stimulus densities, signal-to-noise ratios, natural sounds with or without behavioral meaning attached, etc., and incorporate into the model a decision switch that selects the right combination of STRFs for the given context. It is implicitly understood that this switch in, for example, the primary auditory cortex might be set by higher-order cortex, this offers an interesting challenge for the interaction between several parallel processing channels and their feedback signals. The local flow of activity in cortex can be estimated from directed entropy flow based on a large number (at least from the point of view of the experimenter) of simultaneous recordings, but this number is a pitifully small sample from the point of representation. Local processing clusters appear also to be of small volume, of the order of 1 mm^3, close to the sampling volume of local field potentials. Consequently, modeling will require thoughtful interpolation of the coarse sampling of spiking neurons and potentially integration with the compound activity in the spatially more extended local field potentials. Alternatively, and hardly touched upon in the main text of this chapter, is the use of networks of spiking neurons that can capture some aspects of cortical processing. The more realistic the model neurons become and the more neurons these nets incorporate, the more they may approximate actual cortical processing. But that frontier is still only starting to be explored.

Acknowledgments This work was supported by the Alberta Heritage Foundation for Medical Research, the National Sciences and Engineering Research Council of Canada, a Canadian Institutes of Health – New Emerging Team grant, and the Campbell McLaurin Chair for Hearing Deficiencies.

References

Abeles M (1991) Corticonics: Neural Circuits of the Cerebral Cortex. Cambridge, UK: Cambridge University Press.

Aertsen AM, Johannesma PI (1981) A comparison of the spectro-temporal sensitivity of auditory neurons to tonal and natural stimuli. Biol Cybern 42:145–156.

Aertsen AM, Gerstein GL, Habib MK, Palm G (1989) Dynamics of neuronal firing correlation: modulation of "effective connectivity." J Neurophysiol 61:900–917.

Aertsen A, Erb M, Palm G (1994) Dynamics of functional coupling in the cerebral-cortex – an attempt at a model-based interpretation. Physica D 75:103–128.

Ahveninen J, Jääskelainen IP, Raij T, Bonmassar G, Devore S, Hämäläinen M, Levänen S, Lin FH, Sams M, Shinn-Cunningham BG, Witzel T, Belliveau JW (2006) Task-modulated "what" and "where" pathways in human auditory cortex. Proc Natl Acad Sci U S A 103:14608–14613.

Alonso JM, Usrey WM, Reid RC (2001) Rules of connectivity between geniculate cells and simple cells in cat primary visual cortex. J Neurosci 21:4002–4015.
Altmann CF, Bledowski C, Wibral M, Kaiser J (2007) Processing of location and pattern changes of natural sounds in the human auditory cortex. Neuroimage 35:1192–1200.
Barrett DJ, Hall DA (2006) Response preferences for "what" and "where" in human non-primary auditory cortex. Neuroimage 32:968–977.
Bendor D, Wang X (2005) The neuronal representation of pitch in primate auditory cortex. Nature 436:1161–1165.
Bendor D, Wang X (2006) Cortical representations of pitch in monkeys and humans. Curr Opin Neurobiol 16:391–399.
Bendor D, Wang X (2007) Differential neural coding of acoustic flutter within primate auditory cortex. Nat Neurosci 10:763–771.
Blake DT, Merzenich MM (2002) Changes of AI receptive fields with sound density. J Neurophysiol 88:3409–3420.
Borst A, Theunissen FE (1999) Information theory and neural coding. Nat Neurosci 2:947–957.
Braitenberg V (1974) Thoughts on the cerebral cortex. J Theor Biol 46:421–447.
Brosch M, Schreiner CE (1999) Correlations between neural discharges are related to receptive field properties in cat primary auditory cortex. Eur J Neurosci 11:3517–3530.
Chechik G, Anderson MJ, Bar-Yosef O, Young ED, Tishby N, Nelken I (2006) Reduction of information redundancy in the ascending auditory pathway. Neuron 51:359–368.
Chi T, Gao Y, Guyton MC, Ru P, Shamma S (1999) Spectro-temporal modulation transfer functions and speech intelligibility. J Acoust Soc Am 106:2719–2732.
Chi T, Ru P, Shamma SA (2005) Multiresolution spectrotemporal analysis of complex sounds. J Acoust Soc Am 118:887–906.
Christianson GB, Sahani M, Linden JF (2008) The consequences of response nonlinearities for interpretation of spectrotemporal receptive fields. J Neurosci 28:446–455.
Clarke S, de Ribaupierre F, Rouiller EM, de Ribaupierre Y (1993) Several neuronal and axonal types form long intrinsic connections in the cat primary auditory cortical field (AI). Anat Embryol (Berl) 188:117–138.
Cooke M (1993) Modelling Auditory Processing and Organization. Cambridge, UK: Cambridge University Press.
de Boer E, de Jongh HR (1978) On cochlear encoding: potentialities and limitations of the reverse-correlation technique. J Acoust Soc Am 63:115–135.
deCharms RC, Merzenich MM (1996) Primary cortical representation of sounds by the coordination of action-potential timing. Nature 381:610–613.
de la Mothe LA, Blumell S, Kajikawa Y, Hackett TA (2006) Thalamic connections of the auditory cortex in marmoset monkeys: core and medial belt regions. J Comp Neurol 496:72–96.
Eggermont JJ (1992) Neural interaction in cat primary auditory cortex. Dependence on recording depth, electrode separation, and age. J Neurophysiol 68:1216–1228.
Eggermont JJ (1993) Wiener and Volterra analyses applied to the auditory system. Hear Res 66:177–201.
Eggermont JJ (1994) Neural interaction in cat primary auditory cortex II. Effects of sound stimulation. J Neurophysiol 71:246–270.
Eggermont JJ (1995) Representation of a voice onset time continuum in primary auditory cortex of the cat. J Acoust Soc Am 98:911–920.
Eggermont JJ (2000) Sound-induced synchronization of neural activity between and within three auditory cortical areas. J Neurophysiol 83:2708–2722.
Eggermont JJ (2001) Between sound and perception: reviewing the search for a neural code. Hear Res 157:1–42.
Eggermont JJ (2006) Properties of correlated neural activity clusters in cat auditory cortex resemble those of neural assemblies. J Neurophysiol 96:746–764.
Eggermont JJ, Komiya H (2000) Moderate noise trauma in juvenile cats results in profound cortical topographic map changes in adulthood. Hear Res 142:89–101.

Eggermont JJ, Smith GM (1990) Characterizing auditory neurons using the Wigner and Rihacek distributions: a comparison. J Acoust Soc Am 87:246–259.

Eggermont JJ, Aertsen AM, Hermes DJ, Johannesma PI (1981) Spectro-temporal characterization of auditory neurons: redundant or necessary. Hear Res 5:109–121.

Eggermont JJ, Aertsen AM, Johannesma PI (1983a) Prediction of the responses of auditory neurons in the midbrain of the grass frog based on the spectro-temporal receptive field. Hear Res 10:191–202.

Eggermont JJ, Johannesma PM, Aertsen AM (1983b) Reverse-correlation methods in auditory research. Q Rev Biophys 16:341–414.

Elhilali M, Fritz JB, Klein DJ, Simon JZ, Shamma SA (2004) Dynamics of precise spike timing in primary auditory cortex. J Neurosci 24:1159–1172.

Fishbach A, Nelken I, Yeshurun Y (2001) Auditory edge detection: a neural model for physiological and psychoacoustical responses to amplitude transients. J Neurophysiol 85:2303–2323.

Fishbach A, Yeshurun Y, Nelken I (2003) Neural model for physiological responses to frequency and amplitude transitions uncovers topographical order in the auditory cortex. J Neurophysiol 90:3663–3678.

Friederici AD (2005) Neurophysiological markers of early language acquisition: from syllables to sentences. Trends Cogn Sci 9:481–488.

Gehr DD, Komiya H, Eggermont JJ (2000) Neuronal responses in cat primary auditory cortex to natural and altered species-specific calls. Hear Res 150:27–42.

Gerstein GL, Kirkland KL (2001) Neural assemblies: technical issues, analysis, and modeling. Neural Netw 14:589–598.

Gourévitch B, Eggermont JJ (2007a) Evaluating information transfer between auditory cortical neurons. J Neurophysiol 97:2533–2543.

Gourévitch B, Eggermont JJ (2007b) Spatial representation of neural responses to natural and altered conspecific vocalizations in cat auditory cortex. J Neurophysiol 97:144–158.

Gourévitch B, Eggermont JJ (2008) Spectrotemporal sound density dependent long-term adaptation in cat primary auditory cortex. Eur J Neurosci 27:3310–3321.

Gourévitch B, Noreña A, Shaw G, Eggermont JJ (2009) Spectro-temporal receptive fields in anesthetized cat primary auditory cortex are context dependent. Cereb Cortex 19(6):1448–1461.

Grande LA, Kinney GA, Miracle GL, Spain WJ (2004) Dynamic influences on coincidence detection in neocortical pyramidal neurons. J Neurosci 24:1839–1851.

Grün S, Diesmann M, Aertsen A (2002) Unitary events in multiple single-neuron spiking activity: II. Nonstationary data. Neural Comput 14:81–119.

Hermes DJ, Aertsen AM, Johannesma PI, Eggermont JJ (1981) Spectro-temporal characteristics of single units in the auditory midbrain of the lightly anaesthetised grass frog (Rana temporaria L) investigated with noise stimuli. Hear Res 5:147–178.

Hickok G, Poeppel D (2000) Towards a functional neuroanatomy of speech perception. Trends Cogn Sci 4:131–138.

Holmgren CD, Zilberter Y (2001) Coincident spiking activity induces long-term changes in inhibition of neocortical pyramidal cells. J Neurosci 21:8270–8277.

Hubel DH, Wiesel TN (1962) Receptive fields, binocular interaction and functional architecture in the cat's visual cortex. J Physiol 160:106–154.

Itoh K, Suwazono S, Arao H, Miyazaki K, Nakada T (2005) Electrophysiological correlates of absolute pitch and relative pitch. Cereb Cortex 15:760–769.

Johannesma PIM, Aertsen A, van den Boogaard H, Eggermont JJ, Epping W (1986) From synchrony to harmony: ideas on the function of neural assembles and on the interpretation of neural synchrony. In: Palm G, Aertsen A (eds), Brain Theory, Berlin: Springer, pp. 25–47.

Kaas JH (1987) The organization of neocortex in mammals: implications for theories of brain function. Ann Rev Psychol 38:129–151.

Kadia SC, Wang X (2003) Spectral integration in A1 of awake primates: neurons with single- and multipeaked tuning characteristics. J Neurophysiol 89:1603–1622.

Karmarkar UR, Buonomano DV (2002) A model of spike-timing dependent plasticity: one or two coincidence detectors? J Neurophysiol 88:507–513.

Keenan JP, Thangaraj V, Halpern AR, Schlaug G (2001) Absolute pitch and planum temporale. Neuroimage 14:1402–1408.
Keuroghlian AS, Knudese EI (2007) Adaptive auditory plasticity in developing and adult animals. Prog Neurobiol 32:109–121.
Kimpo RR, Theunissen FE, Doupe AJ (2003) Propagation of correlated activity through multiple stages of a neural circuit. J Neurosci 23:5750–5761.
Kistler WM, Gerstner W (2002) Stable propagation of activity pulses in populations of spiking neurons. Neural Comput 14:987–997.
Klein DJ, Depireux DA, Simon JZ, Shamma SA (2000) Robust spectrotemporal reverse correlation for the auditory system: optimizing stimulus design. J Comput Neurosci 9:85–111.
Klein DJ, Simon JZ, Depireux DA, Shamma SA (2006) Stimulus-invariant processing and spectrotemporal reverse correlation in primary auditory cortex. J Comput Neurosci 20:111–136.
Koelsch S, Siebel WA (2005) Towards a neural basis of music perception. Trends Cogn Sci 9:578–584.
Koenderink JJ (1984) Simultaneous order in nervous nets from a functional standpoint. Biol Cybern 50:35–41.
Kral A, Eggermont JJ (2007) What's to lose and what's to learn: development under auditory deprivation, cochlear implants and limits of cortical plasticity. Brain Res Rev 56:259–269.
Lee CC, Imaizumi K, Schreiner CE, Winer JA (2004a) Concurrent tonotopic processing streams in auditory cortex. Cereb Cortex 14:441–451.
Lee CC, Schreiner CE, Imaizumi K, Winer JA (2004b) Tonotopic and heterotopic projection systems in physiologically defined auditory cortex. Neuroscience 128:871–887.
Lewicki MS (2002) Efficient coding of natural sounds. Nat Neurosci 5:356–363.
Liberman MC (1988) Response properties of cochlear efferent neurons: monaural vs. binaural stimulation and the effects of noise. J Neurophysiol 60:1779–1798.
Liégeois-Chauvel C, de Graaf JB, Laguitton V, Chauvel P (1999) Specialization of left auditory cortex for speech perception in man depends on temporal coding. Cereb Cortex 9:484–496.
Liégeois-Chauvel C, Giraud K, Badier JM, Marquis P, Chauvel P (2001) Intracerebral evoked potentials in pitch perception reveal a functional asymmetry of the human auditory cortex. Ann N Y Acad Sci 930:117–132.
Limb CJ (2006) Structural and functional neural correlates of music perception. Anat Rec A Discov Mol Cell Evol Biol 288:435–446.
Linden JF, Schreiner CE (2003) Columnar transformations in auditory cortex? A comparison to visual and somatosensory cortices. Cereb Cortex 13:83–89.
Lomber SG, Malhotra S (2008) Double dissociation of 'what' and 'where' processing in auditory cortex. Nat Neurosci 11:609–616.
Lomber SG, Malhotra S, Hall AJ (2007) Functional specialization in non-primary auditory cortex of the cat: areal and laminar contributions to sound localization. Hear Res 229:31–45.
London M, Schreibman A, Hausser M, Larkum ME, Segev I (2002) The information efficacy of a synapse. Nat Neurosci 5:332–340.
Lu T, Liang L, Wang X (2001) Temporal and rate representations of time-varying signals in the auditory cortex of awake primates. Nat Neurosci 4:1131–1138.
Luo F, Wang Q, Kashani A, Yan J (2008) Corticofugal modulation of initial sound processing in the brain. J Neurosci 28:11615–11621.
Malhotra S, Hall AJ, Lomber SG (2004) Cortical control of sound localization in the cat: unilateral cooling deactivation of 19 cerebral areas. J Neurophysiol 92:1625–1643.
Martin KAC (2002) Microcircuits in visual cortex. Curr Opin Neurobiol 12:418–425.
Miller LM, Escabi MA, Read HL, Schreiner CE (2001) Functional convergence of response properties in the auditory thalamocortical system. Neuron 32:151–160.
Miller LM, Escabi MA, Read HL, Schreiner CE (2002) Spectrotemporal receptive fields in the lemniscal auditory thalamus and cortex. J Neurophysiol 87:516–527.
Noreña AJ, Eggermont JJ (2005) Enriched acoustic environment after noise trauma reduces hearing loss and prevents cortical map reorganization. J Neurosci 25:699–705.
Noreña AJ, Gourévitch B, Aizawa N, Eggermont JJ (2006) Spectrally enhanced acoustic environment disrupts frequency representation in cat auditory cortex. Nat Neurosci 9:932–939.

Norena AJ, Gourévitch B, Pienkowski M, Shaw G, Eggermont JJ (2008) Increasing spectro-temporal sound density reveals an octave-based organization in cat primary auditory cortex J Neurosci 28:8885–8896.

Oswald AM, Schiff ML, Reyes AD (2006) Synaptic mechanisms underlying auditory processing. Curr Opin Neurobiol 16:371–376.

Palm G, Pöpel B (1985) Volterra representation and Wiener-like identification of nonlinear-systems – scope and limitations. Q Rev Biophys 18:135–164.

Patterson RD, Uppenkamp S, Johnsrude IS, Griffiths TD (2002) The processing of temporal pitch and melody information in auditory cortex. Neuron 36:767–776.

Phillips DP (1993) Representation of acoustic events in the primary auditory cortex. J Exp Psychol Hum Percept Perform 19:203–216.

Picton TW, Alain C, Otten L, Ritter W, Achim A (2000) Mismatch negativity: different water in the same river. Audiol Neurootol 5:111–139.

Pouget A, Dayan P, Zemel R (2000) Information processing with population codes. Nat Rev Neurosci 1:125–132.

Rajan R, Irvine DR, Wise LZ, Heil P (1993) Effect of unilateral partial cochlear lesions in adult cats on the representation of lesioned and unlesioned cochleas in primary auditory cortex. J Comp Neurol 338:17–49.

Rescorla RA, Wagner AR (1972) A theory of Pavlovian conditioning: variations in the effectiveness of reinforcement and nonreinforcement. In: Black AH, Prokasy WF (eds), Classical Conditioning. II. Current Research and Theories. New York: Appleton-Century-Crofts, pp. 64–99.

Reyes AD (2003) Synchrony-dependent propagation of firing rate in iteratively constructed networks in vitro. Nat Neurosci 6:593–599.

Robertson D, Irvine DR (1989) Plasticity of frequency organization in auditory cortex of guinea pigs with partial unilateral deafness. J Comp Neurol 282:456–471.

Rosen S (1992) Temporal information in speech: acoustic, auditory and linguistic aspects. Philos Trans R Soc Lond B Biol Sci 336:367–373.

Ross D, Choi J, Purves D (2007) Musical intervals in speech. Proc Natl Acad Sci U S A. 104:9852–9857.

Rothman JS, Cathala L, Steuber V, Silver RA (2009) Synaptic depression enables neuronal gain control. Nature 457:1015–1018.

Rudolph M, Destexhe A (2003) Tuning neocortical pyramidal neurons between integrators and coincidence detectors. J Comput Neurosci 14:239–251.

Schiff SJ, Jerger K, Duong DH, Chang T, Spano ML, Ditto WL (1994) Controlling chaos in the brain. Nature 370:615–620.

Schneidman E, Berry MJ II, Segev R, Bialek W (2006) Weak pairwise correlations imply strongly correlated network states in a neural population. Nature 440:1007–1012.

Schnupp JW, Hall TM, Kokelaar RF, Ahmed B (2006) Plasticity of temporal pattern codes for vocalization stimuli in primary auditory cortex. J Neurosci 26:4785–4795.

Shamma SA (1996) Auditory cortical representation of complex acoustic spectra as inferred from the ripple analysis method. Network Comp Neural Syst 7:439–476.

Shannon RV, Zeng FG, Kamath V, Wygonski J, Ekelid M (1995) Speech recognition with primarily temporal cues. Science 270:303–304.

Singer W, Gray CM (1995) Visual feature integration and the temporal correlation hypothesis. Annu Rev Neurosci 18:555–586.

Singh NC, Theunissen FE (2003) Modulation spectra of natural sounds and ethological theories of auditory processing. J Acoust Soc Am 114:3394–3411.

Sjöström PJ, Turrigiano GG, Nelson SB (2003) Neocortical LTD via coincident activation of presynaptic NMDA and cannabinoid receptors. Neuron 39:641–654.

Smith PH, Populin LC (2001) Fundamental differences between the thalamocortical recipient layers of the cat auditory and visual cortices. J Comp Neurol 436:508–519.

Spangler KM, Warr WB (1991) The descending auditory system. In: Altschuler RA, Bobbin RP, Clopton BM, Hoffman DW (eds), Neurobiology of Hearing the Central Auditory System. New York: Raven, pp. 27–45.

Stephan KE, Hilgetag CC, Burns GA, O'Neill MA, Young MP, Kötter R. (2000) Computational analysis of functional connectivity between areas of primate cerebral cortex. Philos Trans R Soc Lond B Biol Sci 355:111–126.

Suga N (1989) Principles of auditory information-processing derived from neuroethology. J Exp Biol 146:277–286.

Takeuchi AH, Hulse SH (1993) Absolute pitch. Psychol Bull 113:345–361.

Theunissen FE, Sen K, Doupe AJ (2000) Spectral-temporal receptive fields of nonlinear auditory neurons obtained using natural sounds. J Neurosci 20:2315–2331.

Theunissen FE, Amin N, Shaevitz SS, Woolley SM, Fremouw T, Hauber ME (2004) Song selectivity in the song system and in the auditory forebrain. Ann N Y Acad Sci 1016: 222–245.

Tomita M, Eggermont JJ (2005) Cross-correlation and joint spectro-temporal receptive field properties in auditory cortex. J Neurophysiol 93:378–392.

Ts'o DY, Gilbert CD, Wiesel TN (1986) Relationships between horizontal interactions and functional architecture in cat striate cortex as revealed by cross-correlation analysis. J Neurosci 6:1160–1170.

Vaadia E, Haalman I, Abeles M, Bergman H, Prut Y, Slovin H, Aertsen A (1995) Dynamics of neuronal interactions in monkey cortex in relation to behavioural events. Nature 373:515–518.

Valentine PA, Eggermont JJ (2004) Stimulus dependence of spectro-temporal receptive fields in cat primary auditory cortex. Hear Res 196:119–133.

Villa AE, Rouiller EM, Simm GM, Zurita P, de Ribaupierre Y, de Ribaupierre F (1991) Corticofugal modulation of the information processing in the auditory thalamus of the cat. Exp Brain Res 86:506–517.

von der Malsburg C (1995) Binding in models of perception and brain function. Curr Opin Neurobiol 5:520–526.

Wallace MN, Kitzes LM, Jones EG (1991) Intrinsic inter- and intralaminar connections and their relationship to the tonotopic map in cat primary auditory cortex. Exp Brain Res 86:527–544.

Wang X, Merzenich MM, Beitel R, Schreiner CE (1995) Representation of a species-specific vocalization in the primary auditory cortex of the common marmoset: temporal and spectral characteristics. J Neurophysiol 74:2685–2706.

Wang X, Lu T, Snider RK, Liang L (2005) Sustained firing in auditory cortex evoked by preferred stimuli. Nature 435:341–346.

Watts J, Thomson AM (2005) Excitatory and inhibitory connections show selectivity in the neocortex. J Physiol 562:89–97.

Wehr M, Zador AM (2005) Synaptic mechanisms of forward suppression in rat auditory cortex. Neuron 47:437–445.

Werner G, Mountcastle VB (1965) Neural activity in mechanoreceptive cutaneous afferents: stimulus-response relations, Weber functions, and information Transmission. J Neurophysiol 28:359–397.

Winer JA, Lee CC (2007) The distributed auditory cortex. Hear Res 229:3–13.

Yan J, Ehret G (2001) Corticofugal reorganization of the midbrain tonotopic map in mice. Neuroreport 12:3313–3316.

Yan W, Suga N (1998) Corticofugal modulation of the midbrain frequency map in the bat auditory system. Nat Neurosci 1:54–58.

Yan J, Zhang Y, Ehret G (2005) Corticofugal shaping of frequency tuning curves in the central nucleus of the inferior colliculus of mice. J Neurophysiol 93:71–83.

Zatorre RJ (2001) Neural specializations for tonal processing. Ann N Y Acad Sci 930:193–210.

Zatorre RJ, Belin P, Penhune VB (2002) Structure and function of auditory cortex: music and speech. Trends Cogn Sci 6:37–46.

Zhang Y, Suga N, Yan J (1997) Corticofugal modulation of frequency processing in bat auditory system. Nature 387:900–903.

Zheng XY, Henderson D, McFadden SL, Ding DL, Salvi RJ (1999) Auditory nerve fiber responses following chronic cochlear de-efferentation. J Comp Neurol 406:72–86.

Chapter 6
Computational Models of Inferior Colliculus Neurons

Kevin A. Davis, Kenneth E. Hancock, and Bertrand Delgutte

Abbreviations

AM	Amplitude modulation
AN	Auditory nerve
BD	Best delay
BMF	Best modulation frequency
BF	Best frequency
CD	Characteristic delay
CN	Cochlear nucleus
CP	Characteristic phase
CR	Chopping rate
DCN	Dorsal cochlear nucleus
DNLL	Dorsal nucleus of the lateral lemniscus
EI	Contralaterally excited and ipsilaterally inhibited
EI/F	EI with facilitation
F0	Fundamental frequency
IC	Inferior colliculus
ICC	Central nucleus of the inferior colliculus
ICX	External nucleus of the inferior colliculus
ILD	Interaural level difference
IPD	Interaural phase difference
ITD	Interaural time difference
LSO	Lateral superior olive

K.A. Davis (✉)
Departments of Biomedical Engineering and Neurobiology and Anatomy,
University of Rochester, Rochester, NY 14642, USA
e-mail: kevin_davis@urmc.rochester.edu

R. Meddis et al. (eds.), *Computational Models of the Auditory System*,
Springer Handbook of Auditory Research 35, DOI 10.1007/978-1-4419-5934-8_6,

MNTB	Medial nucleus of the trapezoid body
MSO	Medial superior olive
MTF	Modulation transfer function
PE	Precedence effect
PIR	Postinhibitory rebound
SOC	Superior olivary complex
VNLL	Ventral nucleus of the lateral lemniscus
VCN	Ventral cochlear nucleus

6.1 Introduction

The inferior colliculus (IC), the principal auditory nucleus in the midbrain, occupies a key position in the auditory system. It receives convergent input from most of the auditory nuclei in the brain stem, and in turn, projects to the auditory forebrain (Winer and Schreiner 2005). The IC is therefore a major site for the integration and reorganization of the different types of auditory information conveyed by largely parallel neural pathways in the brain stem, and its neural response properties are accordingly very diverse and complex (Winer and Schreiner 2005). The function of the IC has been hard to pinpoint. The IC has been called the "nexus of the auditory pathway" (Aitkin 1986), a "shunting yard of acoustical information processing" (Ehret and Romand 1997), and the locus of a "transformation [that] adjusts the pace of sensory input to the pace of behavior" (Casseday and Covey 1996). The vague, if not metaphorical, nature of these descriptions partly reflects the lack of a quantitative understanding of IC processing such as that which has emerged from computational studies of the cochlear nucleus (Voigt, Chapter 3) and superior olivary complex (Colburn, Chapter 4).

This chapter reviews computational modeling efforts aimed at understanding how the representations of the acoustic environment conveyed by the ascending inputs are processed and integrated in the IC. Most of these efforts have been directed at the processing of either sound localization cues or amplitude envelope information. To our knowledge, this is the first detailed review of this topic, although previous reviews of the processing of specific stimulus features, for example, binaural processing (Colburn 1996; Palmer and Kuwada 2005) or temporal processing (Langner 1992; Rees and Langner 2005) briefly discussed models for IC neurons. Our focus is primarily on models aimed at explaining biophysical mechanisms leading to emerging physiological response properties of individual IC neurons, although some neural population models for predicting psychophysical performance from patterns of neural activity are also covered. The models are restricted to the mammalian IC, with the exception of one model for the avian homolog of the IC (Rucci and Wray 1999) where the processing may be of general applicability.

6.1.1 *Functional Organization and Afferent Projections to the Inferior Colliculus*

The IC consists of a central nucleus (ICC), characterized by the presence of disk-shaped cells and fibrodendritic laminae, surrounded by layered structures ("cortices") dorsally, laterally, and caudally (Oliver 2005). The tonotopically organized ICC receives the vast majority of afferent inputs from brain stem auditory nuclei and has been the object of most neurophysiological and modeling studies. The surrounding regions receive multisensory and descending inputs as well as ascending projections from the brain stem and ICC, and neurons in these subdivisions tend to have more complex and labile response properties than those in ICC. The main focus of this chapter is therefore on models for ICC neurons.

The main ascending pathways to the ICC are shown in Fig. 6.1 and are described in detail in a recent volume (Cant 2005; Schofield 2005). Two largely monaural,

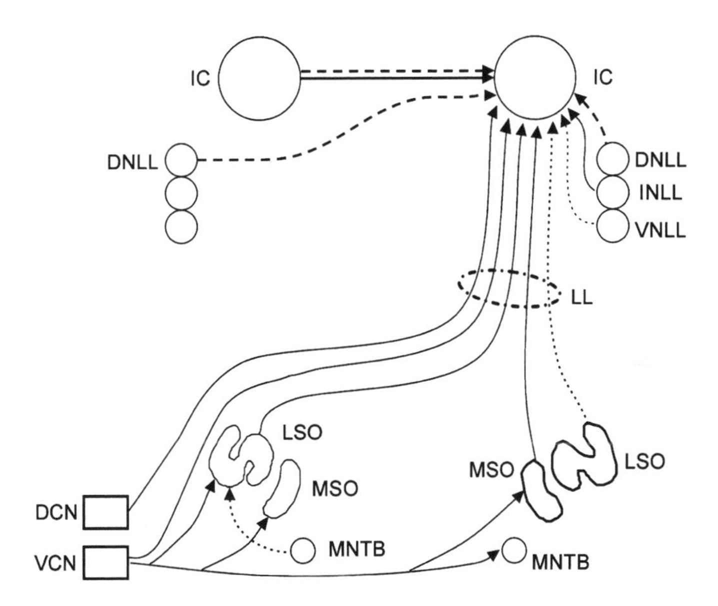

Fig. 6.1 Major ascending projections from the cochlear nucleus to the central nucleus of the inferior colliculus (ICC). Two direct, largely monaural projections arise from the contralateral ventral (VCN) and dorsal cochlear nuclei (DCN). Two projections emerge from binaural nuclei of the superior olivary complex (SOC), one from the medial (MSO) and one from lateral superior olives (LSO). The MSO receives bilateral input from the VCN and projects ipsilaterally to the ICC. The LSO also receives bilateral input from the VCN, via a synapse in the medial nucleus of the trapezoid body (MNTB), and projects bilaterally to the ICC. All of the ascending fibers from CN and SOC run in the lateral lemniscus, where collaterals contact the ventral (VNLL), intermediate (INLL) and dorsal nuclei of the lateral lemniscus (DNLL). The VNLL and INLL project to the ipsilateral ICC, whereas the DNLL projects bilaterally. Putative excitatory pathways are shown with *solid lines*; glycinergic inhibitory pathways are indicated by *dotted lines*; GABAergic inhibitory pathways are represented by *dashed lines*

excitatory projections to the ICC emerge from the stellate cells in the contralateral ventral cochlear nucleus (VCN) and the fusiform and giant cells in the contralateral dorsal cochlear nucleus (DCN) (Cant 2005). The ICC also receives major projections from the two main binaural nuclei of the superior olivary complex (SOC): an ipsilateral excitatory projection from the medial superior olive (MSO) and bilateral projections from the lateral superior olives (LSO) (Schofield 2005). The crossed LSO projection is excitatory, whereas the uncrossed projection is mostly inhibitory and glycinergic (Schofield 2005). LSO neurons show sensitivity to interaural level differences (ILD) arising through interaction of an excitatory input from spherical bushy cells in the ipsilateral VCN, and an inhibitory input from globular bushy cells in the contralateral VCN via an intervening synapse in the medial nucleus of the trapezoid body (MNTB). In contrast, MSO neurons show sensitivity to interaural time differences (ITD) created by convergence of bilateral excitatory inputs from spherical bushy cells in the VCN, with modulation by additional inhibitory inputs.

In addition to ascending projections from the cochlear nuclei (CN) and the SOC, the ICC receives afferent input from three groups of neurons intermingled among the lateral lemniscus fibers ascending to the auditory midbrain: the ventral (VNLL), intermediate (INLL), and dorsal nuclei of the lateral lemniscus (DNLL) (Schofield 2005). The most important of these for our purposes are the DNLL projections, which are bilateral and mostly GABAergic. The DNLL receives most of its inputs from binaural nuclei including MSO, LSO, and the contralateral DNLL, and contains many neurons sensitive to ITD and ILD.

The various brain stem projections to the ICC overlap in some locations, but remain partly segregated in other regions, forming unique synaptic (functional) domains (Oliver and Huerta 1992). The development of physiological criteria for identifying the brain stem cell types from which an ICC neuron receives inputs is an active area of research (Oliver et al. 1997; Ramachandran et al. 1999; Davis 2002; Loftus et al. 2004; Malmierca et al. 2005).

By traditional anatomical criteria, there are only two types of neurons in ICC: disc-shaped and stellate (Oliver and Morest 1984). Disc-shaped cells form the vast majority of ICC neurons, and have planar dendritic fields parallel to the tonotopically-organized afferent fibers from the lateral lemniscus. Stellate cells have omnidirectional dendritic fields spanning several fibrodendritic laminae formed by the lemniscal axons and the dendrites of disc-shaped cells. The physiological correlates of these two anatomical cell types are unknown. In vitro recordings reveal a more complex organization than the traditional disc-shaped/stellate dichotomy. For example, Sivaramakrishnan and Oliver (2001) have identified six physiologically distinct cell types based on their membrane composition in various types of K^+ channels. How the different types of ICC cells defined by intrinsic membrane channels interact with the wide diversity of synaptic domains based on afferent inputs is still an unexplored topic. It is clear, however, that the highly complex organization of the ICC provides a framework for sophisticated processing of acoustic information.

6.1.2 Strategies for Modeling IC Neurons

The great complexity in the functional organization and afferent inputs of the ICC necessarily calls for simplification when modeling the response properties of ICC neurons. To distinguish processing occurring in the IC itself from that inherited from the inputs, an ICC neuron model must include a functional description of the inputs to the IC, that is, it must explicitly or implicitly incorporate a model for processing by the auditory periphery and brain stem nuclei. The models discussed in this chapter illustrate different approaches to this general problem.

The most detailed models explicitly generate action potentials at every stage in the auditory pathway from the auditory nerve (AN) to the IC (Hewitt and Meddis 1994; Cai et al. 1998a, b; Shackleton et al. 2000; Kimball et al. 2003; Voutsas et al. 2005; Dicke et al. 2007; Pecka et al. 2007). This approach has the advantage of directly mimicking the biology, and with a modular software architecture, any improvement in modeling one neuron module can, in principle, be readily propagated to the entire circuit. However, this type of model can be computationally demanding and is limited by the weakest link in the chain of neuron models.

In many cases, the models for individual neurons in the circuit are single-compartment models with explicit biophysical simulation of selected membrane channels; synaptic inputs are implemented as changes in a membrane conductance (Hewitt and Meddis 1994; Cai et al. 1998a, b; Shackleton et al. 2000; Kimball et al. 2003; Pecka et al. 2007). These models generate an action potential when the membrane voltage crosses a threshold (often time-varying). However, some models introduce simplifications in the simulations for at least some of the neurons in the circuit. Some models use point process models rather than biophysical simulations to generate spikes efficiently (Dicke et al. 2007), some replace spiking neuron models by models operating on firing probabilities (Nelson and Carney 2004), and some use algebraic steady-state input–output functions to model the transformations in average firing rates through the circuit (Reed and Blum 1999; Rucci and Wray 1999). These simplifications can increase computational efficiency but also preclude a characterization of certain response properties, for example, temporal discharge patterns when using input–output functions operating on average firing rates, or the influence of firing history (e.g., refractoriness) when using spike probabilities.

Some IC models avoid an explicit modeling of the brain stem circuitry altogether, using signal processing models (such as cross-correlation) to simulate either the immediate inputs to the IC (Rucci and Wray 1999; Borisyuk et al. 2002) or the IC responses themselves (Hancock and Delgutte 2004). Again, this can achieve considerable computational efficiency that is particularly important when modeling a large population of IC neurons (Rucci and Wray 1999; Hancock and Delgutte 2004) but can also limit the set of stimuli to which the model is applicable (e.g., the Borisyuk et al. 2002 model only applies to sinusoidal stimuli, and the cross-correlation models work best for static stimuli).

The models also differ in which subsets of inputs to the ICC they include. Models for binaural processing explicitly include modules for the SOC inputs

(Cai et al. 1998a, b; Shackleton et al. 2000; Pecka et al. 2007), while models for spectral processing (Kimball et al. 2003) focus on inputs from DCN, and models for temporal processing of amplitude modulation (AM) information typically focus on inputs from VCN (Hewitt and Meddis 1994; Nelson and Carney 2004; Voutsas et al. 2005; Dicke et al. 2007). Among the binaural models, those concerned with ITD processing tend to include primarily inputs from MSO and, in some cases, DNLL (Cai et al. 1998a, b; Borisyuk et al. 2002), while models for ILD processing include only inputs from the LSO circuit (Reed and Blum 1999; Pecka et al. 2007). Only one binaural model includes inputs from both MSO and LSO (Shackleton et al. 2000), although even in this case the focus is limited to ITD processing.

6.2 Sound Localization in an Anechoic Environment

Our ability to localize sounds in space is based primarily on three acoustic cues: interaural differences in time and level, and spectral cues (Middlebrooks and Green 1991). The first two cues arise as a result of the spatial separation of the two ears, and are important for left–right (azimuthal) localization. Spectral cues are created by the filtering properties of the head and pinnae, and are critical for accurate localization of elevation in the median plane where interaural cues are minimal. Initial processing of these three cues is done largely in separate brain stem nuclei, with ITD encoded in the MSO (Goldberg and Brown 1969; Yin and Chan 1990), ILD in the LSO (Boudreau and Tsuchitani 1968), and spectral cues (in the form of spectral notches) in the DCN (Young and Davis 2001). Neurons sensitive to binaural and/or monaural sound localization cues are common in the ICC (Palmer and Kuwada 2005). In some respects, the response properties of these ICC neurons are similar to those of their brain stem inputs, but there are also important differences on which modeling efforts have been focused.

6.2.1 Lateralization Based on ITD

One difference between neurons in the brain stem and the midbrain is that ICC unit sensitivity to ITD is often incompatible with a simple coincidence detection mechanism (Jeffress 1948). Most neurons in the MSO respond maximally when their inputs from the two sides arrive simultaneously, that is, they act as interaural coincidence detectors (Goldberg and Brown 1969). These units are termed “peak type” because their pure-tone ITD functions show a maximum response (peak) at the same ITD (called characteristic delay, CD) regardless of stimulus frequency (Yin and Chan 1990). The CD is defined as the slope of the function relating best interaural phase to stimulus frequency (the phase plot), which for simple coincidence detectors is a straight line with a *y*-intercept (the characteristic phase, CP) equal to 0 cycle (dashed line, Fig. 6.2a). Other brain stem neurons, most notably in the LSO, are excited by

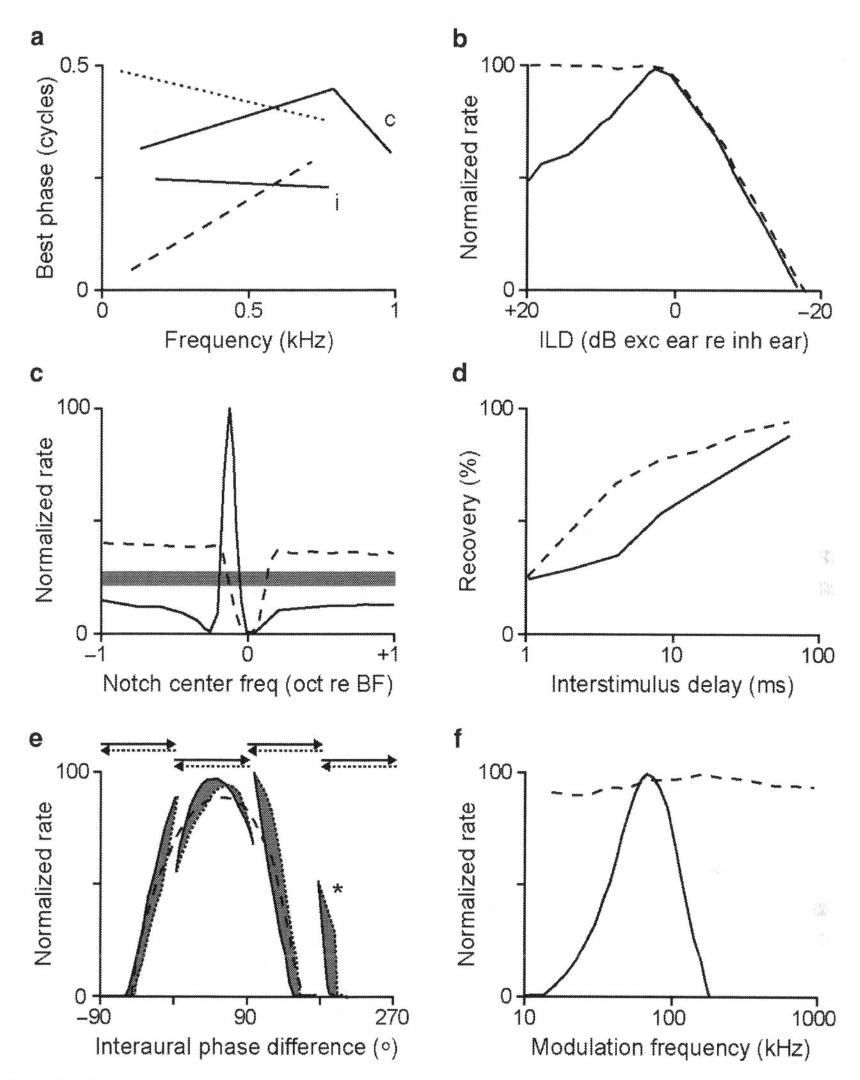

Fig. 6.2 Schematic representations of transformations in binaural and monaural response properties between the brain stem and the midbrain. (**a**) Phase plots derived from responses to binaural pure tones for ITD-sensitive unit(s) in the MSO (peak-type; *dashed line*), the LSO (trough-type; *dotted line*) and the ICC (intermediate (i) and complex-types (c); *solid lines*). (**b**) Rate vs. ILD curves for an excitatory–inhibitory (EI) unit in the LSO (*dashed line*) and an EI/F unit in the ICC (*solid line*). (**c**) Responses of a DCN type IV unit (*dashed line*) and an ICC type O unit (*solid line*) to a notched-noise stimulus whose 3.2-kHz wide notch is swept across the receptive field of the unit. The *gray bar* indicates the range of spontaneous rates for the two units. (**d**) Population response recovery functions to the second of a pair of binaural clicks for CN units (*dashed line*) and ICC units (*solid line*). (**e**) Interaural phase difference (IPD) functions of an MSO (*dashed line*) and an ICC neuron (*solid lines*) derived from responses to simulated sound motion through arcs in the azimuthal plane (sweeps are shown with *arrows* at the *top* of the plot). (**f**) Rate modulation transfer functions for a CN (*dashed line*) and an ICC unit (*solid line*)

sounds in the ipsilateral ear and inhibited by sounds in the contralateral ear (Boudreau and Tsuchitani 1968). Although the spatial receptive fields of LSO units are dominated by ILD cues (Tollin and Yin 2002a, b), many LSO units are also ITD sensitive and are termed "trough type" because they respond minimally when binaural inputs of any frequency are in phase (Caird and Klinke 1983). The phase plots of trough types units are also linear, but with a CP of 0.5 cycle (dotted line, Fig. 6.2a).

In the ICC, many neurons show peak-type and trough-type ITD functions similar to those of MSO and LSO neurons, respectively (Yin and Kuwada 1983). Many other units, however, diverge from the simple coincidence detector model (solid lines, Fig. 6.2a) and show either linear phase plots whose CP is neither 0 nor 0.5 cycles ("intermediate types") or nonlinear phase plots ("complex type") (Palmer and Kuwada 2005). Intermediate and complex type neurons are also observed in the SOC, but they appear to be less prevalent than in ICC (Batra et al. 1997a). It has been hypothesized that the additional complexities of ICC binaural responses may be created through convergence of peak and trough type inputs within the ICC (Yin and Kuwada 1983; McAlpine et al. 1998), or may reflect the effects of ITD-sensitive inhibitory inputs (McAlpine et al. 1998).

Shackleton et al. (2000) tested the hypothesis that convergent inputs from the SOC with different ITD tuning can account for complex and intermediate-types in the ICC. They modeled ICC cells as integrate-and-fire neurons receiving excitatory inputs from either two MSO cells tuned to different ITDs (Fig. 6.3a) or from one MSO cell and one LSO cell. An important assumption in the model is that the two inputs to the ICC cell combine linearly, consistent with physiological observations (McAlpine et al. 1998). This is accomplished by setting the threshold of the membrane model to a low value so that the ICC cell fires when it receives an action potential from either of its two inputs.

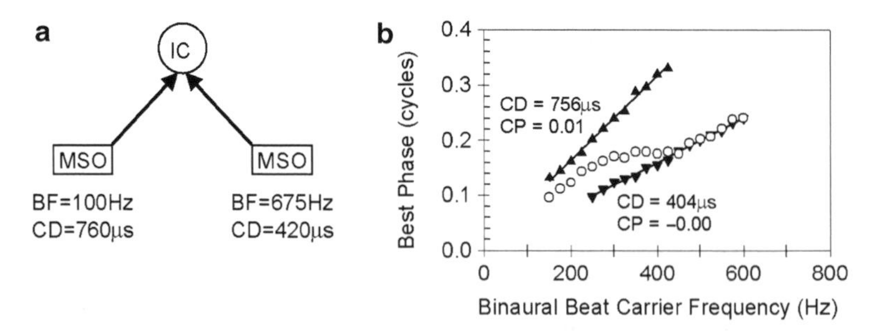

Fig. 6.3 Model for complex phase-frequency plots in ICC (modified from Shackleton et al. 2000). (**a**) An ICC model cell receives input from two MSO cells differing in best frequency (BF) and characteristic delay (CD). (**b**) Best interaural phase difference (IPD) vs. the frequency of a binaural pure tone. *Solid symbols*: Suppressing one input with a tone at its BF and an unfavorable IPD causes the other input to dominate the IC cell's phase-frequency response. *Upward triangles*: high-BF input suppressed. *Downward triangles*: low-BF input suppressed. *Open circles*: when both inputs are active, each dominates the response over a different frequency range, producing an overall phase-frequency plot with complex shape

Figure 6.3b shows a complex phase-frequency plot (open circles) for a model ICC cell that receives two MSO inputs. The important model parameters are the best frequencies (BFs) (100 and 675 Hz) and CDs (760 and 420 μs) of the MSO inputs, chosen to match specific experimental data (McAlpine et al. 1998). The solid symbols show the response when an additional tone stimulus at an unfavorable IPD is introduced to suppress one or the other MSO input, as in the neurophysiological experiments. When the high-BF input is suppressed (upward triangles), the response of the ICC cell is controlled almost entirely by the low-BF input, and vice versa (downward triangles), giving rise to a linear phase plot in either case. When both inputs are active (open circles), the phase plot becomes nonlinear because it is dominated by one input for low frequencies and by the other input for high frequencies, with a transition region in between. This and other examples demonstrate that complex phase-frequency responses could arise in the ICC through simple superposition of excitatory inputs with different best frequencies and/or characteristic delays. As Shackleton et al. (2000) emphasize, these inputs need not arise exclusively from the SOC because intrinsic circuitry within the IC itself could produce equivalent results.

Not all complex and intermediate response types can be accounted for by the Shackleton et al. (2000) model. For example, intermediate types in the ICC and SOC are hard to explain in terms of linear convergence, but could result from a mismatch between the BFs of the ipsilateral and contralateral inputs to SOC cells and the resulting differences in latencies of the cochlear traveling waves on the two sides (Yin and Kuwada 1983; Shackleton et al. 2000). Complex types might also arise through the interaction of inhibition with coincidence detection in the superior olive (Batra et al. 1997b). As Shackleton et al. argue, such nonlinear mechanisms seem to be inconsistent with the simple additive interactions that were the focus of their study. Thus, it is likely that heterogeneous mechanisms underlie the diversity of phase-frequency plots in the ICC.

6.2.2 Lateralization Based on ILD

A second difference between neurons in the brain stem and the ICC is that some ICC units show facilitated responses to a limited range of ILDs and thus exhibit highly focused spatial receptive fields (Pollak et al. 2002). The initial processing of ILD occurs in the LSO, where units are classified as inhibitory–excitatory (IE) because they receive inhibitory input from the contralateral ear and excitatory input from the ipsilateral ear (Boudreau and Tsuchitani 1968). In this way, the coded intensity at one ear is effectively subtracted from the coded intensity in the opposite ear and thus the level difference is represented in the discharge rate of the LSO cell. For most LSO units, firing rate decreases monotonically with increasing stimulation of the inhibitory (contralateral) ear when the excitatory (ipsilateral) input is held constant (dashed line, Fig. 6.2b). Because the excitatory LSO projection to the IC is primarily contralateral, most ILD-sensitive units in ICC are EI and their firing rates decrease with increasing level in the ipsilateral ear. However, a fraction of units (~20%) shows facilitation at intermediate levels of ipsilateral stimulation and inhibition at higher levels (EI/F cells) (solid line, Fig. 6.2b) (Park and Pollak 1993; Davis et al. 1999). These neurons could be the basis for a "place code" of ILD, because they have relatively circumscribed receptive fields. Blocking GABAergic inhibition in the ICC transforms most EI/F cells into conventional EI cells by increasing the spike count elicited by the contralateral sound (Park and Pollak 1993). Hence, the term "facilitation" is somewhat misleading: the essential feature of EI/F cells appears to be GABAergic inhibition evoked by high-level sounds in the contralateral ear. The ipsilateral DNLL has been suggested as the source of this inhibition (Park and Pollak 1993).

This conceptual model was tested computationally by Reed and Blum (1999) in a model created primarily to test connectional hypotheses involving the DNLL. The model ICC cell receives excitatory input from a cell in contralateral LSO and inhibitory input from a cell in ipsilateral DNLL (Fig. 6.4a). In turn, the DNLL cell receives both an excitatory input from the same contralateral LSO cell that projects to the ICC cell and an inhibitory input from a cell in ipsilateral LSO. In Fig. 6.4, the inhibitory connections are introduced one by one to illustrate how the model works. When both inhibitory connection strengths are zero (i.e., DNLL to ICC, and ipsilateral LSO to DNLL), the rate–ILD curves of the DNLL and ICC cells are

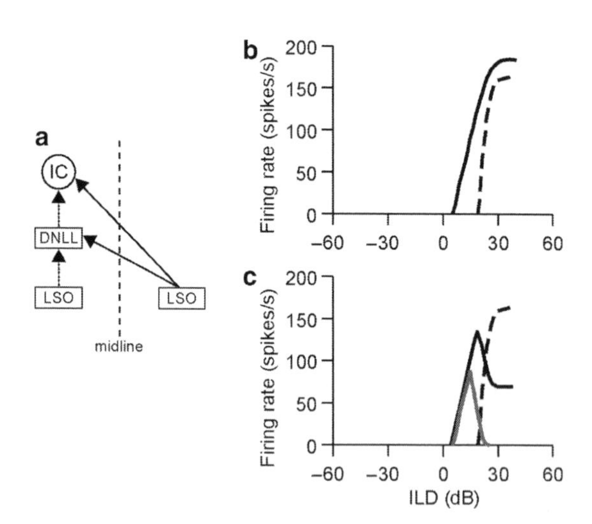

Fig. 6.4 Model for EI/F units in ICC. (Modified from Reed and Blum 1999.) (**a**) Diagram of the neural circuit. Levels below LSO are omitted for clarity. (**b**) *Solid line*: without inhibitory connections, rate–ILD curves of DNLL and ICC model neurons are identical, reflecting common input from contralateral LSO. *Dashed line*: inhibition from ipsilateral LSO shifts the rate–ILD curves of the DNLL to the right. (**c**) *Dashed line*: same as in (**b**). *Solid black line*: moderate inhibition from DNLL reduces ICC response to large contralateral ILDs, creating the EI/F response. *Solid gray line*: Strong inhibition from DNLL completely suppresses ICC response to large ILDs, creating a tuned rate–ILD curve

identical and reflect the IE response of the contralateral LSO cell (Fig. 6.4b). When just the inhibition from DNLL to ICC is added, the response of the ICC cell is reduced at all ILDs (i.e., the curve is compressed, not shown) because the responses of the DNLL and ICC cells exactly align. However, when the ipsilateral LSO to DNLL inhibition is also turned on, the rate–ILD curve of the DNLL cell (and hence its inhibitory effect) shifts to more contralateral ILDs (Fig. 6.4b, c, dashed line), allowing the ICC cell to respond preferentially over a restricted range of intermediate ILDs (Fig. 6.4c, solid line). That is, the ICC cell is disinhibited at intermediate ILDs, where the response appears "facilitated" The gray line in Fig. 6.4c shows that if the DNLL to ICC inhibition is sufficiently strong, the ICC cell responds only to the intermediate ILDs, and thus has completely circumscribed ILD tuning.

Like the Shackleton et al. (2000) model, the Reed and Blum (1999) model stands as a "proof of principle": it demonstrates that the mechanism described is consistent with the observed physiology, without excluding alternative circuits. In the Reed and Blum model, inhibition from DNLL is the critical element underlying the formation of narrow ILD tuning. Reed and Blum suggest that the functional significance of this inhibition may be the transformation of an "edge code," in which sound source ILD is encoded by the location of a boundary between excited and inhibited cells, to a "place code," in which ILD is encoded by the location of maximum activity. That idea awaits further experimental and theoretical evaluation. While place codes can be advantageous for psychophysical acuity, this depends on the sharpness of tuning and the range of stimulus parameters to be encoded (Harper and McAlpine 2004).

6.2.3 Processing of Spectral Cues

A third transformation in the representation of spatial cues that may be directly attributed to processing within the ICC is that some ICC units show more selective responses to spectral notches than do brain stem neurons. Physiological and behavioral evidence in cats suggests that a functional pathway specialized for processing spectral notches is initiated in DCN (Young and Davis 2001). Type O units in the ICC, one of three major unit types defined based on frequency response maps (Ramachandran et al. 1999), are the primary target of ascending DCN projections and are thus thought to represent a midbrain specialization for the processing of spectral cues for sound localization (Davis 2002; Davis et al. 2003). When tested with band-reject noise of varying notch frequency, DCN principal cells (type IV units) show tuned inhibition for notches centered at their BF (dashed line, Fig. 6.2c) (Young et al. 1992). In contrast, type O units in ICC show a tuned excitation for a spectral notch whose rising edge is located just below the BF (solid line, Fig. 6.2c) (Davis et al. 2003). This is the only stimulus known to excite type O units at high stimulus levels. It has been hypothesized that DCN influences are transformed into a more selective representation of sound source location in the ICC through local convergence of frequency-tuned inhibitory and wideband excitatory inputs.

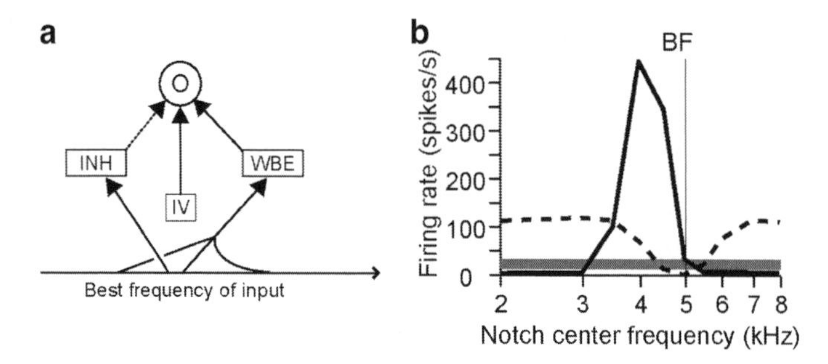

Fig. 6.5 Model for ICC type O responses to spectral notches. (Modified from Kimball et al. 2003.) (**a**) Model type O cell receives on-BF excitatory input from DCN model type IV unit, below-BF inhibition from a narrowly tuned source (INH), and excitation from a broadly tuned source (WBE). (**b**) Responses of model type IV (*dashed line*) and type O (*solid line*) cells as a function of the notch frequency of band-reject noise. The model type O cell shows a tuned excitatory response for a notch located just below BF, consistent with physiological observations

In a preliminary study (Kimball et al. 2003), a computational model of ICC type O units was created to test the idea that the selectivity of these units to spectral edges can be accounted for through interactions between excitatory inputs from DCN type IV units with other inputs. Type O cells in this model have three inputs (Fig. 6.5a): an excitatory input from same-BF type IV cells in the DCN (Hancock and Voigt 1999), a narrowband inhibitory input from "INH" cells centered 0.15 octaves below the BF of the ICC cell, and a wideband excitatory input from "WBE" cells. Although the anatomical identity of the INH and WBE cells is unknown, in the model they both receive excitatory input from model AN fibers (Carney 1993). INH cells receive strong excitatory input from same-BF AN fibers and thus are narrowly tuned and respond well to both tones and noise. WBE cells, on the other hand, receive weak excitatory input from AN fibers covering a wide range of BFs and thus respond weakly to tones but strongly to noise. Figure 6.5b plots the firing rate as a function of notch center frequency for a model type O unit, and shows an excitatory response tuned to notches located just below the BF (vertical line), consistent with the physiology. The results suggest that the conceptual model can account for the unique spectral integration properties of ICC type O units.

An obvious task for future experimental and modeling work is to associate the INH and WBE inputs with specific brain stem sources. Such identification may reveal connections to other IC models. For example, the DNLL is a putative source of inhibition in several models, and is also a suitable candidate for the INH input to type O cells (Davis et al. 2003). The time course of inhibition is often critically important. Although spectral coding by auditory neurons is typically described in the steady state, a characterization of temporal response properties may help identify the inhibitory inputs and facilitate comparison to other IC models. The across-BF convergence occurring in the WBE neuron is an important feature of population models for ITD processing and AM coding (Sects. 6.5 and 6.7) and models for Onset neurons in the cochlear nucleus (Voigt, Chapter 3). Evaluation of this mechanism in the context of spectral processing may reveal another unifying element among IC models.

6.3 Sound Localization in Reverberant Environments: The Precedence Effect

The precedence effect (PE, a.k.a. "law of the first wavefront") refers to a group of perceptual phenomena relevant to sound localization in reverberant environments (for review, see Litovsky et al. 1999). The PE is experienced when two (or more) sounds originating from different spatial locations reach the ears closely spaced in time. Typically, the leading sound represents the direct wavefront and the lagging sound a single acoustic reflection. If the delay between the two sounds is less than approximately 1 ms, the listener hears a single auditory event located between the two sources, but biased toward the leading source. This phenomenon is called summing localization. For longer delays, the listener still hears only one auditory event but localizes the event near the location of the leading source. This phenomenon is called localization dominance. Finally, if the interstimulus delay exceeds an "echo threshold," then the two sounds are perceived as separate entities each with its own spatial location. The echo threshold depends on the characteristics of the sound source, ranging from 5 to 10 ms for transient sounds to 30 to 50 ms for continuous speech or music (for review, see Litovsky et al. 1999). It is important to note that, while the spatial location of the lagging sound is suppressed during the period of localization dominance, its presence nonetheless affects other aspects of the percept including its timbre, loudness, and spatial extent.

Neural correlates of the PE have been described at virtually all levels of the auditory system, including the auditory nerve, cochlear nucleus, SOC, ICC and auditory cortex (reviewed by Litovsky et al. 1999; Palmer and Kuwada 2005). At each level, when a pair of successive sounds (e.g., clicks) is presented to the ears with a short delay, the response to the lagging stimulus is (almost always) suppressed for short delays and recovers for long delays. The rate of recovery depends on the stage of the auditory system under study. At the level of the AN and CN (dashed line, Fig. 6.2d), neurons typically respond to lag stimuli for delays as short as 1–2 ms, whereas in the ICC (solid line, Fig. 6.2d) and auditory cortex, recovery takes 10–100 ms, although there is considerable variability across neurons. The recovery time in ICC depends on anesthetic state, and is about half as long in awake animals compared to anesthetized animals (Tollin and Yin 2004).

Neural echo suppression observed with pairs of transient stimuli is often interpreted as a correlate of localization dominance in the PE (Litovsky et al. 1999). Consistent with this interpretation, echo suppression in the ICC is directional and depends on binaural cues (Litovsky and Yin 1998). For example, in some neurons, a leading stimulus at the neuron's best ITD evokes stronger suppression than a stimulus at the worst ITD; in other neurons, the converse is true. On the other hand, neural echo suppression is observed with similar properties in binaural and monaural neurons, so that some forms of suppression may represent a more general correlate of context dependent perceptual phenomena (such as forward masking) rather than being specifically linked to the PE. The long duration of echo suppression in most ICC neurons (10–50 ms) compared to psychophysical echo thresholds for transient stimuli (5–10 ms) poses a challenge to the view that the two phenomena are linked.

Regardless of the exact function of neural suppression, it is likely to play an important role in signal processing in the IC.

Echo suppression in the IC has been primarily attributed to synaptic inhibition from a binaural source such as the DNLL (Yin 1994), consistent with earlier models for the PE (Lindemann 1986; Zurek 1987) that postulated a central echo suppression mechanism. A role for central mechanisms is supported by the observation that, although some suppression is observed in the AN, it is too brief to account for the suppression observed in the ICC (Fitzpatrick et al. 1999). Moreover, suppression in the ICC can be observed even when the leading sound evokes no spikes, thus ruling out intrinsic mechanisms such as refractoriness or recurrent inhibition (Yin 1994). Nevertheless peripheral mechanisms such as amplitude compression and cochlear filtering are likely to play a role in echo suppression by making the basilar membrane responses to the lead and the lag interact (Hartung and Trahiotis 2001; Paterson and McAlpine 2005). While peripheral mechanisms alone can account for some psychophysical PE results with transient stimuli and short interstimulus delays, a mechanism operating over longer time scales seems necessary to account for the PE observed with longer stimuli and longer delays (Braasch and Blauert 2003).

Two computational modeling studies have investigated the mechanisms underlying neural echo suppression in the IC. The model of Cai et al. (1998a) focuses on the binaural response properties of low-frequency ICC neurons that are sensitive to ITDs. The model incorporates a model of auditory nerve fibers (Carney 1993), models of VCN spherical and globular bushy cells (Rothman et al. 1993), and a model of MSO cells (Brughera et al. 1996). The ICC model neuron is excited by an ipsilateral MSO model neuron and inhibited by a contralateral MSO neuron via an inhibitory interneuron, presumed to be in the DNLL, which is not explicitly modeled (Fig. 6.6a). The rate–ITD curve of the model ICC cell for a single binaural click stimulus peaks at the characteristic delay of the ipsilateral MSO input (0 μs, Fig. 6.6b). This response is controlled entirely by the ipsilateral MSO because the contralateral inhibition arrives too late to have an effect with transient stimuli. The effects of the inhibitory input on the ICC cell are relatively long-lasting compared to those of the excitatory input, thus the inhibitory inputs activated by the leading sound suppress the response to the lagging sound over a range of delays. This is illustrated in Fig. 6.6c, d, which plot responses to the lag click as a function of the interclick delay. The lag click is held at the best ITD, while the lead click is either at the best (+) or worst (–) ITD. If the excitatory and inhibitory MSO cells have the same best ITD, then the response of the model ICC neuron to the lagging sound is most strongly suppressed for a leading stimulus at the best ITD of the ICC cell (Fig. 6.6c). If, on the other hand, the best ITD of the inhibitory MSO cell is set at the worst ITD of the excitatory MSO neuron, then a leading stimulus at the worst ITD of the ICC cell creates the maximum suppression (Fig. 6.6d).

The model of Pecka et al. (2007) simulates a different (and much smaller) population of ICC neurons for which the response to a lagging stimulus is *enhanced* when the lead stimulus arises from a favorable location (Burger and Pollak 2001). Experiments in which DNLL activity is blocked pharmacologically suggest that these ICC neurons do not inherit their ILD sensitivity from the LSO, but rather create

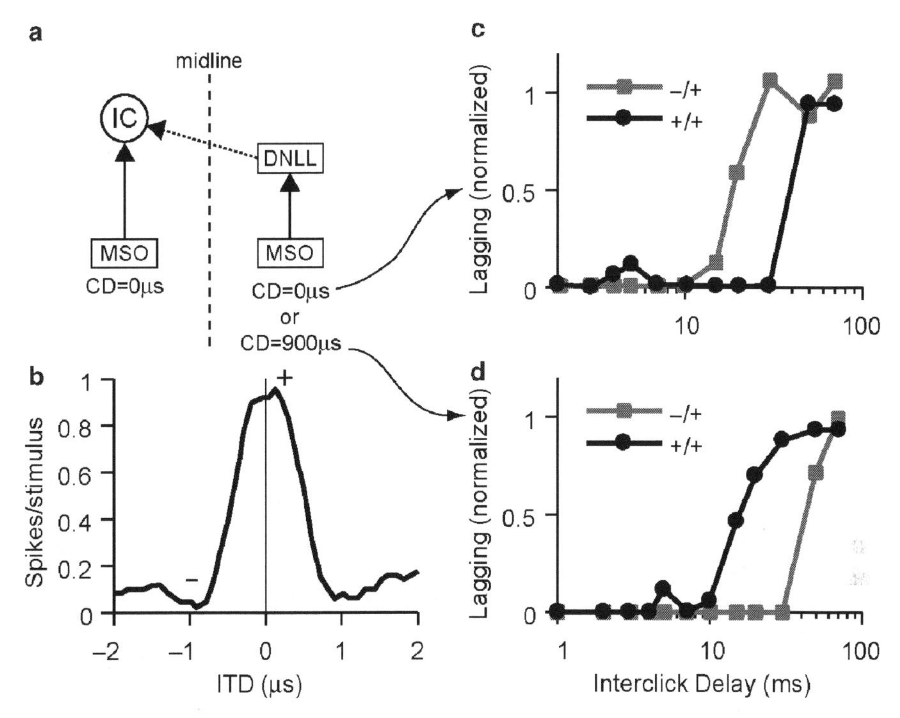

Fig. 6.6 Model of responses to precedence effect stimuli (Cai et al. 1998a). (**a**) Neural circuit that suppresses responses of ITD-sensitive ICC neuron to the lagging click in a pair of binaural clicks presented in succession. Circuit below MSO is omitted for clarity. (**b**) Rate–ITD curve for model ICC cell in response to a single binaural click. (**c, d**) Recovery curves show the response to the lagging click as a function of the interclick interval. The two curves in each *panel* correspond to leading clicks at favorable (+) and unfavorable (–) ITDs. The lagging click always has a favorable ITD. Suppression of the response to the lag is greatest when the leading click is near the characteristic delay (CD) of the "inhibitory" MSO neuron. (**c**) Inhibitory CD=0 μs. (**d**) Inhibitory CD=–900 μs

it de novo by combining a monaural excitatory input (assumed to be from VCN in the model) with an inhibitory input from contralateral DNLL (Li and Kelly 1992; Burger and Pollak 2001). Thus, the mechanism for ILD sensitivity of ICC neurons in the Pecka et al. (2007) model fundamentally differs from that in the Reed and Blum (1999) model. The model DNLL cells receive an excitatory input from the contralateral LSO and an inhibitory input from the contralateral DNLL (Fig. 6.7a). A critical property of the model is that the inhibition from one DNLL to the other is *persistent*: the DNLL that prefers the leading sound suppresses the response of the opposite DNLL for tens of milliseconds after the lead sound ends (Yang and Pollak 1998). This persistent inhibition is implemented as a slowly decaying hyperpolarization with a time constant of 12 ms, as determined from in vitro measurements (Pecka

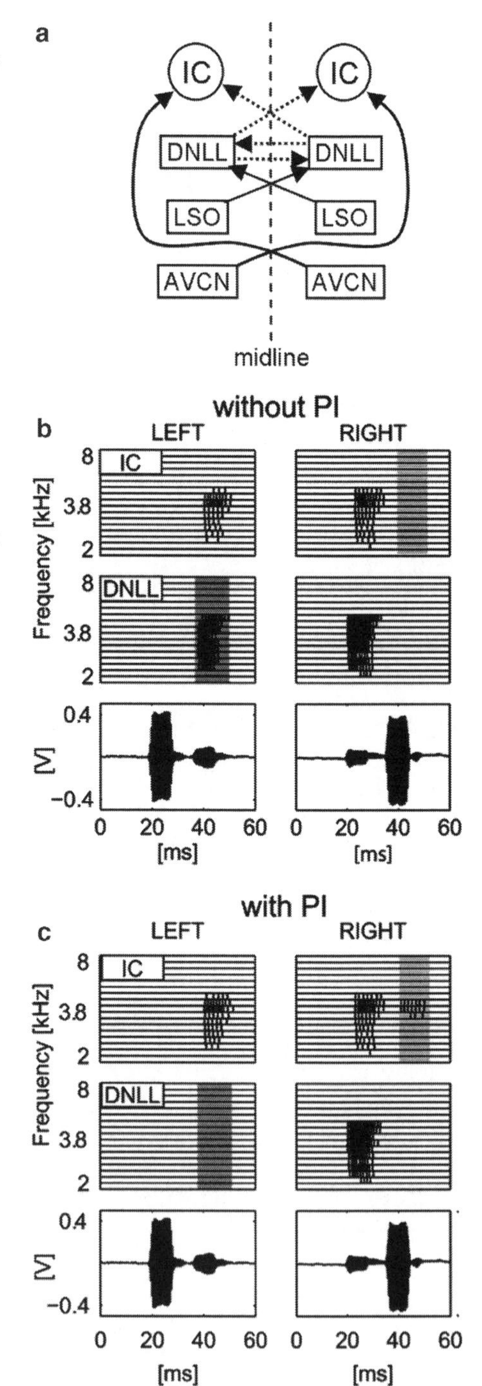

Fig. 6.7 (**a**) Neural circuit that enhances responses of an ILD-sensitive ICC neuron to a lagging stimulus in a pair of tones. (Modified from Pecka et al. 2007.) ILD sensitivity is created de novo in the ICC by interaction between a monaural excitatory input from AVCN and an inhibitory input from contralateral DNLL. (**b**, **c**) Responses to a pair of 10-ms binaural tone bursts as a function of tone frequency. The *bottom panel* shows the stimulus waveform in each ear. (**b**) Without persistent inhibition between the two DNLLs, each DNLL and IC cell only responds to the binaural tone which is more intense in the contralateral ear. (**c**) With persistent inhibition, the right DNLL shuts down the left DNLL, which can no longer inhibit the right IC in response to the lagging tone pip. Thus, the right IC responds to a lagging stimulus at an unfavorable ILD

et al. 2007). Figure 6.7b, c show responses of the DNLL and ICC model cells that demonstrate the effect of persistent inhibition. The stimulus is a pair of 10-ms binaural tone bursts such that the first ILD favors the left ear, and the second favors the right ear. Without persistent inhibition (Fig. 6.7b), the right DNLL and ICC cells are excited by the leading sound, while the left DNLL and ICC cells are excited by the lag. With inhibition, however, the left DNLL is persistently inhibited by the lead sound in the left ear, so that it cannot inhibit the response of the right ICC to the lagging sound (Fig. 6.7c). Therefore, ICC cell responses to the lagging sound depend on the ILD of the leading sound for a range of delays between the two sounds. In particular, persistent inhibition evoked by the lead stimulus causes an ICC cell to respond to a lagging stimulus to which it would not respond if presented in isolation. This altered neural representation of the lagging sound may partially underlie the poor localization of trailing sounds by human listeners, which is a characteristic of the PE (Zurek 1987).

Persistent inhibition arising from the DNLL is a common element of the models of Cai et al. (1998a) and Pecka et al. (2007), but the models diverge in other respects. In the Cai et al. model inhibition occurs in the ICC, where it directly inhibits the response to a lagging sound. In the Pecka et al. model, the persistent inhibition occurs between the two DNLLs and *disinhibits* the IC response to the lagging sound. The Cai et al. model focuses on the ITD-sensitive pathway while Pecka et al. are concerned with ILD-sensitive responses. Such differences highlight the fact that the PE is not a unitary phenomenon and is not likely to be mediated by a single neural mechanism. Moreover, the relationship between neural echo suppression observed with a simple lead-lag pair and the practical challenge of hearing in reverberant environments comprising a large number of reflections is not clear (Devore et al. 2009). Exploration of these models using realistic reverberant stimuli is an important next step. Another important task is to clearly distinguish the contributions of peripheral and central mechanisms to neural echo suppression phenomena in the ICC (Hartung and Trahiotis 2001). Finally, it will be important to develop neural population models to explore the psychophysical consequences of echo suppression, not only for the PE, but also for other phenomena such as forward masking and sensitivity to motion cues (Xia et al. 2009).

6.4 Localization of a Moving Sound Source

In everyday listening environments, the acoustic cues for sound localization often vary over time as a result of the motion of a sound source relative to the listener's head. Several psychophysical (Grantham 1998; Dong et al. 2000) and neuroimaging (Griffiths et al. 1998) studies suggest there may exist specialized detectors and brain regions that are selectively activated by sound motion cues such as time-varying interaural phase differences (IPDs). However, other studies have concluded that the auditory system responds sluggishly to changing localization cues and that acoustic motion generally impairs the accuracy of sound localization (e.g., Grantham and

Wightman 1978). The two points of view are not mutually exclusive as there could be specialized detectors that detect motion regardless of the instantaneous location of the sound source.

Studies of neuronal sensitivity to temporal variations in IPD suggest that a major transformation in the coding of this motion cue occurs between the SOC and the IC. Neurons in the MSO, where IPD is initially encoded, are generally insensitive to motion cues (Spitzer and Semple 1998). That is, the responses of MSO neurons to time-varying IPD stimuli resemble their responses to static IPD stimuli for each time instant, that is, these neurons track the instantaneous IPD (dashed line, Fig. 6.2e). In contrast, many low-frequency neurons in the ICC are sensitive to dynamic changes in IPD (Spitzer and Semple 1993, 1998; McAlpine and Palmer 2002). As shown in Fig. 6.2e (solid lines), the responses of ICC neurons to dynamic stimuli swept across complementary narrow ranges of IPDs are not continuous with each other, although the entire set of partial response profiles tends to follow the general shape of the static IPD function measured over a wide IPD range. The responses to dynamic IPD arcs are always steeper than the static IPD function, with motion toward the peak of the static function resulting in overshoot and motion away from the peak resulting in undershoot. Over a restricted range of IPDs, responses to opposite directions of motion thus form hysteresis loops (gray shaded areas). Importantly, dynamic IPD stimuli can evoke a strong excitatory response for instantaneous IPDs for which there is no response to static IPDs (asterisk), a phenomenon that has been termed "rise-from-nowhere."

The difference between the responses of ICC and MSO neurons to dynamic IPD stimuli suggests that sensitivity to sound-motion cues emerges at the level of the ICC. One explanation for sensitivity to motion cues is a nonlinear interaction between IPD-tuned excitatory and inhibitory inputs to ICC (Spitzer and Semple 1998). Alternatively, motion sensitivity could reflect "adaptation of excitation" (Spitzer and Semple 1998; McAlpine et al. 2000), in which the firing rates of ICC neurons depend on both the instantaneous value of IPD and their recent firing history (not the history of the IPD cues per se). In support of the adaptation hypothesis over the inhibition hypothesis, McAlpine and Palmer (2002) found that sensitivity to apparent motion cues in ICC is actually decreased by the inhibitory transmitter γ-aminobutyric acid (GABA) and increased by the GABA-antagonist bicuculline.

The roles of adaptation and inhibition in shaping ICC unit sensitivity to motion cues have been examined in two computational modeling studies (Cai et al. 1998a, b; Borisyuk et al. 2001, 2002). Both support the proposal that the primary mechanism responsible for sensitivity to motion cues is adaptation-of-excitation. Cai et al. (1998a) first tested their model ICC cell discussed in Sect. 6.3, which receives IPD-tuned excitatory and inhibitory inputs from the ipsilateral and contralateral MSO, respectively. They found that this model version does not produce differential responses to stimuli with dynamic and static IPDs. Subsequently, they added an adaptation mechanism that causes the model to exhibit sharpened dynamic IPD tuning, and sensitivity to the direction of IPD change (hysteresis). The adaptation mechanism in their model is a calcium-activated potassium channel. After each spike of the ICC cell, the conductance of this channel increases by a small amount, and then decreases exponentially with a time constant of 500 ms. The effects of this channel are minimal for transient stimuli that evoke just a few spikes, but increase with stimulus duration through superposition of the effects of each spike. As a result, dynamic-IPD stimuli evoke higher firing rates than static-IPD stimuli, and responses to dynamic stimuli depend on the recent response-history of the cell, giving rise to sensitivity to the direction of motion. Consistent with pharmacological data obtained later (McAlpine and Palmer 2002), Cai et al. (1998b) showed that model sensitivity to dynamic IPD stimuli decreases with increasing inhibition. Inhibition reduces the discharge rate of the ICC model cell, thus reducing the amount of adaptation of excitation experienced by the cell. However, this model does not predict the rise-from-nowhere phenomenon. The Borisyuk et al. (2002) model has many similarities with the Cai et al. (1998b) model, but further introduces a postinhibitory rebound (PIR) mechanism that allows the model to predict the rise-from-nowhere phenomenon. Both models have membrane conductances giving rise to firing rate adaptation and also include both an excitatory input from MSO and an inhibitory input from DNLL. In the Borisyuk et al. (2002) model, adaptation is produced by a slowly-activating voltage-gated potassium current with an inactivation time constant of 150 ms. As in the Cai et al. (1998b) model, this adaptation mechanism simulates many features of the responses to dynamic IPD stimuli including sharper tuning and hysteresis (Fig. 6.8a). In addition, a PIR mechanism is implemented as a transient inward current with fast activation and slow inactivation. When the membrane is abruptly released from a prolonged state of hyperpolarization, the PIR current activates instantaneously and depolarizes the membrane for tens of milliseconds. This PIR current is activated when a dynamic IPD stimulus coming from an unfavorable IPD where inhibition dominates moves toward a more favorable IPD, thereby producing the rise-from-nowhere phenomenon (Fig. 6.8b, arrow).

The models of Cai et al. (1998b) and Borisyuk et al. (2002) suggest that intrinsic membrane properties underlie the sensitivity of IC neurons to dynamic IPDs. In both models, hyperpolarizing conductances with time constants of a few hundred milliseconds produce long-lasting spike rate adaptation. The channel used by Cai et al. is consistent with the calcium-gated, voltage-insensitive potassium current associated with rate adaptation in IC neurons in vitro (Sivaramakrishnan and Oliver 2001). The PIR mechanism adopted by Borisyuk et al. is consistent with observation that over half of ICC neurons exhibit calcium-dependent post-inhibitory rebound in vitro (Sivaramakrishnan and Oliver 2001).[1]

It is unclear whether the dynamic IPD sensitivity of ICC neurons represents a specific encoding of motion cues or a more general mechanism for processing dynamic aspects of sound. An important question is the extent to which the membrane properties influence the neural representation of other dynamic stimulus features. A similar history dependence appears to hold for changes in ILD (Sanes et al.

[1]Borisyuk et al. (2002) created hyperpolarization using a voltage-gated potassium channel because their model does not explicitly generate action potentials, which are needed to activate the calcium-gated channels used by Cai et al. (1998b). This difference does not appear significant; the two models similarly replicate the hysteresis effects observed in the physiological data.

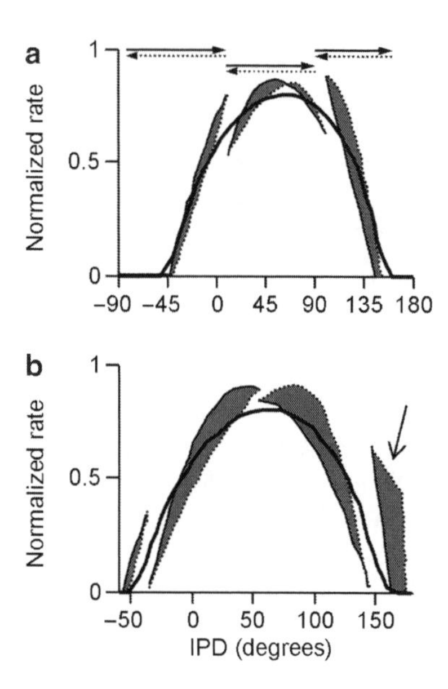

Fig. 6.8 Model of responses to stimuli with time-varying IPD (Borisyuk et al. 2002). (**a**) Rate–IPD curves for model ICC cell with a membrane channel producing strong spike rate adaptation. Response to static stimuli (*thick solid line*) is superimposed on responses to three different dynamic stimuli swept over different ranges of IPD. *Arrows* at the *top* indicate the sweep ranges. *Dashed lines* correspond to sweeps toward negative IPDs, *thin solid lines* sweeps toward positive IPDs. Rate adaptation is responsible for the hysteresis in the responses to dynamic IPD stimuli. (**b**) Rate–IPD curves for model IC cell with an additional channel producing postinhibitory rebound (PIR). With PIR channel, model produces "rise-from-nowhere" (*arrow*)

1998). However, most of the models described in other sections of this chapter consider only either static or brief stimuli that would minimally activate the conductances used by Cai et al. and Borisyuk et al. An exception is amplitude modulated (AM) stimuli; yet surprisingly, none of the proposed models for AM tuning in ICC (Sect. 6.6) considers intrinsic membrane channels as a possible mechanism. The slow adaptive effects described in this section may also contribute to PE phenomena evoked with long-duration stimuli, but this has yet to be explored.

6.5 Population Models of ITD Processing

The coincidence detection mechanism that underlies ITD encoding in the MSO can be viewed over long time scales as a cross-correlation of the inputs from the two ears after taking into account cochlear filtering (Yin and Chan 1990). Rate responses of ITD-sensitive ICC neurons to simple low-frequency binaural stimuli are quantitatively consistent with a cross-correlation operation (Yin et al. 1987; Devore et al. 2009). More complex stimuli encountered in everyday acoustic environments, however, reveal discrepancies between IC behavior and this model. Responses in cat ICC show greater sensitivity to ITD in the presence of reverberation than predicted by long-term cross-correlation (Devore et al. 2009). Similarly, ITD sensitivity in the external nucleus of the inferior colliculus (ICX) of the barn owl is more robust to noise than the model predicts (Keller and Takahashi 1996). Such observations suggest that specialized processing may occur in the IC to maintain the potency of spatial cues in the presence of degrading influences.

6.5.1 Robust Coding of ITD in Noise in Barn Owl IC

Rucci and Wray (1999) used a model to investigate the role of IC intrinsic circuitry in creating robust ITD coding in noise in the barn owl ICX. The ICX in this species contains a map of auditory space where neurons are broadly tuned to frequency, but sharply tuned to both ITD and ILD (which is primarily an elevation cue in the barn owl due to the asymmetry of the external ears). The ICX receives inputs from ICC where neurons are sharply tuned to both frequency and ITD. The model comprises two layers of ICC neurons, one excitatory and one inhibitory (Fig. 6.9b). Each layer is a grid with best ITD systematically varied in one dimension and best frequency (BF) varied in the other. The input to the ICC comes from the ipsilateral nucleus laminaris (NL, the avian homolog of MSO), where cells are arranged in a corresponding grid making point-to-point (i.e., topographic) connections to the excitatory layer in ICC. Each NL neuron is modeled by bandpass filtering the acoustic inputs around the BF to approximate cochlear processing, delaying the filter output on one side by the best delay, and multiplying and integrating (i.e., cross-correlating) the filter outputs from the two sides (Fig. 6.9a). A memoryless nonlinear function converts the resulting correlation value into a firing rate. The excitatory layer in ICC is topographically connected to the inhibitory layer. Specifically, each cell in the inhibitory layer receives input from one excitatory-layer neuron with the same BF and best ITD, and projects back to all excitatory cells having the same BF but different best ITDs (i.e., the inhibition is lateral but not recurrent). The ICX model is a one-dimensional array of neurons organized by best ITD. Each ICX cell receives convergent input from all excitatory ICC cells that have the same best ITD, regardless of BF. The ICX also contains its own lateral inhibitory connections.

The results show that sharp ITD tuning is maintained in ICX in the presence of noise only when the lateral connections are included. The stimulus was a broadband target sound contaminated by a band of uncorrelated noise. Figure 6.9c plots model rate–ITD curves with and without the lateral inhibitory connections, and shows that the lateral inhibition across BD sharpens tuning in ICC and, especially, ICX without altering the best ITD. From the perspective of a single model neuron, the sharpening due to lateral inhibition is relatively modest, but when pooled across the entire population, the improvement of ITD coding in noise is substantial. Rucci and Wray further demonstrate that lateral inhibition and convergence across BF effectively implement a generalized cross-correlation, in which the cross-correlation function in

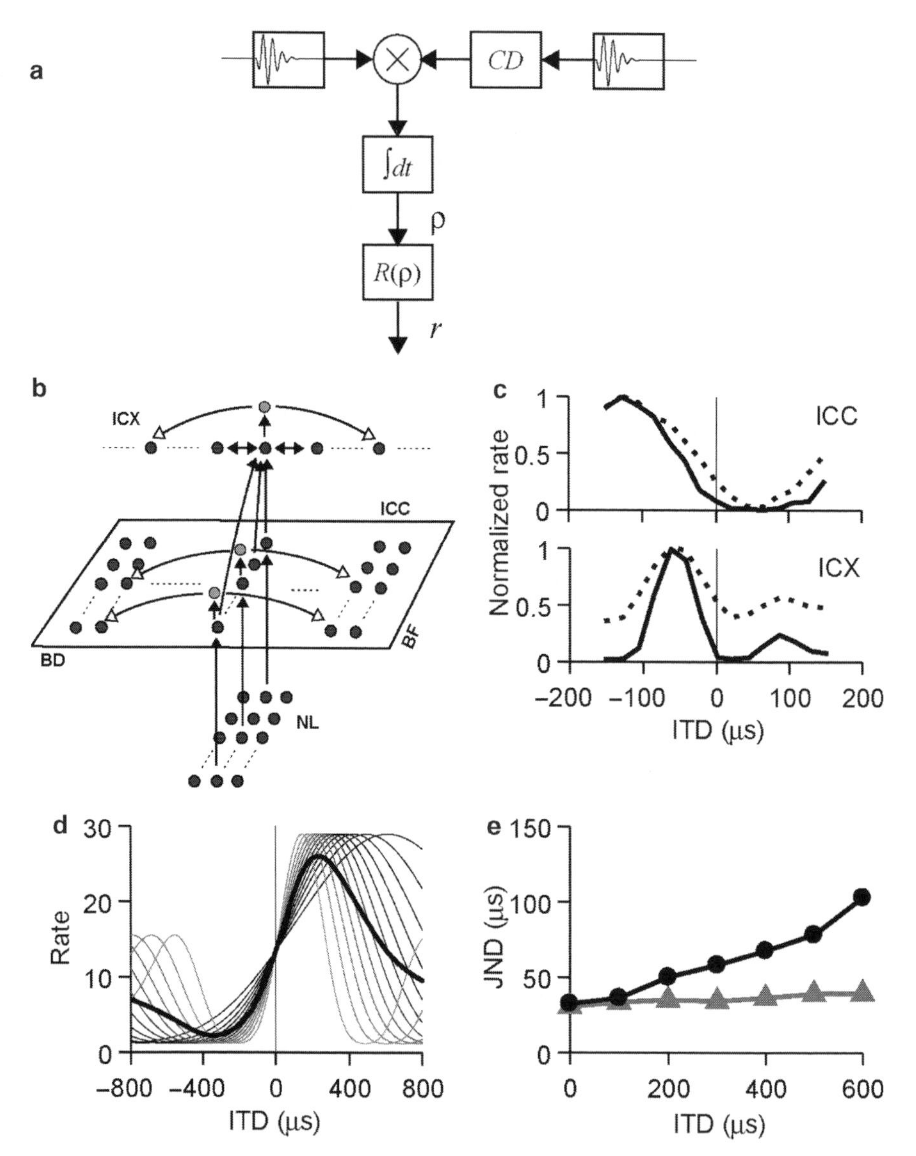

Fig. 6.9 Population models of ITD processing. (**a**) Cross-correlation model for rate responses of single ICC neurons. (From Hancock and Delgutte 2004.) Acoustic inputs are bandpass filtered at the BF, delayed by the characteristic delay (CD) on one side, then multiplied and integrated over time (cross-correlation). A nonlinear function $R(\rho)$ converts correlation coefficient into average firing rate. (**b**) Barn owl network that robustly codes ITD in noise. (From Rucci and Wray 1999.) Nucleus laminaris (NL) and ICC cells are arranged in two-dimensional grids with best ITD (BD) on one axis and best frequency (BF) on the other. The ICC has both an excitatory layer (*solid circles*) and an inhibitory layer (*hatched circles*) which projects back to the excitatory layer. The excitatory layer of ICC forms a convergent projection across BF to the external nucleus (ICX). Additional details of connections are described in the text. (**c**) Lateral inhibition sharpens model rate–ITD curves in ICC and ICX. (From Rucci and Wray 1999.) *Dashed lines*: no lateral inhibition. *Solid lines*: with lateral

each frequency band is weighted by the statistical reliability of the IPD estimate. Thus, careful analysis of the underlying anatomy and physiology can enhance a signal-processing approach to modeling sound localization.

An important consideration is the extent to which this mechanism applies to mammalian spatial hearing, which differs from that of the barn owl in significant ways. Barn owls use ITD to determine the azimuth of a sound source, and ILD to determine elevation. The two cues are combined multiplicatively to create a topographic space map in ICX (Peña and Konishi 2001); no comparable map has been found in mammals. The Rucci and Wray (1999) model deals only with ITD processing and the existence of a space map is not an essential assumption. To this extent, the results may have general applicability to robust ITD coding in mammals, although the inhibitory connections need not be located within the ICC. Indeed across-ITD inhibitory connections similar to those of the Rucci and Wray model are a key feature of the Lindemann (1986) model for binaural processing in humans.

6.5.2 Predicting Psychophysical ITD Acuity

Models for populations of neurons such as the Rucci and Wray (1999) model can be used not only to understand neural processing but also to predict psychophysical performance from patterns of neural activity across the population. This approach is particularly appropriate for the IC, where sufficient information for all auditory behavioral tasks must be available because the IC is the site of an (almost) obligatory synapse in the ascending auditory pathway.

An example of this approach is the Hancock and Delgutte (2004) model, which aimed at explaining trends in ITD discrimination performance using a population model of the ICC structurally similar to that of Rucci and Wray (1999). For human listeners, just noticeable differences (JNDs) in ITD for broadband noise stimuli are smallest when measured about a reference ITD of 0 (i.e., on the midline), and systematically increase as the reference ITD increases (Mossop and Culling 1998). ICC neurons are modeled using the cross-correlation computation of Fig. 6.9a, and are arranged in a grid in which BF varies in one dimension, and the "best IPD" (the product of the BF and the best ITD) in the other dimension. Best IPD is used instead of best ITD for consistency with the experimental observation in mammals

Fig. 6.9 (continued) inhibition. *Top*: ICC. *Bottom*: ICX. (**d**) Population model for ITD discrimination based on ICC responses (Hancock and Delgutte 2004). Rate–ITD curves of model ICC cells with a common best IPD but different BFs (*gray lines*) align only near the midline. Thus, the across-BF average rate–ITD curve (*black line*) is steepest at this point. (**e**) Model predicts increasing just-noticeable-difference (JND) in ITD with increasing reference ITD when rates are summed across BF before performing signal detection (*black line*), but predicts constant acuity without across-BF integration (*gray line*)

that best IPD and BF are more nearly orthogonal than best ITD and BF (McAlpine et al. 2001; Hancock and Delgutte 2004). Unlike the models considered so far, the Hancock and Delgutte model explicitly models the internal noise resulting from variability in firing rates in order to compute the expected performance for an ideal observer of the population neural activity. ITD JNDs are computed using standard signal detection techniques (Green and Swets 1988) from patterns of model firing rates evoked by pairs of stimuli with different ITDs. Figure 6.9e compares model-generated ITD JNDs to psychophysical data (Mossop and Culling 1998) as a function of reference ITD. The model predicts the degradation in ITD acuity with increasing reference ITD when information is pooled across BF (black line), but predicts constant ITD acuity when there is no across-BF convergence (gray line). This property occurs because the slopes of rate–ITD curves tend to align across BF near the midline, but not for more lateral locations (Fig. 6.9d) (McAlpine et al. 2001). Consequently, convergence across BF reinforces sensitivity to changes in ITD near the midline, but blurs it more laterally. An extension of this model (Hancock 2007) predicts the lateralization of stimuli with independently applied interaural time and phase shifts (Stern et al. 1988).

Convergence across BF is an important element of both the Rucci and Wray (1999) and Hancock and Delgutte (2004) models and is also a feature of some functional models for predicting human binaural performance (Stern et al. 1988; Shackleton et al. 1992). Together, the results from the IC models suggest that pooling of information across BF helps create a robust neural ITD code on the midline at the expense of ITD acuity on the periphery. The Hancock and Delgutte (2004) model considered only single sound sources in anechoic space and hence did not need the lateral inhibitory connections of Rucci and Wray (1999). Application of the model in the presence of competing sources or reverberation may require such connections.

6.6 Sensitivity to Amplitude Modulation

Many natural sounds contain prominent fluctuations in their amplitude envelope (Singh and Theunissen 2003). Amplitude modulation (AM) information is used by the auditory system in a variety of tasks, including speech perception (Steeneken and Houtgast 1980; Rosen 1992), pitch perception (Plack and Oxenham 2005) and auditory scene analysis (Darwin and Carlyon 1995). Envelope information appears to be primarily encoded in the temporal response patterns of neurons in the early stages of the auditory system, and then partly transformed into a rate-based code at higher stages of processing (Joris et al. 2004).

Neural responses to AM stimuli, most often sinusoidally amplitude-modulated (SAM) tones, are typically quantified based on average firing rate and synchrony to the modulation period. These metrics are plotted against modulation frequency to obtain rate and temporal modulation transfer functions (MTFs), respectively. A temporal MTF has both a magnitude (the modulation gain) and a phase representing the delay between the modulation in the acoustic stimulus and the modulation of the neural response. Unfortunately, the MTF phase is rarely reported in neurophysiological studies, so the focus of modeling has been primarily on the magnitude, which we will abbreviate sync-MTF. Rate-MTFs are not true transfer functions in the sense of linear systems analysis, but the acronym is nevertheless widely used by analogy with temporal MTFs.

At the level of the AN and CN, rate-MTFs are usually flat (dashed line, Fig. 6.2f), whereas sync-MTFs can have lowpass or bandpass shapes (Joris et al. 2004). Encoding of AM in the CN is therefore assumed to be primarily based on a temporal code, although there is also a place code at high modulation frequencies when each component of an AM tone is peripherally resolved. In contrast, rate-MTFs of ICC neurons are typically bandpass (Fig. 6.2f) or lowpass in shape, meaning that many ICC neurons are selective to a range of modulation frequencies and respond with a maximum spike rate for a specific best modulation frequency (BMF) (Joris et al. 2004). Furthermore, in cat, there is some evidence for a coarse map of BMFs running orthogonal to the tonotopic map in the ICC (Schreiner and Langner 1988).

Four models have been proposed for the transformation from a temporal representation of AM in the CN to a rate-base representation in the ICC. Two of these models rely on coincidence detection (Langner 1981; Hewitt and Meddis 1994), while the other two involve inhibition (Nelson and Carney 2004; Dicke et al. 2007).

6.6.1 Models Based on Coincidence Detection

The basic structure of the model suggested by Langner (1981) (see also Langner 1985; Langner and Schreiner 1988; Voutsas et al. 2005) consists of a trigger neuron, an oscillator circuit and an integrator circuit, each associated with different unit types in the CN, and a coincidence detector in the ICC (Fig. 6.10a). The trigger neuron fires once on each cycle of the modulation of an AM stimulus. In turn, the trigger neuron triggers the oscillator circuit to fire a small number of spikes at an intrinsic (chopping) interval independent of the envelope period of the signal. Simultaneously, the trigger neuron activates the integrator circuit, which then fires a single spike delayed in time by the integration period of the circuit. Importantly, the integrator circuit also receives AN inputs that are phase-locked to the carrier frequency of the AM signal and requires a specific number of carrier-locked input spikes before firing; thus, the integration period is always an integer multiple of the carrier period. The coincidence detector in the ICC receives excitatory inputs from both the oscillator circuit and the integrator circuit, and fires when spikes from both sources coincide in time. This occurs when the delay of the integrator equals the modulation period of the AM signal (e.g., when a spike produced by the integrator in response to one modulation cycle coincides in time with an oscillator spike triggered by the next modulation cycle). The resulting rate-MTF of the ICC cell is bandpass in shape with a BMF equal to

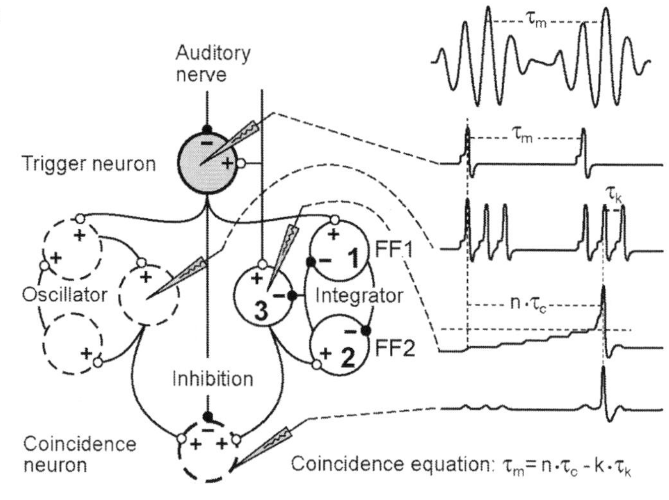

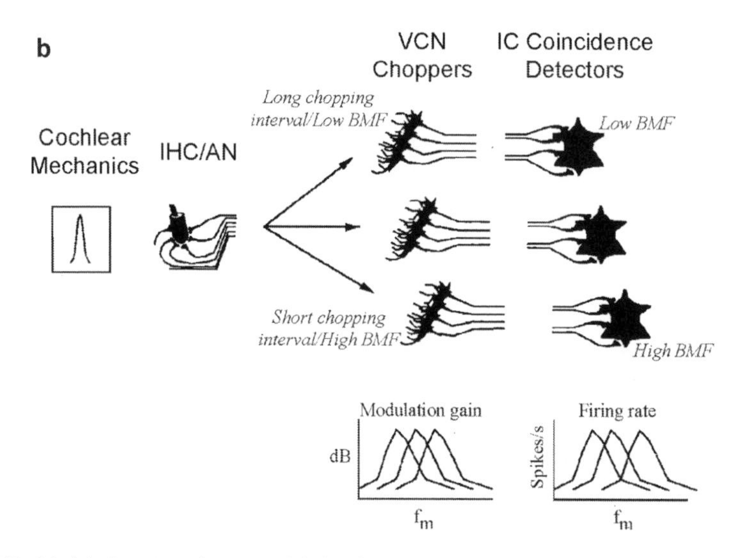

Fig. 6.10 Models for rate tuning to modulation frequency based on coincidence detection. (**a**) Block diagram of the Langner (1981) model. (From Voutsas et al. 2005.) The trigger neuron located in the CN receives inputs from the auditory nerve (AN) and produces one spike per modulation cycle τ_m of an AM tone. This neuron triggers both an oscillator circuit and an integrator circuit. The oscillator fires a small number k of spikes at regular intervals τ_k in response to the trigger. The integrator circuit receives inputs phase locked to the carrier frequency of the AM stimulus from the AN and fires one spike a predetermined number n of carrier cycles τ_c after each trigger. The coincidence detector located in ICC fires when it receives simultaneous spikes from the integrator and oscillator circuits, thereby converting a temporal code for periodicity into a rate code. The coincidence detector neuron has a bandpass rate MTF, with a BMF approximately equal to $1/n\tau_c$. The circuit as a whole implements the "coincidence" or "periodicity" equation (*bottom right*) which accounts for the pitch shift of harmonic complex tones. (**b**) Block diagram of the Hewitt and Meddis (1994) model for rate tuning to AM in ICC. Auditory nerve fibers innervating a given cochlear place project to an array of chopper neurons in VCN that show bandpass temporal MTFs and differ in their best modulation frequency (BMF). A set of chopper neurons having the same BMF projects to one coincidence detector neuron in ICC, endowing this neuron with a bandpass rate MTF. Coincidence detection converts a temporal code for AM in the VCN into a rate code in the IC

the inverse of the integration period of the integrator.[2] A key prediction of this model is that the BMF period must be linearly related to the stimulus carrier period (the so-called "periodicity" or "coincidence" equation; see Fig. 6.10a). While such a linear relationship holds approximately for some IC neurons in both cat (Langner and Schreiner 1988) and guinea fowl (Langner 1983), it does not appear to hold in general (Krishna and Semple 2000). An additional drawback is that the model only works at low carrier frequencies (<5 kHz), where the input to the integrator circuit input retains synchrony to the carrier. Thus, another mechanism is required to explain results at higher carrier frequencies despite the basic similarity of rate-MTF shapes throughout the tonotopic axis of the ICC.

The second coincidence detector model for the formation of bandpass AM tuning in the ICC (Hewitt and Meddis 1994) is based on an array of VCN chopper cells that fire spikes at fairly regular intervals in response to AM tones (Fig. 6.10b). The chopping rate of these model choppers (Hewitt et al. 1992) is typically unrelated to either the carrier or the modulation frequency of the AM stimulus, that is, their rate-MTFs are flat. Importantly, with unmodulated stimuli, two chopper units with the same chopping rate gradually lose temporal synchrony with one another over time because of random fluctuations in their interspike intervals. On the other hand, when the intrinsic chopping rate of a group of VCN chopper cells matches the modulation frequency of an AM stimulus, then the chopper cells remains tightly synchronized to the AM stimulus and to each other throughout the stimulus duration. Thus, chopper units exhibit bandpass sync-MTFs with BMFs equal to their intrinsic chopping rate. The BMF is controlled by adjusting the strength of a K^+ membrane conductance in the chopper model neuron (Hewitt et al. 1992). A group of VCN chopper units with the same intrinsic chopping rate converge onto an ICC cell acting as a coincidence detector, that is, the ICC cell fires only when it receives a sufficient number of synchronous VCN inputs. Such coincidences are most likely to occur when the modulation frequency of the stimulus matches the chopping rate of the VCN inputs. The Hewitt and Meddis (1994) model therefore transforms the bandpass sync-MTFs of VCN chopper units into bandpass rate-MTFs in the ICC. This transformation from a temporal code to a rate code is achieved under the strong constraint that all the VCN inputs to a given ICC cell have identical chopping rates. Guerin et al. (2006) found that if this constraint is relaxed, then the model ICC cell still exhibits a bandpass shaped rate-MTF but the sync-MTF becomes more noisy and less bandpass in shape than those of real ICC units. The strength of the synchronous response can be enhanced by adding a second layer of coincidence detectors in the ICC.

The Langner (1981) and Hewitt and Meddis (1994) models have several circuit elements in common. Both use coincidence detectors located in ICC to transform a temporal code for modulation frequency into a rate code, and both include oscillator

[2] One problem with this model structure is that the ICC cell responds equally well to harmonics of the BMF as to the BMF itself. Voutsas et al. (2005) avoid this problem by having the trigger neuron inhibit the ICC cell, via an interneuron putatively located in the VNLL, for a predefined period of time equal to at least one-half of the integration period.

neurons (choppers) putatively located in the CN. However, whereas bandpass AM tuning in the Hewitt and Meddis model is inherited from the tuning of the VCN chopper neurons, it is controlled by the integration time of the integrator circuit in the Langner model. One outstanding issue for both models is the disparity in the ranges of VCN chopper cell synchrony BMFs (150–700 Hz; (Frisina et al. 1990; Rhode and Greenberg 1994) and ICC unit rate BMFs (1–150 Hz; Krishna and Semple 2000).[3] Moreover, intracellular recordings from ICC neurons with bandpass rate-MTFs fail to show strong stimulus-locked modulations of the membrane potential when the modulation frequency of an AM tone stimulus matches the BMF, posing a challenge to the idea that these cells act as coincidence detectors (Geis and Borst 2009).

6.6.2 Models Based on Inhibition

Nelson and Carney (2004) proposed a model for bandpass rate tuning to AM that differs from the models discussed in the preceding text in that it relies on interaction between excitation and inhibition rather than coincidence detection to produce tuning. In this model, ICC cells receive same-BF inhibitory and excitatory (SFIE) input from a VCN bushy cell which, in turn, receives SFIE input from auditory nerve fibers (Fig. 6.11a). Both excitation and inhibition are phase locked to each modulation cycle, but with different delays. At the VCN level, the excitatory inputs are stronger and faster than the inhibitory inputs. As a result, the bushy cell's response to SAM tones largely resembles that of its AN inputs: that is, the bushy cell has a flat rate-MTF and a lowpass sync-MTF (Fig. 6.11b). In contrast, at the level of the ICC, the inhibitory inputs are stronger (but still slower) than the excitatory

[3]Studies in both anesthetized cat (Langner and Schreiner 1988) and awake chinchilla (Langner et al. 2002) have reported IC units with rate-BMFs as high at 1,000 Hz, but most of these were multi-units (as opposed to single units) and could be recordings from axons of incoming lemniscal inputs rather than from cell bodies of ICC neurons (see Joris et al. 2004 for discussion).

Fig. 6.11 Models for rate tuning to modulation in ICC based on inhibition. (**a**) Block diagram of the Nelson and Carney (2004) model. An ICC model neuron receives both a fast excitatory input and a delayed inhibitory input from model VCN neurons having the same BF. The VCN model neuron receives structurally similar excitatory and inhibitory inputs from the AN. (**b**) Typical rate-MTFs and sync-MTFs for AN, VCN and ICC model neurons. (From Nelson and Carney 2004.) (**c**) Block diagram of the Dicke et al. (2007) model. The model includes two neuron types in VCN: entraining (EN) neurons similar to octopus cells with rate MTFs linearly increasing up to 800 Hz, and constant-rate (CR) neurons with flat rate MTFs. Lowpass (LP) neurons in ICC receive excitatory inputs from an adjustable number of CR neurons and an inhibitory input from one EN neuron, giving them a lowpass rate MTF whose cutoff increases with the number of CR inputs. Bandpass (BP) neurons in ICC (BP) receive an excitatory input from one LP neuron and an inhibitory neuron from another LP neuron with a slightly lower cutoff frequency, giving them bandpass rate MTFs. (**d**) Rate MTFs of the four neuron types in the Dicke et al. (2007) model

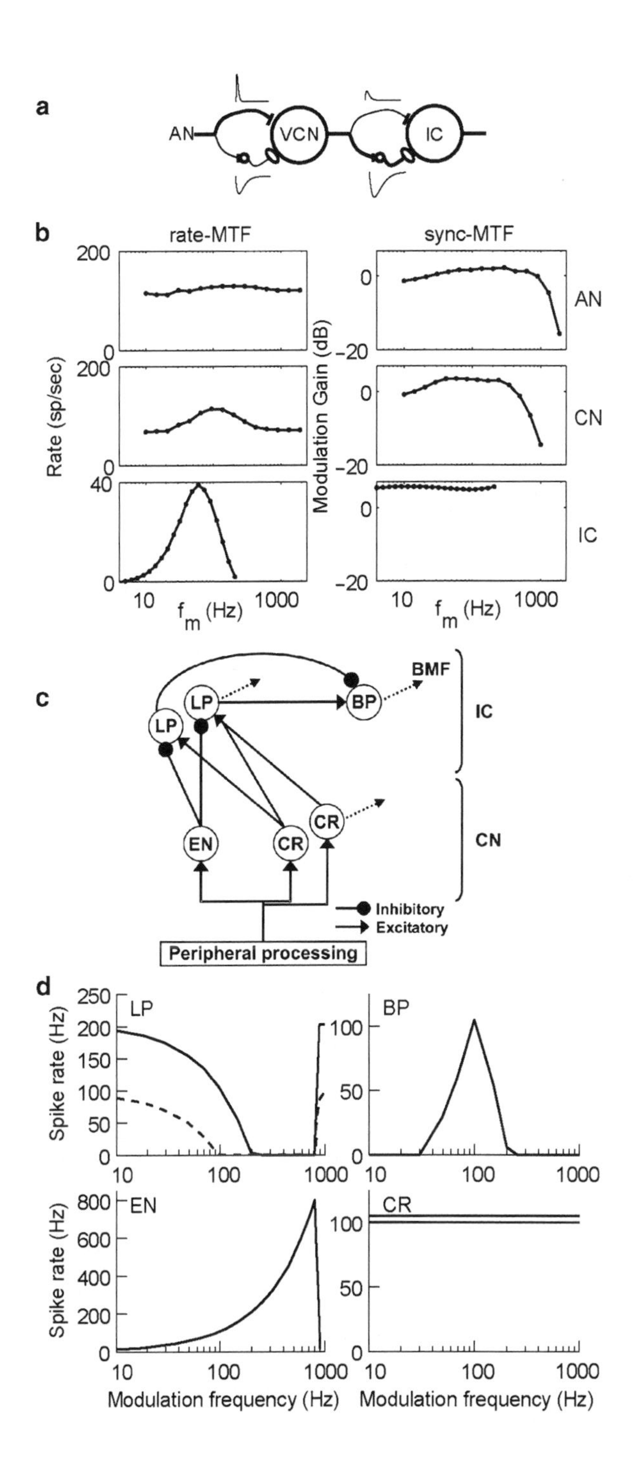
a
AN
VCN
IC
b
rate-MTF
sync-MTF
Rate (sp/sec)
Modulation Gain (dB)
AN
CN
IC
f_m (Hz)
c
BMF
LP
BP
EN
CR
IC
CN
Inhibitory
Excitatory
Peripheral processing
d
Spike rate (Hz)
LP
BP
EN
CR
Modulation frequency (Hz)

inputs. For AM stimuli with low modulation frequencies, excitation and inhibition occur largely in phase, giving rise to a weak rate response in the ICC neuron. For modulation frequencies near the cell's BMF, the stimulus-locked excitatory peaks line up in time with the "valleys" in the phase-locked inhibition, giving rise to a robust response. At modulation frequencies above the BMF, the inhibition is no longer phase locked (due its slow time constant) and completely blocks the excitation. The ICC model cell thus has a bandpass rate-MTF (Fig. 6.11b), with a BMF determined mainly by the time constants of its inhibitory and excitatory inputs: short time constants give rise to high BMFs and long time constants to lower BMFs. Interestingly, the Nelson and Carney (2004) model is the cascade of two structurally identical SFIE circuits, either of which could, by itself, implement the transformation from a synchrony to a rate code. Nonetheless, the model has the advantage of simulating physiological data from both VCN and ICC. However, interpretation of the model structure is complicated by the fact that VCN bushy cells do not project directly to the ICC, but rather to LSO and MSO. Neurons in the LSO show lowpass rate-MTFs (Joris and Yin 1998) suggesting that the transformation into a bandpass MTF must occur at a later stage such as the ICC. MSO neurons in the big brown bat show lowpass or all-pass rate-MTFs (Grothe et al. 2001), also suggesting that a major transformation must occur at a later stage.

Dicke et al. (2007) proposed a model for bandpass rate tuning to AM which, like the Nelson and Carney (2004) model uses inhibition, but has a different principle of operation. In this model, ICC cells with bandpass rate-MTFs receive inputs from other ICC cells with lowpass rate-MTFs which, in turn, receive inputs from neurons in the CN (Fig. 6.11c). The CN stage of the model contains two unit types: (1) entraining (EN) neurons, which fire exactly one spike per modulation cycle of SAM stimuli and (2) constant-rate (CR) neurons, which fire a series of regularly spaced spikes at a rate unrelated to the modulation frequency (i.e., the CR neurons are choppers and have flat rate-MTFs). The EN model neurons resemble PVCN units with ideal onset (On-I) response patterns to pure tones. Because they entrain to low-frequency AM tones, their rate-MTFs increase linearly with modulation frequency up to an upper limit at about 800 Hz, above which they fire only at the onset of SAM stimuli (Fig. 6.11d). The cells in the first layer of the ICC, called lowpass (LP) units, are excited by a variable number of CR units and inhibited by EN unit activity. These cells exhibit basically lowpass[4] rate-MTFs, and their cutoff frequency increases with increasing number of CR inputs (Fig. 6.11d). The bandpass (BP) cells in the second ICC layer receive a strong inhibitory input from one LP unit and weak excitatory input from another LP unit with a slightly higher cutoff frequency. The interaction of excitatory and inhibitory inputs results in a bandpass rate-MTF, with a BMF between the cutoff frequencies of the two LP units that comprise its inputs (Fig. 6.11d). This model thus exploits the fact that some ICC cells exhibit lowpass rate-MTFs to generate

[4]At modulation frequencies above approximately 800 Hz, the EN neurons providing inhibitory inputs to the LP neurons no longer respond in a sustained manner, causing the firing rate of the LP neurons to increase rapidly. Thus, strictly speaking, the LP neurons have band-reject, not lowpass, rate-MTFs.

bandpass rate tuning in the ICC. It is not known whether the orderly lateral inhibition among neighboring cells assumed in the circuit is biologically realistic.

While the timing between excitation and inhibition is critical in the Nelson and Carney (2004) model to create bandpass rate tuning to AM, the Dicke et al. (2007) model is much less dependent on precise timing. The critical element in the Dicke et al. model is the entraining neuron (EN), which fires exactly one spike per modulation cycle, thereby coding the modulation frequency in both its firing rate (which equals the modulation frequency) and its temporal discharge patterns. The temporal information available in EN neurons is largely ignored at the IC stage, where first lowpass and then bandpass AM tuning are created through slow inhibitory interactions. That the model does not require precise temporal processing is consistent with the general sluggishness of IC neurons, as shown by their poor phase locking to both pure tones and AM tones with frequencies above a few hundred Hz (Krishna and Semple 2000; Liu et al. 2006). However, the robustness of the rate code for modulation frequency to variations in modulation depth, background noise and reverberation has not been investigated in either CN Onset cells or model EN neurons. An additional issue for both the Dicke et al. (2007) model and the Nelson and Carney (2004) model is that intracellular recording from ICC neurons with bandpass rate MTFs do not consistently show evidence for inhibition (Geis and Borst 2009), suggesting that, if inhibition plays a role, it may operate at levels below the ICC.

6.6.3 Conclusion

In conclusion, a wide variety of neural circuits and mechanisms have been proposed to account for rate tuning to modulation frequency in ICC. All four models discussed can produce the most commonly found bandpass rate MTFs, and some of the models can also produce lowpass MTFs, but none accounts for the rate-MTFs showing two peaks flanking a suppression region occasionally observed in the ICC (Krishna and Semple 2000). Given the wide variety of afferent inputs to ICC neurons, it is unlikely that the same model structure can account for all rate-MTF shapes observed in ICC. None of the models explicitly includes inputs from MSO, LSO, and DNLL, which constitute a significant fraction of the afferent inputs to the ICC. While the focus of the models has been on the transformation from a temporal code to a rate code for AM frequency, temporal codes remain important in the ICC. In particular, modulation frequencies in the 3–7 Hz range critical for speech reception are primarily coded temporally at this stage. Moreover, Nelson and Carney (2007) have argued that temporal information in IC neurons better accounts for psychophysical performance in modulation detection than rate information. For the most part, the proposed models have not attempted to account for the phase of the MTF and the detailed temporal discharge patterns of ICC neurons in response to AM stimuli. Yet, these very features are likely to provide valuable information for testing and refining the models. Clearly, much additional experimental and theoretical work is needed to develop a comprehensive model of modulation processing in ICC.

6.7 Pitch of Harmonic Complex Tones

Many natural sounds such as human voice, animal vocalizations, and the sounds produced by musical instruments contain harmonic complex tones in which all the frequency components are multiples of a common fundamental, resulting in a periodic waveform. Such harmonic sounds produce a strong pitch sensation at their fundamental frequency (F0), even if they contain no energy at the fundamental. Pitch plays an important role in music perception, in speech perception (particularly for tonal languages), and in auditory scene analysis (Darwin and Carlyon 1995).

Both the Langner (1981) model and the Hewitt and Meddis (1994) model discussed in Sect. 6.6.1 in the context of amplitude modulation processing are basic building blocks for neural population models intended to account for the pitch of harmonic complex tones. This section revisits these models with a focus on how well they predict psychophysical data on pitch perception, rather than the physiology of brain stem and midbrain neurons. We begin with a brief review of perceptual pitch phenomena, peripheral representations of pitch, and the autocorrelation model of pitch.

6.7.1 Basic Psychophysics of Pitch

Harmonic complex tones have two mathematically equivalent properties: periodic waveforms and harmonic spectra (in which all frequency components are integer multiples of a common fundamental). These two properties are converted into two kinds of neural pitch cues through peripheral auditory processing. On the one hand, the frequency selectivity and frequency-to-place mapping in the cochlea transforms a harmonic spectrum into a tonotopic representation that is maintained (and elaborated upon) up to at least the primary auditory cortex. On the other hand, neural phase locking provides a representation of the waveform periodicity in the temporal discharge patterns of auditory neurons, particularly in the interspike intervals. Contrasting hypotheses about the central processing of these two pitch cues available in the auditory periphery lead to "place" (a.k.a. spectral) and "temporal" models for pitch perception, respectively.

An important factor in pitch perception of harmonic complex tones is whether the stimulus contains peripherally resolved harmonics (see Plack and Oxenham 2005 for review). An individual frequency component in a complex tone is said to be peripherally resolved if it produces a local maximum in the spatial pattern of neural activity along the tonotopic axis of the cochlea. In general, a component will be resolved if the bandwidth of the cochlear filter tuned to its frequency is narrower than the frequency spacing between adjacent components. In contrast, if the bandwidth of a cochlear filter encompasses several harmonics, these harmonics are said to be unresolved and are not individually associated with place cues along the tonotopic axis. Because the bandwidths of auditory filters increase roughly proportionately to their center frequency (except at very low frequencies), the low-order harmonics of a complex tone are peripherally resolved, while high-order harmonics are unresolved. The first six to ten harmonics in a tone complex are thought to be resolved in humans (Plomp 1964; Bernstein and Oxenham 2003).

While periodicity pitch sensations can be produced by both resolved and unresolved harmonics, in general sounds containing resolved harmonics give rise to stronger pitches that are less dependent on the relative phases of the components than sounds consisting entirely of unresolved harmonics (see Plack and Oxenham 2005 for review). Because most natural sounds contain low-order, resolved harmonics that are dominant for pitch, pitch perception by normal-hearing listeners outside of the psychophysics laboratory is likely to be based almost entirely on resolved harmonics.

6.7.2 Peripheral Representations of Pitch

Tones with resolved harmonics give rise to three distinct pitch cues in the activity patterns of the auditory nerve: spatial, temporal, and spatio-temporal (Fig. 6.12). First, each resolved harmonic produces a local maximum in firing rate at its place along the tonotopic axis for low and moderate stimulus levels (Cedolin and Delgutte 2005) (Fig. 6.12f, solid line). In principle, the pitch of a tone complex could be estimated by matching harmonic templates to this pattern of local maxima (de Cheveigné 2005). However, these rate-place cues may not be very robust to variations in stimulus level and signal-to-noise ratio due to the limited dynamic range of most AN fibers (Sachs and Young 1979). Second, the waveform of the basilar membrane motion at the cochlear place tuned to a resolved harmonic is nearly sinusoidal. For harmonic frequencies below 4–5 kHz, AN fibers innervating that place phase lock to this sinusoidal waveform (Fig. 6.12c), thereby allowing a precise identification of the harmonic frequency from interspike interval statistics. In principle, the temporal information about each resolved harmonic can then be combined in various ways across the tonotopic axis to identify the pitch of the tone complex (Srulovicz and Goldstein 1983; Meddis and Hewitt 1991). Third, the cochlear traveling wave slows down markedly near the place of each resolved harmonic, giving rise to a rapid phase transition along the cochlear axis which is reflected in the local spatio-temporal pattern of auditory nerve activity (Fig. 6.12b) (Shamma 1985; Cedolin and Delgutte 2007). These spatio-temporal pitch cues might be extracted by a central mechanism sensitive to the coincidence of spike activity in AN fibers innervating neighboring cochlear places (Fig. 6.12f, dashed line).

In contrast to the rich set of pitch cues provided by resolved harmonics, unresolved harmonics give rise to temporal cues only in peripheral neural responses. These temporal cues can be either in the envelope or in the fine time structure of the temporal discharge patterns. Envelope cues are always available with unresolved harmonics, and are the only cues available if the harmonics lie above the 4–5 kHz upper frequency limit of phase locking (Fig. 6.12e). If a set of partials are

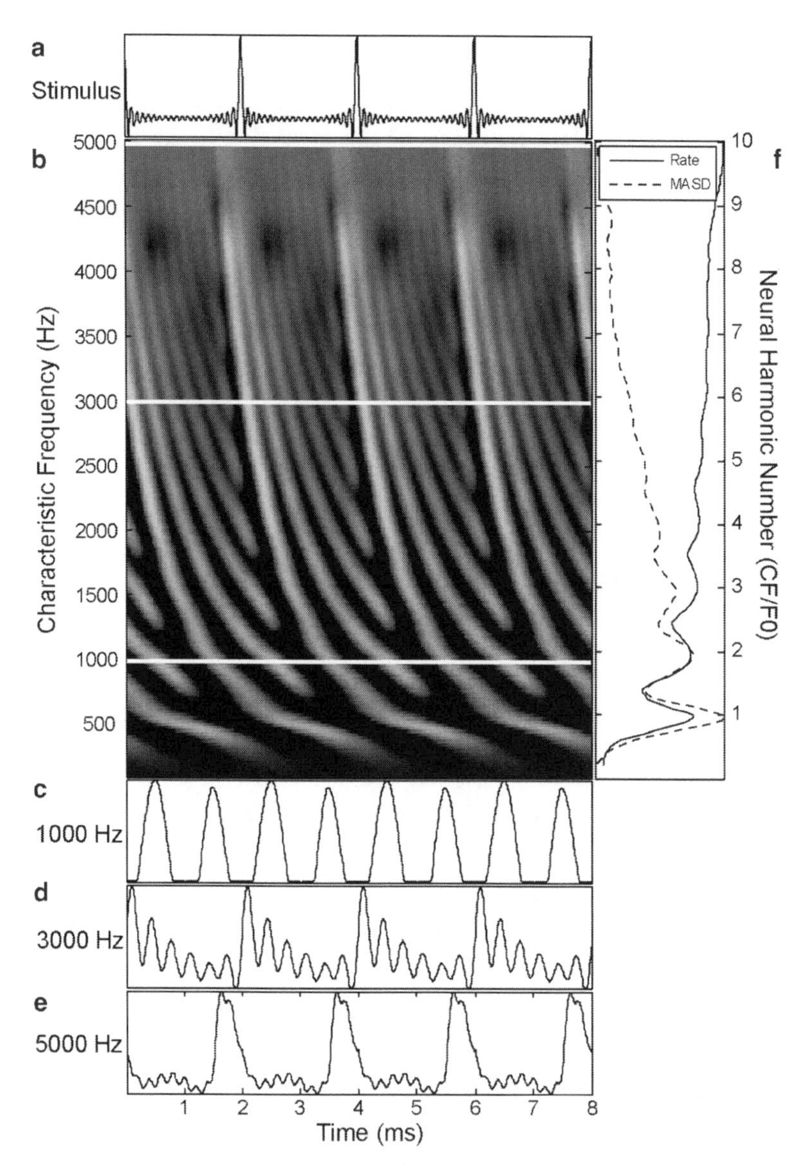

Fig. 6.12 Pitch cues available in the spatio-temporal response pattern of the auditory nerve. (**a**) Waveform of harmonic complex tone (F0 = 500 Hz) with equal-amplitude harmonics in cosine phase. (**b**) Spatio-temporal response pattern of the Zilany and Bruce (2006) model of peripheral auditory processing in cat to the complex tone stimulus at 60 dB SPL. Characteristic frequency (CF) of AN fibers is on the *y*-axis and time on the *x*-axis. Note the rapid changes in response phase (latency) at the places of Harmonics 1–4 (500, 1,000, 1,500, and 2,000 Hz) and the more gradual phase changes in between these harmonics. (**c**–**e**) Temporal response patterns of three model fibers to the harmonic complex tone. *White horizontal lines* in (**b**) indicate the CFs of the three fibers. (**c**) Response of fiber with CF near the second harmonic (1,000 Hz) of the F0 resembles half-wave rectified 1,000-Hz sinusoid, indicating that this harmonic is resolved. (**d**) Response of 3,000-Hz fiber shows phase locking to both the envelope and the fine structure of the complex tone.

high enough in frequency to be unresolved, but still below the upper limit of phase locking, their common periodicity is also represented in the fine time structure of the AN fiber responses (Fig. 6.12d) (e.g., Young and Sachs 1979; Cariani and Delgutte 1996). For harmonic stimuli, the envelope and fine structure periodicities are congruent and directly reflect the perceived pitch. However, envelope periodicity can be dissociated from fine-structure periodicities using inharmonic stimuli in which all the frequency components are shifted from their harmonic values by a constant offset. With such stimuli, the perceived pitch generally differs from the envelope frequency and can be predicted from fine structure periodicities using a simple formula called "de Boer's rule" (de Boer 1956). Thus, fine structure pitch cues usually dominate envelope cues when both are available.

6.7.3 Autocorrelation Model of Pitch

Two broad classes of mathematical models have been successful in accounting for a wide variety of pitch phenomena: spectral pattern recognition models and temporal models based on autocorrelation (see de Cheveigné 2005 for review). Both types of models take into account peripheral frequency analysis and describe the computations needed to estimate the pitch of a stimulus without specifying a detailed physiological implementation of these computations. Spectral pattern recognition models estimate pitch by matching internal harmonic templates to maxima in activity associated with resolved harmonics along a tonotopic representation of the stimulus spectrum. While these models account for most pitch phenomena with resolved harmonics, they are not further discussed here because they have not yet led to physiologically realistic implementations involving the IC.

Autocorrelation is a general mathematical technique for detecting periodicities in a signal. For binary signals such as neural spike trains, the autocorrelation is equivalent to the all-order interspike interval distribution (Perkel et al. 1967). Licklider (1951) proposed a two-dimensional ("duplex") representation of sound obtained by computing the autocorrelation of the spike train in every tonotopic channel. The resulting representation is organized by both cochlear place (mapping frequency) and time delay or interspike interval. Licklider pointed out that the operations required for computing an autocorrelation (delay, multiplication, and temporal integration) can be implemented by known neural mechanisms (axonal conduction delays, coincidence

Fig. 6.12 (continued) (**e**) Response of 5,000-Hz fiber only phase locks to the stimulus envelope which is periodic at F0. (**f**) Profiles of average firing rate (*solid line*) and spatio-temporal phase cue ("MASD," *dashed line*) against neural harmonic number CF/F0. Both profiles show peaks at small integer values (1–4) of CF/F0, indicating that these harmonics are resolved. The spatio-temporal phase cue is obtained by taking the first spatial derivative (with respect to CF) of the spatio-temporal pattern in (**b**), then full-wave rectifying and integrating over time (Cedolin and Delgutte 2007)

detection, and lowpass filtering by the membrane capacitance, respectively). Slaney and Lyon (1990) and Meddis and Hewitt (1991) implemented Licklider's duplex representation using a computational peripheral auditory model as input, and added a stage in which the autocorrelations are summed across all frequency channels to obtain a one-dimensional "summary" or "pooled" autocorrelation. The pitch period is then estimated from the delay of the most prominent maximum (or maxima) in the pooled autocorrelation. The summation across all frequency channels discards all place information and makes the autocorrelation model a purely temporal model invariant to a random shuffling of the tonotopic axis.

The autocorrelation model accounts for the pitches produced by both resolved and unresolved harmonics using the same computations (Meddis and Hewitt 1991). Whereas the autocorrelation model is sensitive to periodicities in both the envelope and the fine structure, the autocorrelation peaks associated with fine structure periodicities tend to be sharper than those produced by envelope periodicities, and thus receive more weight in the pooled autocorrelation (Cariani and Delgutte 1996). This is how the autocorrelation model accounts for the pitch shift of inharmonic tones (for which fine structure and envelope periodicities conflict) and for the dominance of low-frequency harmonics that give rise to fine structure cues over high-frequency harmonics that give rise to only envelope cues. However, the autocorrelation model is not sensitive to whether temporal fine structure cues originate from resolved or unresolved harmonics. This conflicts with psychophysical observations showing that resolved harmonics give rise to stronger pitches than unresolved harmonics even when both sets of harmonics occupy the same frequency region (Carlyon and Shackleton 1994; Bernstein and Oxenham 2004). Also, the poor pitch discrimination performance of cochlear implant listeners (Shannon 1983; Carlyon et al. 2002) raises issues for the autocorrelation model because appropriate electrical stimulation creates precise temporal fine structure cues in the auditory nerve (Dynes and Delgutte 1992) that should be detectable by an autocorrelation mechanism. Despite these shortcomings, the autocorrelation model parsimoniously accounts for a large number of psychophysical phenomena and predicts the perceived pitch for a vast majority of the stimuli experienced in everyday listening.

6.7.4 Physiologically Based Models of Pitch Processing

Despite the remarkable effectiveness of the autocorrelation model of pitch, there is no compelling physiological or anatomical evidence for the orderly neural architecture of delay lines and coincidence detectors postulated by Licklider (1951). The delays are particularly problematic since detecting the periodicity at the lower limit of pitch near 30 Hz (Pressnitzer et al. 2001) would require an approximately 30 ms delay, which is much longer than the neural latencies typically observed in the brain stem and midbrain. The models presented in this section represent efforts at implementing autocorrelation-like processing using neural mechanisms more consistent with the anatomy and physiology of the auditory brain stem.

The Langner (1981) model circuit discussed in Sect. 6.6.1 is tuned to AM tones with specific modulation and carrier frequencies through coincidence between a delayed trigger signal locked to the modulation frequency and an integrator signal locked to an integer submultiple of the carrier frequency. Thus, a two-dimensional array of such circuits mapped according to both their best carrier frequency and their best modulation frequency provides a periodicity map similar to Licklider's (1951) duplex representation. This structure is consistent with evidence for roughly orthogonal maps of best frequency and best modulation frequency in the ICC (Schreiner and Langner 1988). For a SAM tone, the pitch can be estimated from the location of the maximum of activity along the "periodotopy" (modulation frequency) axis of the neural map (Voutsas et al. 2005). For broadband stimuli containing many harmonics, an additional processing stage is required to estimate the pitch from the two-dimensional neural map, either by integrating activity across the tonotopic axis as in the autocorrelation model, or by computing pair wise cross-correlations between all tonotopic channels (Borst et al. 2004).

By requiring coincidence between carrier-locked and modulation-locked signals, the Langner (1981) model can, in principle, account for the pitch of inharmonic stimuli in which the temporal cues in the envelope and the fine structure conflict. Indeed, the periodicity equation describing the dependence of BMF on carrier frequency in the Langner model is formally similar[5] to the de Boer's rule for the pitch shift of inharmonic tones (Langner 1985). While this property holds for a neural circuit made up of ideal trigger and integrator neurons receiving inputs that perfectly entrain to the modulation and the carrier, respectively, the robustness of this model for more realistic inputs is an issue, especially in the presence of noise and reverberation. For example, strong pitches can be evoked by stimuli that have very little amplitude modulation, such as iterated ripple noise or harmonic complex tones in random phase. The model trigger neuron may have trouble reliably firing once per modulation cycle with such stimuli. Also, while the function of the integrator neuron is to count a specific number of carrier cycles before firing, auditory nerve fibers do not fire exactly once per carrier cycle (i.e., they do not entrain). Even if the volley principle is invoked to get a neural signal more directly representing the carrier frequency, the robustness of the integrator with intrinsically noisy neural signals as inputs remains an issue. These issues were partly addressed in two implementations of the Langner model that use fairly realistic integrate-and-fire model neurons and receive inputs from a simplified model of the auditory periphery (Borst et al. 2004; Voutsas et al. 2005). So far, both implementations have been tested only with harmonic complex tones having pronounced amplitude modulations. More thorough tests are needed to evaluate the efficacy of this model as a general pitch processor.

[5]The periodicity equation gives a linear relationship between best modulation *period* and carrier *period* (in which the slope is a ratio of small integers), while de Boers' rule gives a linear relationship between perceived pitch *frequency* and carrier *frequency*. The two relationships are very similar when examined over a limited range of carrier frequencies (Langner 1985).

The "chopper model of pitch" (Wiegrebe and Meddis 2004), performs an autocorrelation-like operation using physiologically realistic elements based on the Hewitt and Meddis (1994) model for rate tuning to AM frequency in the ICC. In the Hewitt and Meddis (1994) model, rate tuning to AM arises through coincidence between many inputs from VCN chopper neurons that have the same intrinsic chopping rate (CR) and best frequency. As in the Licklider (1951) and Langner (1981) models, a two-dimensional temporal representation of sound is first obtained at the VCN stage through a large ensemble of chopper neurons arranged based on both their best frequency and their chopping rate (Fig. 6.13a). This temporal representation of periodicity information is transformed into a two-dimensional rate-based representation at the IC stage through coincidence detection between chopper inputs having the same best frequency and chopping rates. At a final stage of the model (presumably located above the IC), the firing rate of IC neurons is summed across the tonotopic axis for each CR to obtain a one dimensional representation of pitch that plays the same role as the pooled autocorrelation in the Meddis and Hewitt (1991) model (Fig. 6.13a). The model in its present form does not specify how the pitch frequency is estimated from the model output profile. Rather, changes in the shape of the output profile with stimulus manipulations are related to changes in pitch in psychophysical experiments.

Because the model's effectiveness depends critically on the tuning of chopper neurons to a specific periodicity, the VCN stage of the model has been analyzed in detail (Wiegrebe and Meddis 2004). Model VCN neurons, like real sustained chopper (Chop-S) neurons, show enhanced representation of the stimulus periodicity in their first-order interspike intervals (i.e., they fire more regularly) when the stimulus F0 matches the neuron's intrinsic chopping rate (Winter et al. 2001). Importantly, periodicity enhancement in interspike intervals can also occur when the stimulus F0 is a small integer multiple (2 or 4) of the CR. Wiegrebe and Meddis (2004) interpret this result as evidence that chopper neurons act as temporal harmonic templates in the sense that a given chopper neuron would fire highly regularly when a resolved harmonic of a complex tone is an integer multiple of its CR as well as when the fundamental matches its CR. However, the physiological data reported so far only show that a chopper can fire regularly in response to a stimulus whose *fundamental* (not a resolved harmonic) is a small integer multiple of its CR, which is not quite the same thing since the regular firing to the fundamental could be based on envelope cues. Additional physiological data are needed on this important point.

The complete model including the ICC stage (Meddis and O'Mard 2006) shows pitch cues in the activity profile of the output layer as a function of CR but there is no prominent activity peak at the pitch period as in the autocorrelation model (Fig. 6.13b). For stimuli consisting entirely of unresolved harmonics, the profile is basically highpass (Fig. 6.13b, right), with little activity for low CRs, a steep rise to a shallow maximum when the CR matches the stimulus F0, and then a broad plateau at higher CRs. This plateau arises because chopper neurons with CRs above the F0 fire in synchrony to the stimulus envelope (the F0) rather than to their intrinsic CR. For complex tones with resolved harmonics (Fig. 6.13b, left), the model output pattern shows a steep rise for CRs near the stimulus F0 as in the unresolved case, but then shows a second, shallower

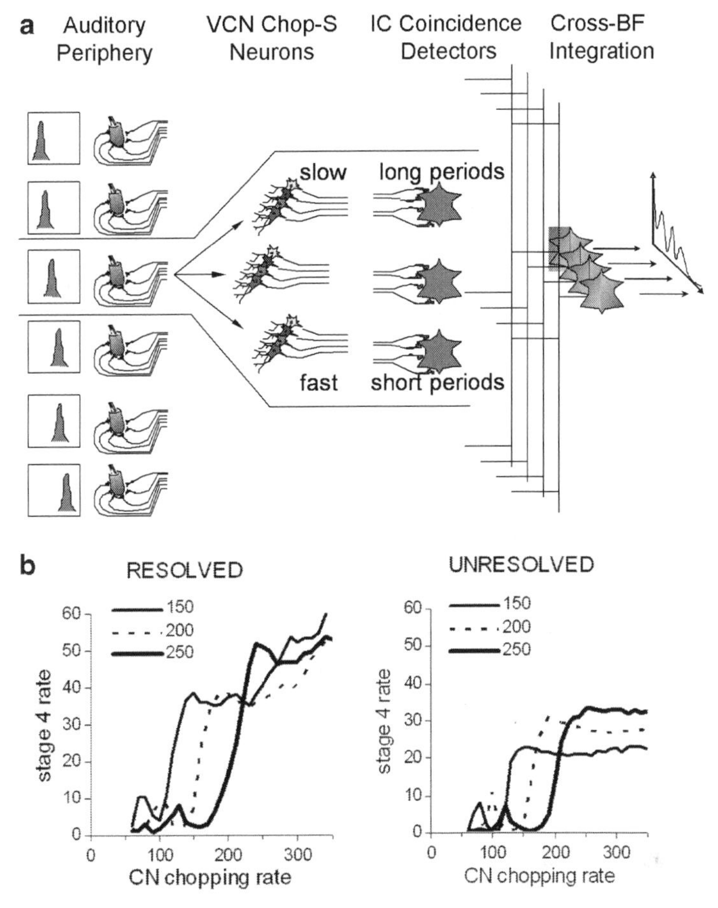

Fig. 6.13 Chopper model for pitch processing. (From Meddis and O'Mard 2006.) (**a**) Block diagram of the model. The model has four stages: auditory periphery, VCN, ICC, and cross-BF integrator located beyond the ICC. At the ICC stage, the model is a two-dimensional grid of neurons arranged by both BF and rate BMF. Each ICC neuron is modeled after Hewitt and Meddis (1994; see Fig. 6.10b), and receives inputs from a set of model VCN chopper neurons that all have the same BF and chopping rate (CR) (which determines the BMF of the ICC neuron). At the integrator stage, the firing rates of all ICC neurons having the same CR is summed across the tonotopic (BF) axis, giving rise to a one-dimensional representation of periodicity as a function of CR. (**b**) Response profiles at the output of the model as a function of CR for two sets of harmonic complex tones with F0s of 150, 200, and 250 Hz presented at 70 dB SPL. *Left*: Harmonics 3–8, which are peripherally resolved. *Right*: Harmonics 13–18, which are unresolved

rise rather than a broad plateau at higher CRs. This second rise is attributed to the ability of chopper neurons to fire regularly in response to resolved harmonics whose frequency is an integer multiple of the CR. The overall model activity with resolved harmonics exceeds that in the unresolved case, consistent with the greater pitch strength of resolved harmonics. The model also predicts the pitch shift of inharmonic tones when the harmonics are resolved but not when they are unresolved, broadly consistent with psychophysical data (Moore and Moore 2003).

The chopper model of pitch represents the most comprehensive and thoroughly tested effort at developing a physiologically realistic model for pitch processing to date. There are nevertheless a number of issues with this model. First, the model has mostly been tested with a narrow range of F0s near 200 Hz. With such stimuli, resolved harmonics are typically within the frequency range (<1,000–1,500 Hz) where VCN choppers phase lock to the fine structure (Blackburn and Sachs 1989), while unresolved harmonics are always above the upper limit of phase locking. Thus, the differences in model behavior attributed to differences in harmonic resolvability may actually reflect differences between pitch cues available in the temporal fine structure vs. the envelope of the responses of chopper neurons. Harmonics located in the 1,800- to 3,000-Hz region (above the upper limit of phase locking in choppers) can give rise to strong pitch sensations in human listeners, and also show sharp perceptual differences between resolved and unresolved harmonics (Carlyon and Shackleton 1994; Bernstein and Oxenham 2004). It will be important to test this condition with the model by varying F0 over a wider range. A second issue is that the model output profile for harmonic complex tones is sensitive not only to changes in F0 that affect the pitch, but also to changes in harmonic composition (timbre) and stimulus level that produce no change in pitch. The present strategy of looking at changes in model output profiles associated with changes in pitch may not work for psychophysical experiments where stimulus level and harmonic composition are randomized. To account for performance in such experiments, it may be necessary to identify a specific feature associated with pitch in the model output profile. Because small changes in F0 can produce widespread changes in the model output that are not restricted to CRs near the F0 (Fig. 5a in Meddis and O'Mard 2006), identifying a such a feature may prove challenging. A third and more fundamental issue with the model is that the 100- to 500-Hz range of intrinsic CRs typically observed in VCN neurons (Frisina et al. 1990) covers only a portion of the 30- to 1,400-Hz range of F0 that produce pitch percepts with missing fundamental stimuli in human listeners As mentioned in the preceding text, if the coincidence detector is located in ICC (as assumed by the model), there is a further mismatch between the range of VCN chopping rates and the 10- to 150-Hz range where most IC neurons have their best modulation frequencies (Krishna and Semple 2000).

6.8 Summary

We have reviewed a wide variety of models that aim at accounting for emerging response properties of IC neurons in terms of interactions between various inputs to the IC and circuitry within the IC itself. These models assume specific neural circuits to model specific aspects of spatial hearing or temporal processing. While these models have advanced our understanding of how response properties may be formed, they leave us yet far from a comprehensive understanding of IC function.

Such a comprehensive view might be furthered by a focus on the integration of stimulus features, and a clearer identification of essential commonalities across models.

Despite the wide range of modeling strategies and phenomena being modeled, the models use a small number of basic circuit modules much like bricks in a Lego game. Prominent among these is coincidence detection which sharpens temporal response patterns and can also transform a temporal code into a rate or rate-place code. Coincidence detection, first proposed by Jeffress (1948), is used in models for ITD processing in the MSO (Colburn, Chapter 4), in models for generating Onset response patterns in the cochlear nucleus (Voigt, Chapter 3), and in models for rate tuning to modulation frequency and periodicity detection in the IC (Langner 1981; Hewitt and Meddis 1994; Meddis and O'Mard 2006). Contrasting with coincidence detection, which is a highly nonlinear interaction between excitatory inputs, is the additive summation of excitatory inputs used in the Shackleton et al. (2000) model. Also widely used is the interplay between excitatory and inhibitory inputs which can perform a wide variety of signal processing operations depending on the characteristics of the inhibition. Inhibition can be either temporally precise (phase locked) as in models for ITD sensitivity in MSO (Brand et al. 2002) and AM tuning in MSO and IC (Grothe 1994; Nelson and Carney 2004), or be delayed and persistent to create a long-term dependence on previous inputs (Cai et al. 1998a; Pecka et al. 2007). Inhibition can also be lateral to sharpen tuning to a stimulus feature along a neural map (Cai et al. 1998a; Rucci and Wray 1999; Kimball et al. 2003; Dicke et al. 2007).

Whereas interactions between synaptic inputs have been the main workhorse of computational IC models, the role of intrinsic membrane channels, which play an important role in models of cochlear nucleus neurons (Voigt, Chapter 3), has been relatively unexplored. Only two models (Cai et al. 1998b; Borisyuk et al. 2002) explicitly introduced membrane conductances for dynamic processing, and there has been no effort to exploit the signal processing capabilities of the arsenal of membrane channels documented in IC neurons (Sivaramakrishnan and Oliver 2001; Wu et al. 2002). This omission is particularly surprising for models of AM tuning, where it would seem to be natural. Another omission is that all IC models use a strictly feed-forward architecture and thus do not take advantage of the signal processing capabilities of recurrent connections (Cariani 2001; Friedel et al. 2007). Recurrent connections are widespread in the central nervous system, and the existence of intricate circuitry within the IC, commissural connections between the two ICs (Saldaña and Merchán 2005) and descending projections from the thalamus and cortex (Winer 2005) suggest that feedback may play an important role in IC processing (Suga and Ma 2003).

Perhaps the main limitation of IC models so far has been their focus on a rather narrow set of physiological phenomena. Binaural IC models focus on either ITD or ILD processing, and consider only inputs from MSO and LSO, respectively, while temporal processing models focus on the transformation from a temporal code to a rate code for AM tuning, and consider only direct, monaural inputs from the CN. In reality, the vast majority of IC neurons are sensitive to both binaural cues and amplitude modulation, and physiological studies reveal interactions between binaural and temporal processing (Sterbing et al. 2003; Lane and Delgutte 2005).

Also, a majority of IC neurons are sensitive to both ITD and ILD (Chase and Young 2005), particularly when tested with broadband stimuli. An important next step will be to integrate models of binaural, temporal, and spectral processing to develop a more comprehensive view of IC function. A major stumbling block toward this goal is the lack of physiological criteria for identifying the types of brain stem inputs that a given IC neuron receives. Such criteria are well established for the cochlear nucleus where there is a clear correspondence between anatomical cell types and physiological types defined from tone-burst response patterns and frequency response maps, but this type of understanding has been slow to emerge for the IC. Recent progress in this important area is encouraging (Ramachandran et al. 1999; Davis 2002; Loftus et al. 2004; Malmierca et al. 2005). Another stumbling block is that physiological studies typically report only a fairly narrow set of response properties from each neuron, even when the neuron was actually tested with a wide range of stimulus conditions. Perhaps the increased availability of publicly accessible databases of physiological data can remedy this problem.

Despite the limitations of present IC models, the very fact that this topic could be reviewed indicates the enormous progress that has been made since the first edition of this book in 1996. We may be at the threshold of developing a more integrative understanding of IC function in which computational modeling will undoubtedly play a key role.

Acknowledgment We thank M. Slama, G.I. Wang, L. Wang, B. Wen, and Y Zhou for their comments on the manuscript, and G.I. Wang for help with figure preparation. Preparation of this chapter was supported by NIH grants DC002258 (Delgutte and Hancock) and DC05161 (Davis).

References

Aitkin LM (1986) The Auditory Midbrain: Structure and Function of the Central Auditory Pathway. Clifton, NJ: Humana.

Batra R, Kuwada S, Fitzpatrick DC (1997a) Sensitivity to interaural temporal disparities of low- and high-frequency neurons in the superior olivary complex. I. Heterogeneity of responses. J Neurophysiol 78:1222–1236.

Batra R, Kuwada S, Fitzpatrick DC (1997b) Sensitivity to interaural temporal disparities of low- and high-frequency neurons in the superior olivary complex. II. Coincidence detection. J Neurophysiol 78:1237–1247.

Bernstein JG, Oxenham AJ (2003) Pitch discrimination of diotic and dichotic tone complexes: harmonic resolvability or harmonic number? J Acoust Soc Am 113:3323–3334.

Bernstein JG, Oxenham AJ (2004) Effect of stimulus level and harmonic resolvability on pitch discrimination. J Acoust Soc Am 115:2389.

Blackburn CC, Sachs MB (1989) Classification of unit types in the anteroventral cochlear nucleus: PST histograms and regularity analysis. J Neurophysiol 62:1303–1329.

Borisyuk A, Semple MN, Rinzel J (2001) Computational model for the dynamic aspects of sound processing in the auditory midbrain. Neurocomputing 38–40:1127–1134.

Borisyuk A, Semple MN, Rinzel J (2002) Adaptation and inhibition underlie responses to time-varying interaural phase cues in a model of inferior colliculus neurons. J Neurophysiol 88: 2134–2146.

Borst M, Langner G, Palm G (2004) A biologically motivated neural network for phase extraction from complex sounds. Biol Cybern 90:98–104.
Boudreau JC, Tsuchitani C (1968) Binaural interaction in the cat superior olive S segment. J Neurophysiol 31:442–454.
Braasch J, Blauert J (2003) The precedence effect for noise bursts of different bandwidths. II. Comparison of model algorithms. Acoust Sci Technol 24:293–303.
Brand A, Behrend O, Marquardt T, McAlpine D, Grothe B (2002) Precise inhibition is essential for microsecond interaural time difference coding. Nature 417:543–547.
Brughera A, Stutman E, Carney LH, Colburn HS (1996) A model for excitation and inhibition for cells in the medial superior olive. Audit Neurosci 2:219–233.
Burger RM, Pollak GD (2001) Reversible inactivation of the dorsal nucleus of the lateral lemniscus reveals its role in the processing of multiple sound sources in the inferior colliculus of bats. J Neurosci 21:4830–4843.
Cai H, Carney LH, Colburn HS (1998a) A model for binaural response properties of inferior colliculus neurons. I. A model with interaural time difference-sensitive excitatory and inhibitory inputs. J Acoust Soc Am 103:475–493.
Cai H, Carney LH, Colburn HS (1998b) A model for binaural response properties of inferior colliculus neurons. II. A model with interaural time difference-sensitive excitatory and inhibitory inputs and an adaptation mechanism. J Acoust Soc Am 103:494–506.
Caird D, Klinke R (1983) Processing of binaural stimuli by cat superior olivary complex neurons. Exp Brain Res 52:385–399.
Cant NB (2005) Projections from the cochlear nuclear complex to the inferior colliculus. In: Winer JA, Schreiner CE (eds), The Inferior Colliculus. New York: Springer, pp 115–131.
Cariani PA (2001) Neural timing nets. Neural Netw 14:737–753.
Cariani PA, Delgutte B (1996) Neural correlates of the pitch of complex tones. I. Pitch and pitch salience. J Neurophysiol 76:1698–1716.
Carlyon RP, Shackleton TM (1994) Comparing the fundamental frequencies of resolved and unresolved harmonics: evidence for two pitch mechanisms? J Acoust Soc Am 95:3541–3554.
Carlyon RP, van Wieringen A, Long CJ, Deeks JM, Wouters J (2002) Temporal pitch mechanisms in acoustic and electric hearing. J Acoust Soc Am 112:621–633.
Carney LH (1993) A model for the responses of low-frequency auditory-nerve fibers in cat. J Acoust Soc Am 93:401–417.
Casseday JH, Covey E (1996) A neuroethological theory of the operation of the inferior colliculus. Brain Behav Evol 47:311–336.
Cedolin L, Delgutte B (2005) Pitch of complex tones: rate-place and interspike interval representations in the auditory nerve. J Neurophysiol 94:347–362.
Cedolin L, Delgutte B (2007) Spatio-temporal representation of the pitch of complex tones in the auditory nerve. In: Kollmeier B, Klump G, Hohmann V, Langemann U, Mauermann M, Uppenkamp S, Verhey J (eds), Hearing – From Sensory Processing to Perception Heidelberg: Springer, pp 61–70.
Chase SM, Young ED (2005) Limited segregation of different types of sound localization information among classes of units in the inferior colliculus. J Neurosci 25:7575–7585.
Colburn HS (1996) Computational models of binaural processing. In: Hawkins HH, McMullen TA, Popper AN, Fay RR (eds), Auditory Computation. New York: Springer, pp 332–401.
Darwin CJ, Carlyon RP (1995) Auditory grouping. In: Moore BCJ (ed), The Handbook of Perception and Cognition. New York: Academic, pp 387–424.
Davis KA (2002) Evidence of a functionally segregated pathway from dorsal cochlear nucleus to inferior colliculus. J Neurophysiol 87:1824–1835.
Davis KA, Ramachandran R, May BJ (1999) Single-unit responses in the inferior colliculus of decerebrate cats. II. Sensitivity to interaural level differences. J Neurophysiol 82:164–175.
Davis KA, Ramachandran R, May BJ (2003) Auditory processing of spectral cues for sound localization in the inferior colliculus. J Assoc Res Otolaryngol 4:148–163.
de Boer E (1956) On the “Residue” in Hearing (PhD thesis) Amsterdam: University of Amsterdam.

de Cheveigné A (2005) Pitch perception models. In: Plack CJ, Oxenham AJ, Fay RR, Popper AN (eds), Pitch Neural Coding and Perception. New York: Springer, pp 169–233.

Devore S, Ihlefeld A, Hancock K, Shinn-Cunningham B, Delgutte B (2009) Accurate sound localization in reverberant environments is mediated by robust encoding of spatial cues in the auditory midbrain. Neuron 62:123–134.

Dicke U, Ewert SD, Dau T, Kollmeier B (2007) A neural circuit transforming temporal periodicity information into a rate-based representation in the mammalian auditory system. J Acoust Soc Am 121:310–326.

Dong CJ, Swindale NV, Zakarauskas P, Hayward V, Cynader MS (2000) The auditory motion aftereffect: its tuning and specificity in the spatial and frequency domains. Percept Psychophys 62:1099–1111.

Dynes SB, Delgutte B (1992) Phase-locking of auditory-nerve discharges to sinusoidal electric stimulation of the cochlea. Hear Res 58:79–90.

Ehret G, Romand R, eds (1997) The Central Auditory System. New York: Oxford University Press.

Fitzpatrick DC, Kuwada S, Kim DO, Parham K, Batra R (1999) Responses of neurons to click-pairs as simulated echoes: auditory nerve to auditory cortex. J Acoust Soc Am 106:3460–3472.

Friedel P, Burck M, Leo van Hemmen J (2007) Neuronal identification of acoustic signal periodicity. Biol Cybern 97:247–260.

Frisina RD, Smith RL, Chamberlain SC (1990) Encoding of amplitude modulation in the gerbil cochlear nucleus: I. A hierarchy of enhancement. Hear Res 44:99–122.

Geis HR, Borst JG (2009) Intracellular responses of neurons in the mouse inferior colliculus to sinusoidal amplitude-modulated tones. J Neurophysiol 101:2002–2016.

Goldberg JM, Brown PB (1969) Response of binaural neurons of dog superior olivary complex to dichotic tonal stimuli: some physiological mechanisms of sound localization. J Neurophysiol 32:613–636.

Grantham DW (1998) Auditory motion aftereffects in the horizontal plane: the effects of spectral region, spatial sector, and spatial richness. Acustica 84:337–347.

Grantham DW, Wightman FL (1978) Detectability of varying interaural temporal differences. J Acoust Soc Am 63:511–523.

Green DM, Swets JA (1988) Signal Detection Theory and Psychophysics. Los Altos, CA: Peninsula.

Griffiths TD, Rees G, Rees A, Green GG, Witton C, Rowe D, Buchel C, Turner R, Frackowiak RS (1998) Right parietal cortex is involved in the perception of sound movement in humans. Nat Neurosci 1:74–79.

Grothe B (1994) Interaction of excitation and inhibition in processing of pure tone and amplitude-modulated stimuli in the medial superior olive of the mustached bat. J Neurophysiol 71:706–721.

Grothe B, Covey E, Casseday JH (2001) Medial superior olive of the big brown bat: neuronal responses to pure tones, amplitude modulations, and pulse trains. J Neurophysiol 86: 2219–2230.

Guerin A, Jeannes Rle B, Bes J, Faucon G, Lorenzi C (2006) Evaluation of two computational models of amplitude modulation coding in the inferior colliculus. Hear Res 211:54–62.

Hancock KE (2007) A physiologically-based population rate code for interaural time differences (ITDs) predicts bandwidth-dependent lateralization. In: Kollmeier B, Klump G, Hohmann V, Langemann U, Mauermann M, Uppenkamp S, Verhey J (eds), Hearing – From Sensory Processing to Perception. Heidelberg: Springer, pp 389–397.

Hancock KE, Delgutte B (2004) A physiologically based model of interaural time difference discrimination. J Neurosci 24:7110–7117.

Hancock KE, Voigt HF (1999) Wideband inhibition of dorsal cochlear nucleus type IV units in cat: a computational model. Ann Biomed Eng 27:73–87.

Harper NS, McAlpine D (2004) Optimal neural population coding of an auditory spatial cue. Nature 430:682–686.

Hartung K, Trahiotis C (2001) Peripheral auditory processing and investigations of the "precedence effect" which utilize successive transient stimuli. J Acoust Soc Am 110:1505–1513.

Hewitt MJ, Meddis R (1994) A computer model of amplitude-modulation sensitivity of single units in the inferior colliculus. J Acoust Soc Am 95:2145–2159.

Hewitt MJ, Meddis R, Shackleton TM (1992) A computer model of a cochlear-nucleus stellate cell: responses to amplitude-modulated and pure-tone stimuli. J Acoust Soc Am 91:2096–2109.

Jeffress LA (1948) A place theory of sound localization. J Comp Physiol Psychol 41:35–39.

Joris PX, Yin TC (1998) Envelope coding in the lateral superior olive. III. Comparison with afferent pathways. J Neurophysiol 79:253–269.

Joris PX, Schreiner CE, Rees A (2004) Neural processing of amplitude-modulated sounds. Physiol Rev 84:541–577.

Keller CH, Takahashi TT (1996) Binaural cross-correlation predicts the responses of neurons in the owl's auditory space map under conditions simulating summing localization. J Neurosci 16:4300–4309.

Kimball J, Lomakin O, Davis KA (2003) A computational model of spectral processing pathway from dorsal cochlear nucleus to inferior colliculus. In: BME Annual Fall Meeting, Nashville, TN, p 173.

Krishna BS, Semple MN (2000) Auditory temporal processing: responses to sinusoidally amplitude-modulated tones in the inferior colliculus. J Neurophysiol 84:255–273.

Lane CC, Delgutte B (2005) Neural correlates and mechanisms of spatial release from masking: single-unit and population responses in the inferior colliculus. J Neurophysiol 94:1180–1198.

Langner G (1981) Neuronal mechanisms for pitch analysis in the time domain. Exp Brain Res 44:450–454.

Langner G (1983) Evidence for neuronal periodicity detection in the auditory system of the Guinea fowl: implications for pitch analysis in the time domain. Exp Brain Res 52:333–355.

Langner G (1985) Time coding and periodicity pitch. In: Michelsen A (ed), Time Resolution in Auditory Systems. Heidelberg: Springer, pp 108–121.

Langner G (1992) Periodicity coding in the auditory system. Hear Res 60:115–142.

Langner G, Schreiner CE (1988) Periodicity coding in the inferior colliculus of the cat. I. Neuronal mechanisms. J Neurophysiol 60:1799–1822.

Langner G, Albert M, Briede T (2002) Temporal and spatial coding of periodicity information in the inferior colliculus of awake chinchilla (*Chinchilla laniger*). Hear Res 168:110–130.

Li L, Kelly JB (1992) Inhibitory influence of the dorsal nucleus of the lateral lemniscus on binaural responses in the rat's inferior colliculus. J Neurosci 12:4530–4539.

Licklider JC (1951) A duplex theory of pitch perception. Experientia 7:128–134.

Lindemann W (1986) Extension of a binaural cross-correlation model by contralateral inhibition. II. The law of the first wave front. J Acoust Soc Am 80:1623–1630.

Litovsky RY, Yin TC (1998) Physiological studies of the precedence effect in the inferior colliculus of the cat. II. Neural mechanisms. J Neurophysiol 80:1302–1316.

Litovsky RY, Colburn HS, Yost WA, Guzman SJ (1999) The precedence effect. J Acoust Soc Am 106:1633–1654.

Liu LF, Palmer AR, Wallace MN (2006) Phase-locked responses to pure tones in the inferior colliculus. J Neurophysiol 95:1926–1935.

Loftus WC, Bishop DC, Saint Marie RL, Oliver DL (2004) Organization of binaural excitatory and inhibitory inputs to the inferior colliculus from the superior olive. J Comp Neurol 472:330–344.

Malmierca MS, Saint Marie RL, Merchán MA, Oliver DL (2005) Laminar inputs from dorsal cochlear nucleus and ventral cochlear nucleus to the central nucleus of the inferior colliculus: two patterns of convergence. Neuroscience 136:883–894.

McAlpine D, Palmer AR (2002) Blocking GABAergic inhibition increases sensitivity to sound motion cues in the inferior colliculus. J Neurosci 22:1443–1453.

McAlpine D, Jiang D, Shackleton TM, Palmer AR (1998) Convergent input from brainstem coincidence detectors onto delay-sensitive neurons in the inferior colliculus. J Neurosci 18:6026–6039.

McAlpine D, Jiang D, Shackleton TM, Palmer AR (2000) Responses of neurons in the inferior colliculus to dynamic interaural phase cues: evidence for a mechanism of binaural adaptation. J Neurophysiol 83:1356–1365.

McAlpine D, Jiang D, Palmer AR (2001) A neural code for low-frequency sound localization in mammals. Nat Neurosci 4:396–401.
Meddis R, Hewitt MJ (1991) Virtual pitch and phase sensitivity of a computer model of the auditory periphery. I. Pitch identification. J Acoust Soc Am 89:2866–2882.
Meddis R, O'Mard L (2006) Virtual pitch in a computational physiological model. J Acoust Soc Am 120:3861–3869.
Middlebrooks JC, Green DM (1991) Sound localization by human listeners. Annu Rev Psychol 42:135–159.
Moore GA, Moore BC (2003) Perception of the low pitch of frequency-shifted complexes. J Acoust Soc Am 113:977–985.
Mossop JE, Culling JF (1998) Lateralization of large interaural delays. J Acoust Soc Am 104:1574–1579.
Nelson PC, Carney LH (2004) A phenomenological model of peripheral and central neural responses to amplitude-modulated tones. J Acoust Soc Am 116:2173–2186.
Nelson PC, Carney LH (2007) Neural rate and timing cues for detection and discrimination of amplitude-modulated tones in the awake rabbit inferior colliculus. J Neurophysiol 97:522–539.
Oliver DL (2005) Neuronal organization in the inferior colliculus. In: Winer JA, Schreiner CE (eds), The Inferior Colliculus. New York: Springer, pp 69–114.
Oliver DL, Huerta MF (1992) Inferior and superior colliculi. In: Webster DB, Popper AN, Fay RR (eds), The Mammalian Auditory Pathway: Neuroanatomy. New York: Springer, pp 168–222.
Oliver DL, Morest DK (1984) The central nucleus of the inferior colliculus in the cat. J Comp Neurol 222:237–264.
Oliver DL, Beckius GE, Bishop DC, Kuwada S (1997) Simultaneous anterograde labeling of axonal layers from lateral superior olive and dorsal cochlear nucleus in the inferior colliculus of cat. J Comp Neurol 382:215–229.
Palmer AR, Kuwada S (2005) Binaural and spatial coding in the inferior colliculus. In: Winer JA, Schreiner CE (eds), The Inferior Colliculus. New York: Springer, pp 377–410.
Park TJ, Pollak GD (1993) GABA shapes sensitivity to interaural intensity disparities in the mustache bat's inferior colliculus: implications for encoding sound location. J Neurosci 13:2050–2067.
Paterson M, McAlpine D (2005) Effects of peripheral processing are evident in neural correlates of precedence effect. Assoc Res Otolaryn Abstr 28:1414.
Pecka M, Zahn TP, Saunier-Rebori B, Siveke I, Felmy F, Wiegrebe L, Klug A, Pollak GD, Grothe B (2007) Inhibiting the inhibition: a neuronal network for sound localization in reverberant environments. J Neurosci 27:1782–1790.
Peña JL, Konishi M (2001) Auditory spatial receptive fields created by multiplication. Science 292:249–252.
Perkel D, Gerstein G, Moore G (1967) Neuronal spike trains and stochastic point processes: I. The single spike train. Biophys J 7:391–418.
Plack CJ, Oxenham AJ (2005) The psychophysics of pitch. In: Plack CJ, Oxenham AJ, Fay RR, Popper AN (eds), Pitch Neural Coding and Perception. New York: Springer, pp 7–55.
Plomp R (1964) The ear as a frequency analyzer. J Acoust Soc Am 36:1628–1636.
Pollak GD, Burger RM, Park TJ, Klug A, Bauer EE (2002) Roles of inhibition for transforming binaural properties in the brainstem auditory system. Hear Res 168:60–78.
Pressnitzer D, Patterson RD, Krumbholz K (2001) The lower limit of melodic pitch. J Acoust Soc Am 109:2074–2084.
Ramachandran R, Davis KA, May BJ (1999) Single-unit responses in the inferior colliculus of decerebrate cats. I. Classification based on frequency response maps. J Neurophysiol 82:152–163.
Reed MC, Blum JJ (1999) Model calculations of steady state responses to binaural stimuli in the dorsal nucleus of the lateral lemniscus. Hear Res 136:13–28.
Rees A, Langner G (2005) Temporal coding in the auditory midbrain. In: Winer JA, Schreiner CE (eds), The Inferior Colliculus. New York: Springer, pp 346–376.
Rhode WS, Greenberg S (1994) Encoding of amplitude modulation in the cochlear nucleus of the cat. J Neurophysiol 71:1797–1825.

Rosen S (1992) Temporal information in speech: acoustic, auditory and linguistic aspects. Philos Trans R Soc Lond B Biol Sci 336:367–373.
Rothman JS, Young ED, Manis PB (1993) Convergence of auditory nerve fibers onto bushy cells in the ventral cochlear nucleus: implications of a computational model. J Neurophysiol 70:2562–2583.
Rucci M, Wray J (1999) Binaural cross-correlation and auditory localization in the barn owl: a theoretical study. Neural Netw 12:31–42.
Sachs MB, Young ED (1979) Encoding of steady-state vowels in the auditory nerve: representation in terms of discharge rate. J Acoust Soc Am 66:470–479.
Saldaña E, Merchán MA (2005) Intrinsic and commissural connections of the inferior colliculus. In: Winer JA, Schreiner CE (eds), The Inferior Colliculus. New York: Springer, pp 155–181.
Sanes DH, Malone BJ, Semple MN (1998) Role of synaptic inhibition in processing of dynamic binaural level stimuli. J Neurosci 18:794–803.
Schofield BR (2005) Superior olivary complex and lateral lemniscal connections of the auditory midbrain. In: Winer JA, Schreiner CE (eds), The Inferior Colliculus. New York: Springer, pp 132–154.
Schreiner CE, Langner G (1988) Periodicity coding in the inferior colliculus of the cat. II. Topographical organization. J Neurophysiol 60:1823–1840.
Shackleton TM, Meddis R, Hewitt MJ (1992) Across frequency integration in a model of lateralization. J Acoust Soc Am 91:2276–2279.
Shackleton TM, McAlpine D, Palmer AR (2000) Modelling convergent input onto interaural-delay-sensitive inferior colliculus neurones. Hear Res 149:199–215.
Shamma SA (1985) Speech processing in the auditory system. I: The representation of speech sounds in the responses of the auditory nerve. J Acoust Soc Am 78:1612–1621.
Shannon RV (1983) Multichannel electrical stimulation of the auditory nerve in man. I. Basic psychophysics. Hear Res 11:157–189.
Singh NC, Theunissen FE (2003) Modulation spectra of natural sounds and ethological theories of auditory processing. J Acoust Soc Am 114:3394–3411.
Sivaramakrishnan S, Oliver DL (2001) Distinct K currents result in physiologically distinct cell types in the inferior colliculus of the rat. J Neurosci 21:2861–2877.
Slaney M, Lyon RF (1990) A perceptual pitch detector. Proc IEEE-ICASSP-90:357–360.
Spitzer MW, Semple MN (1993) Responses of inferior colliculus neurons to time-varying interaural phase disparity: effects of shifting the locus of virtual motion. J Neurophysiol 69:1245–1263.
Spitzer MW, Semple MN (1998) Transformation of binaural response properties in the ascending auditory pathway: influence of time-varying interaural phase disparity. J Neurophysiol 80:3062–3076.
Srulovicz P, Goldstein JL (1983) A central spectrum model: a synthesis of auditory-nerve timing and place cues in monaural communication of frequency spectrum. J Acoust Soc Am 73:1266–1276.
Steeneken HJ, Houtgast T (1980) A physical method for measuring speech-transmission quality. J Acoust Soc Am 67:318–326.
Sterbing SJ, D'Angelo WR, Ostapoff EM, Kuwada S (2003) Effects of amplitude modulation on the coding of interaural time differences of low-frequency sounds in the inferior colliculus. I. Response properties. J Neurophysiol 90:2818–2826.
Stern RM, Zeiberg AS, Trahiotis C (1988) Lateralization of complex binaural stimuli: a weighted-image model. J Acoust Soc Am 84:156–165.
Suga N, Ma X (2003) Multiparametric corticofugal modulation and plasticity in the auditory system. Nat Rev 4:783–794.
Tollin DJ, Yin TC (2002a) The coding of spatial location by single units in the lateral superior olive of the cat. I. Spatial receptive fields in azimuth. J Neurosci 22:1454–1467.
Tollin DJ, Yin TC (2002b) The coding of spatial location by single units in the lateral superior olive of the cat. II. The determinants of spatial receptive fields in azimuth. J Neurosci 22:1468–1479.

Tollin DJ, Populin LC, Yin TC (2004) Neural correlates of the precedence effect in the inferior colliculus of behaving cats. J Neurophysiol 92:3286–3297.

Voutsas K, Langner G, Adamy J, Ochse M (2005) A brain-like neural network for periodicity analysis. IEEE Trans Syst Man Cybern B Cybern 35:12–22.

Wiegrebe L, Meddis R (2004) The representation of periodic sounds in simulated sustained chopper units of the ventral cochlear nucleus. J Acoust Soc Am 115:1207–1218.

Winer JA (2005) Three systems of descending projections to the inferior colliculus. In: Winer JA, Schreiner CE (eds), The Inferior Colliculus. New York: Springer, pp 231–247.

Winer JA, Schreiner CE, eds (2005) The Inferior Colliculus. New York: Springer.

Winter IM, Wiegrebe L, Patterson RD (2001) The temporal representation of the delay of iterated rippled noise in the ventral cochlear nucleus of the guinea-pig. J Physiol 537:553–566.

Wu SH, Ma CL, Sivaramakrishnan S, Oliver DL (2002) Synaptic modification in neurons of the central nucleus of the inferior colliculus. Hear Res 168:43–54.

Xia J, Brughera A, Colburn HS, Shinn-Cunningham BG (2009) Modeling physiological and psychophysical responses to precedence effect stimuli. Presented at 15th International Symposium on Hearing. Salamanca Spain.

Yang L, Pollak GD (1998) Features of ipsilaterally evoked inhibition in the dorsal nucleus of the lateral lemniscus. Hear Res 122:125–141.

Yin TC (1994) Physiological correlates of the precedence effect and summing localization in the inferior colliculus of the cat. J Neurosci 14:5170–5186.

Yin TC, Chan JC (1990) Interaural time sensitivity in medial superior olive of cat. J Neurophysiol 64:465–488.

Yin TC, Kuwada S (1983) Binaural interaction in low-frequency neurons in inferior colliculus of the cat. III. Effects of changing frequency. J Neurophysiol 50:1020–1042.

Yin TC, Chan JC, Carney LH (1987) Effects of interaural time delays of noise stimuli on low-frequency cells in the cat's inferior colliculus. III. Evidence for cross-correlation. J Neurophysiol 58:562–583.

Young ED, Davis KA (2001) Circuitry and function of the dorsal cochlear nucleus. In: Oertel D, Popper AN, Fay RR (eds), Integrative Functions in the Mammalian Auditory Pathway. New York: Springer 160–206.

Young ED, Sachs MB (1979) Representation of steady-state vowels in the temporal aspects of the discharge patterns of populations of auditory-nerve fibers. J Acoust Soc Am 66:1381–1403.

Young ED, Spirou GA, Rice JJ, Voigt HF (1992) Neural organization and responses to complex stimuli in the dorsal cochlear nucleus. Philos Trans R Soc Lond B Biol Sci 336:407–413.

Zilany MS, Bruce IC (2006) Modeling auditory-nerve responses for high sound pressure levels in the normal and impaired auditory periphery. J Acoust Soc Am 120:1446–1466.

Zurek PM (1987) The precedence effect. In: Yost WA, Gourevitch G (eds), Directional Hearing. New York: Springer 85–105.

Chapter 7
Computational Modeling of Sensorineural Hearing Loss

Michael G. Heinz

7.1 Introduction

Models of auditory signal processing have been used for many years to provide parsimonious explanations for how the normal auditory system functions at the physiological and perceptual levels, as well as to provide useful insight into the physiological bases for perception. Such models have also been used in many applications, such as automatic speech recognition, audio coding, and signal-processing strategies for hearing aids and cochlear implants. Here, an application framework is considered that is motivated by the long-term goal of using computational models to maximize the ability to apply physiological knowledge to overcome auditory dysfunction in individual patients. This long-term goal motivates the present chapter's focus, which is on physiologically based computational models of auditory signal processing that have been used to explore issues related to sensorineural hearing loss (SNHL). Specifically, this chapter considers phenomenological signal processing models (rather than biophysical models) that predict normal and impaired peripheral responses to complex stimuli. These types of models are likely to be most useful in the specific applications needed to achieve the stated long-term goal, such as explaining the physiological bases for perceptual effects of SNHL, diagnosing the underlying physiological cochlear status of individual patients, and fitting and designing hearing-aid algorithms in a quantitative physiological framework.

Much insight into perceptual effects of SNHL has come from simple and complex models of basilar-membrane (BM) responses and the effects of outer-hair-cell (OHC) dysfunction on those responses. However, recent studies suggest that inner-hair-cell (IHC) dysfunction can also significantly affect perceptually relevant response properties in the auditory nerve (AN) related to intensity and speech coding (Bruce et al. 2003; Heinz and Young 2004). Physiological evidence suggests

M.G. Heinz (✉)
Department of Speech, Language, and Hearing Sciences & Weldon School of Biomedical Engineering, Purdue University, West Lafayette, IN 47907, USA
e-mail: mheinz@purdue.edu

R. Meddis et al. (eds.), *Computational Models of the Auditory System*, Springer Handbook of Auditory Research 35, DOI 10.1007/978-1-4419-5934-8_7,

that many common forms of SNHL are likely to involve mixed OHC and IHC damage, including noise-induced and age-related hearing loss (Liberman and Dodds 1984b; Schmiedt et al. 2002). Thus, the application of models to investigate SNHL effects in individual patients requires a more general modeling framework that includes accurate descriptions of physiological response properties associated with both OHC and IHC dysfunction.

The majority of this chapter (Sect. 7.2) describes the significant progress that has been made in modeling physiological effects of SNHL in AN responses. The level of the AN is important to consider because it includes the effects of both OHC and IHC dysfunction, as well as providing the obligatory input to the central nervous system. Thus, this chapter focuses on the contributions of *sensory* (cochlear) dysfunction to sensorineural loss, rather than *neural* dysfunction, for which there has been less physiological modeling. Modeling approaches are described that account for effects of SNHL at low to moderate sound levels, for which the normal-hearing system benefits from cochlear amplification associated with OHC function. A separate section describes modeling effects of SNHL at high sound levels, which are particularly important because hearing aids often operate at high sound levels. A brief overview (Sect. 7.3) is given of the wide range of application areas for which physiologically based models of SNHL have been used. Finally, the potential to develop quantitative links between physiological and psychoacoustical approaches to modeling SNHL is considered (Sect. 7.4). These links are critical to realize fully the potential for computational models to contribute to the translation of basic-science knowledge to applications that will improve the daily lives of people suffering from SNHL.

7.2 Physiologically Based Models of Sensorineural Hearing Loss

Numerous physiological models have included some aspects of SNHL; however, few have been tested directly and rigorously against physiological data in terms of the effects of both OHC and IHC damage. A recent model by Zilany and Bruce (2006, 2007b) represents the most thorough modeling study to date in terms of the inclusion of both OHC and IHC effects. This model is also the most rigorously tested against physiological data from both normal-hearing and hearing-impaired animals. Thus, this model is used here as a framework for discussing the multitude of issues that are involved with predicting physiological AN responses to complex stimuli in cases with mixed hair cell losses. Other computational models that have provided useful insight to our understanding of the effects of SNHL on physiological responses are reviewed within this framework.

A schematic diagram of the Zilany and Bruce (2006, 2007b) model is shown in Fig. 7.1 and provides an illustration of many of the typical components of physiologically based computational models of peripheral auditory processing. This model is a composite signal-processing model in that it has modules that represent

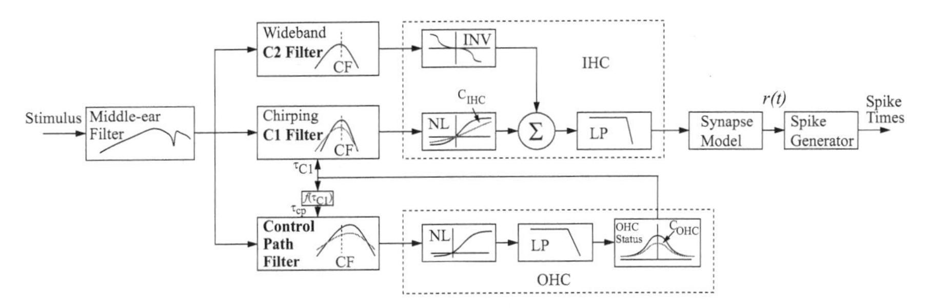

Fig. 7.1 Schematic diagram of a physiologically based auditory-nerve (AN) model that accounts for a wide range of response properties from both normal and hearing-impaired cats for stimuli ranging from simple to complex (e.g., tones, noise, and speech). The model takes as input an arbitrary acoustic waveform and produces an output in terms of either the time-varying discharge rate waveform $r(t)$ or a set of spike times for a single AN fiber with a specified characteristic frequency (CF). The model has two parallel signal paths, a component-1 (C1) path that accounts primarily for low and moderate sound levels and a component-2 (C2) path that accounts for high-level responses. An inner-hair-cell (IHC) module combines the two signal-path outputs as input to a synapse model and spike generator. The control path contains an outer-hair-cell (OHC) module that accounts for nonlinear peripheral response properties associated with the cochlear-amplifier mechanism. Sensorineural hearing loss is simulated with the parameters C_{OHC} and C_{IHC}, which provide independent control of OHC and IHC function and range from 1 (normal) to 0 (fully dysfunctional). *Other abbreviations*: *LP* low-pass filter; τ filter time constant (Reprinted with permission from Zilany and Bruce 2006. Copyright 2006, American Institute of Physics.)

various stages of cochlear signal processing, such as middle-ear filtering, OHC function, BM tuning, IHC transduction, IHC synapse properties, and spike generation. Numerous nonlinear composite models have been developed for normal hearing (reviewed by Lopez-Poveda 2005; Chapter 2). These models are phenomenological by nature in that their ultimate focus is on predicting responses at the level of the BM or AN, rather than on the biophysical mechanisms that produced those responses. However, effort is often made when possible to represent as closely as possible the true anatomical and physiological structure of cochlear processing. This approach has significant benefits for generalizing the models, in particular toward including the various effects of SNHL that arise from dysfunction of specific anatomical structures, such as OHCs and IHCs. As shown in Fig. 7.1, composite models typically take an arbitrary stimulus and produce an output designed to replicate the main physiological response properties at a given location within the periphery (e.g., BM or AN). Auditory-nerve models either predict a time-varying waveform representing the discharge rate as a function of time, $r(t)$, or spike times from a spike-generating module to match the type of data recorded in neurophysiological experiments. The choice of a waveform or spike-based output depends on the specific application, but ideally models would have the potential to provide either type of output.

The Zilany and Bruce (2006, 2007b) model represents the latest extension of the model originally developed by Carney (Carney 1993; Heinz et al. 2001c; Zhang et al. 2001; Bruce et al. 2003; Tan and Carney 2003). The historical development of this model has been reviewed elsewhere (Bruce and Zilany 2007) and thus is not explicitly reviewed here except through discussion of the relevant physiological properties that have been incorporated over the years. The original motivation for the development of this model was to provide a computational description of the effects of cochlear nonlinearity on temporal (as well as average-rate) response properties in the AN for arbitrary stimuli. A computational model was desired so that the implications of cochlear nonlinearity could be studied for issues related to the neural coding of complex sounds. Throughout the extensions of this model, these overriding goals have remained and have been extended to include neural coding issues related to SNHL.

7.2.1 Low and Moderate Sound Level Effects

The common description of the physiological effects associated with the cochlear amplifier (e.g., compression, level-dependent frequency selectivity, and suppression) is based on data from BM or AN responses at low to moderate sound levels (for reviews, see Ruggero 1992; Robles and Ruggero 2001). Thus, most models of SNHL have focused on nonlinear cochlear effects for low and moderate sound levels and have ignored high-level irregularities that occur in AN responses above 80–100 dB SPL. Models that include high-level effects are discussed in Sect. 7.2.2 after the basic issues of modeling OHC and IHC damage are reviewed.

7.2.1.1 Modeling Outer Hair Cell Damage

The key insight that allowed the original Carney model to be successfully extended to model the effects of SNHL was the way in which many of the nonlinear cochlear response properties were attributed to a single mechanism. Damage to OHCs has been shown to result in increased thresholds, broadened tuning, reduced cochlear compression, shifted best frequency (BF), reduced two-tone suppression, and reduced level dependence in phase responses (Dallos and Harris 1978; Schmiedt et al. 1980; Ruggero and Rich 1991; Heinz et al. 2005b). Thus, each of these properties are believed to be associated with a single mechanism, the cochlear amplifier, for which OHCs play a major role (Patuzzi 1996). Despite knowing the molecular protein responsible for OHC electromotility and that this protein is necessary and sufficient for providing cochlear amplification (Zheng et al. 2000; Liberman et al. 2002), the exact biophysical details of how this amplification occurs remain unknown. However, from a signal-processing perspective, the key observation for modeling SNHL has been that some single mechanism related to OHC function is responsible for each of these properties. The relation between BM compression (reduced gain at characteristic frequency (CF) with increased level) and nonlinear frequency selectivity (broadened bandwidth and phase response with increased level) was implemented by specifying OHC-mediated control of BM tuning in terms of a single filter parameter (the time constant τ in Fig. 7.1, which is inversely related to filter bandwidth). This implementation allowed the natural filter gain/bandwidth tradeoff to link these two properties. The original Carney (1993) model had a control path that adjusted the filter time constant based on how much signal energy passed through the BM filter, which produced compression and level-dependent bandwidth (*BW*) and phase responses. A similar approach was taken by Kates (1991a, b), except that the filter *Q* value (a measure of tuning sharpness, equal to *CF*/*BW*) was adjusted for both the tip and tail filters in his model. Zhang et al. (2001) demonstrated that suppression also fit into this phenomenological framework by increasing the bandwidth of the control-path filter to account for the observation that stimulus energy well away from CF could influence peripheral gain and bandwidth (Kiang and Moxon 1974).

The schematic in Fig. 7.1 shows an implementation of this framework, in which the output of the middle-ear filter is passed through both a narrowband filter representing the excitatory BM motion at low to moderate sound levels (middle path, component-1 (C1) response) and a wide-band control-path filter representing the stimulus-based control of cochlear tuning in the C1 path. For narrowband stimuli (e.g., tones at CF), energy passes through both the sharp C1 filter and the control-path filter. At low sound levels, the control path does not change cochlear gain. As stimulus level increases above roughly 30 dB SPL, the energy passing through the control-path filter becomes large enough to activate OHC control, such that the control path begins to turn down the gain of the C1 filter by decreasing τ. Such a reduction in gain results in a compressive input–output function at CF, increased bandwidth, and shallower phase response for frequencies near CF. The nonlinear control of BM tuning does not affect the frequency response well away from CF, accounting for the common description that off-CF responses are linear (Ruggero et al. 1997). When off-CF stimulus energy is added to an excitatory CF tone (e.g., Sachs and Kiang 1968), the off-frequency energy does not pass through the narrow C1 filter, but does pass through the control-path filter and thus suppresses the at-CF response by reducing the gain and increasing the bandwidth of the C1 filter. Thus, this phenomenological approach parsimoniously captures the main signal processing properties associated with OHC function that have been hypothesized to be perceptually relevant (Moore 1995; Oxenham and Bacon 2003; Bacon and Oxenham 2004) for normal and hearing-impaired listeners (i.e., those properties likely to be most important to restore with hearing aids).

Such a phenomenological framework for the role of OHCs in nonlinear cochlear tuning allowed Bruce et al. (2003) to extend the Zhang et al. (2001) model in a parsimonious way to account for the effects of OHC damage on each of these properties. Physiological AN responses to a vowel from normal-hearing cats and from cats with a noise-induced hearing loss (Miller et al. 1997; Wong et al. 1998) provided an excellent data set to develop and test their implementation of hearing impairment (Bruce et al. 2003). The effects of OHC damage were implemented by including a single parameter (C_{OHC}) to the control path (Fig. 7.1) that controls the maximum amount of low-level cochlear gain provided by the OHCs (Fig. 7.2). A value of $C_{OHC} = 1$ represents normal OHC function with sharp tuning at low sound levels (Fig. 7.2a) and full cochlear-amplifier gain. The BM input–output function (Fig. 7.2b) shows linear response growth at low levels and

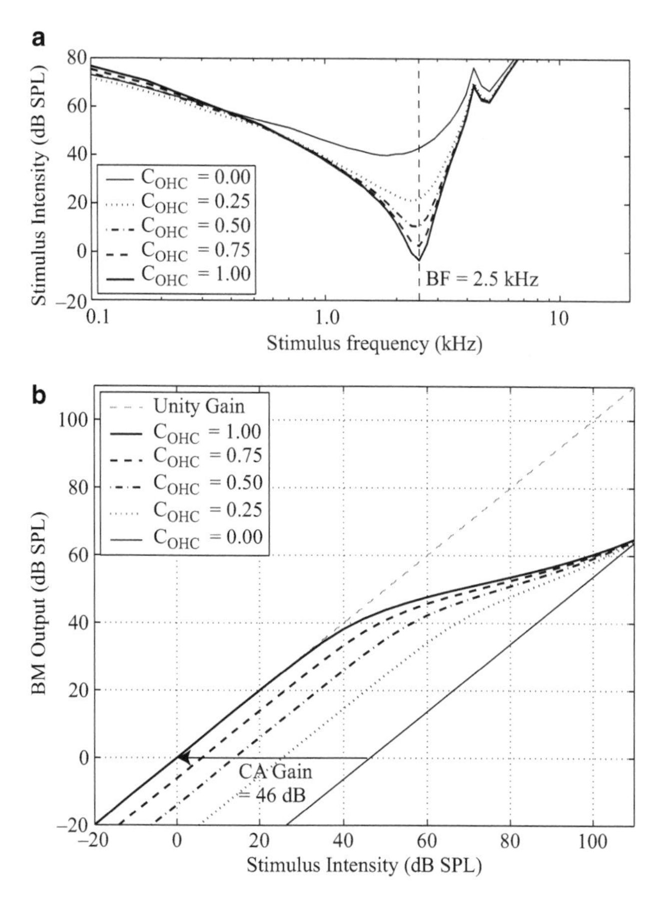

Fig. 7.2 The effect of outer-hair-cell (OHC) dysfunction on model responses. (**a**) Tuning curves. (**b**) Basilar-membrane (BM) input–output functions at best frequency (BF). The parameter C_{OHC} represents the degree of OHC dysfunction, ranging from normal hearing ($C_{OHC}=1$) to complete OHC loss ($C_{OHC}=0$). Normal OHC function produces sharp tuning and full cochlear-amplifier (CA) gain at low sound levels and compressive response growth at moderate levels. Complete OHC loss produces broad tuning, no CA gain, and linear response growth at all levels. A systematic reduction in tuning and cochlear nonlinearity is produced as the degree of OHC dysfunction increases (i.e., as C_{OHC} is reduced from 1 to 0) (Reprinted with permission from Bruce et al. 2003. Copyright 2003, Acoustical Society of America.)

compressive growth at moderate levels because the cochlear-amplifier gain is reduced to 0 dB with increases in sound level, as expected for normal hearing. A value of $C_{OHC}=0$ represents total OHC loss, where no cochlear gain is provided at any sound level. Thus, thresholds are elevated in dB by the amount of cochlear-amplifier gain normally provided (Fig. 7.2b), and tuning is as broad as for normal hearing at high levels (Fig. 7.2a). Values of C_{OHC} between 0 and 1 represent partial OHC damage, where cochlear gain still exists at low levels, but is reduced from the normal-hearing gain. In this case, the nonlinear properties associated with

OHC function still exist, but are reduced in size because the OHCs still acts to reduce the cochlear gain to 0 dB as the sound level increases, there is simply less gain to reduce (Fig. 7.2b). A reduction in nonlinear properties, rather than a complete abolition, is consistent with numerous physiological studies where partial OHC damage is likely to have occurred. For example, compression was reduced as the effect of furosemide increased (Ruggero and Rich 1991), AN tuning-curve bandwidth increased as the effect of furosemide increased (Sewell 1984a), suppression was reduced but not eliminated after acoustic trauma (Miller et al. 1997), and the degree of nonlinear phase shifts was reduced with mild noise-induced hearing loss (Heinz et al. 2005b). Thus, it appears reasonable to model OHC damage as spanning a continuum of OHC function ranging from fully functional ($C_{OHC} = 1$) to completely dysfunctional ($C_{OHC} = 0$), with each of the non-linear properties associated with the cochlear amplifier scaled similarly based on the degree of OHC dysfunction.

The approach taken by Bruce et al. (2003) to modeling the continuum of OHC-damage effects on numerous response properties is similar to several previous models. Kates (1991a, b) limited the control of filter sharpness (Q) provided by the OHCs, which resulted in the compressive and suppressive properties of the model scaling with OHC damage. However, the implications of adjusting both the tip and tail filters in this way were not fully explored. A BM model described by Giguère and Woodland (1994a, b) also included compression, suppression, and the downward shift in BF and increase in bandwidth with level. OHC damage was included in this BM transmission-line model by decreasing the maximum gain provided by the OHCs. Although this model did not include the effects of IHC damage, the fact that the major effects of OHC damage were included via a single mechanism allowed this model to provide useful insight into the limitations imposed by the cochlear amplifier on the ability of hearing-aid amplification to restore normal response patterns across the BM to a variety of stimuli (Giguère and Smoorenburg 1998; see also Sect. 7.3.4). Several models have evaluated the physiological effects of gain reductions and increases in bandwidth associated with OHC damage without including mechanisms to link these response properties (Geisler 1989; Schoonhoven et al. 1994). These effects were accounted for by fitting linear BM filters to measured AN-fiber tuning curves. By doing so, the change in gain was not constrained by the change in bandwidth, which makes it difficult to isolate the effects of OHC damage from the effects of IHC damage that are also likely to be present following acoustic trauma. The justification for the use of linear filters in both these models of SNHL was based on the argument that cochlear filtering becomes linear after SNHL (Geisler 1989; Schoonhoven et al. 1994); however, as described above, cochlear nonlinearity is completely removed only after total OHC loss and the majority of tuning curves considered in those studies were W-shaped, indicating only partial OHC damage (Liberman and Dodds 1984b). It is likely that IHC damage contributed to the hearing loss, although IHC damage was not explicitly considered in either model. Despite these limitations, useful insight was gained by demonstrating which changes in temporal responses to speech and clicks could be accounted for simply by changes in cochlear frequency selectivity.

Another property related to OHC function is the BF shift with level or impairment, which is important for modeling degraded tonotopic representations after SNHL due to enhanced upward spread of masking. The gain/bandwidth tradeoff that was successfully used by Zhang et al. (2001) with gammatone filters can also be applied to filters designed based on pole and zero locations in the complex plane (Tan and Carney 2003). Careful selection of the configuration of poles and zeros can create asymmetric filters that have instantaneous frequency glides in their impulse response, which have been shown to be responsible for BF shifts with level (Carney et al. 1999). Moving the poles away from the imaginary axis as sound level increases creates a broader filter with reduced gain, and thus produces the same phenomenological representation of OHC function (Tan and Carney 2003). Zilany and Bruce (2006) updated this model to include the effects of OHC damage in much the same way as was done in the Bruce et al. (2003) model (see Fig. 7.1). This implementation added BF shifts with level to the list of nonlinear properties that are affected by OHC damage in the model. With impairment, filters are broader at threshold and may have a shifted BF relative to the CF of the cochlear location they innervate (Liberman 1984), depending on the CF and the corresponding direction of the frequency glide in derived filter impulse responses (Carney et al. 1999). This property provides another example of the successful strategy of modeling OHC damage as a reduction in the maximum cochlear gain to capture the phenomenon that nonlinearity in each of the properties associated with OHC function is reduced after OHC damage.

7.2.1.2 Modeling Inner Hair Cell Damage

It has often been assumed that the primary effects of SNHL on perception are related to OHC damage (e.g., Kates 1993; Moore 1995; Edwards 2004); however, physiological and anatomical evidence suggests that IHC damage often occurs to a comparable extent (Liberman and Dodds 1984a, b). Perceptual consequences of IHC loss have been considered in relation to dead regions (Moore 2004); however, recent physiological evidence suggests that partial IHC damage significantly affects AN response properties relevant for speech and intensity coding (Bruce et al. 2003; Heinz and Young 2004). Although partial IHC damage does not significantly affect frequency selectivity, there are other consistent effects on AN response properties following IHC damage. These effects include elevated thresholds, reduced spontaneous and driven rates, and reduced slopes of rate-level functions (Liberman and Dodds 1984a, b; Liberman and Kiang 1984; Wang et al. 1997; Heinz and Young 2004). Potential degradations in temporal coding associated with IHC damage or disrupted AN activity have been implicated in perceptual effects of auditory neuropathy (Zeng et al. 2005); however, these potential degradations typically have not been incorporated into computational models due to the lack of thorough physiological data characterizing these effects.

The signal processing associated with IHC transduction is typically represented in composite models by an instantaneous saturating and rectifying nonlinearity followed by a low-pass filter (reviewed by Lopez-Poveda 2005; Chapter 2). In the model illustrated in Fig. 7.1, the middle path (C1) includes the IHC transduction function that is responsible for low-moderate sound levels (the component-2 (C2) path is discussed in Sect. 7.2.2). Dysfunction in the IHC is modeled using the parameter C_{IHC} to reduce the slope of the IHC transducer function, with $C_{IHC}=1$ representing normal function and $C_{IHC}=0$ representing total C1 dysfunction (Zilany and Bruce 2006). Such an implementation serves to produce most of the major characteristics of IHC damage, specifically elevated thresholds, no significant change in frequency selectivity, and shallower rate-level functions. However, it does not produce a reduction in spontaneous rate or a reduction in the maximum driven rate.

An alternative approach to modeling IHC damage has been described by Kates (1991a, b), where the resistance associated with his IHC transduction model was increased to represent the closure of transduction channels associated with stereocilia damage. For example, an increase in the resistance by a factor of 10 can be thought to represent damage to 90% of the tallest row of stereocilia, which acts to reduce the spontaneous rate and the maximum discharge rate. The continuum of IHC damage represented in both approaches is intuitive in terms of reduced IHC transduction current that would be expected to occur in cases of either noise-induced hearing loss (with damage to some of the stereocilia) or strial damage (with reduced endocochlear potential to drive IHC transduction). Since both implementations capture the primary effects of IHC damage, the differences between these two implementations may by subtle. Nonetheless, the most appropriate way to model the detailed effects of IHC damage remains an open area for research.

7.2.1.3 Predicting the Relative Contributions of OHC and IHC Damage

Computational AN models that constrain the relation between changes in bandwidth and gain due to OHC damage have been used to make quantitative predictions of the relative contributions of OHC and IHC damage to threshold shifts observed in AN data (Miller et al. 1997; Bruce et al. 2003; Zilany and Bruce 2006). Values of C_{OHC} were chosen as a function of CF in order to produce model Q_{10} estimates that matched the population of Q_{10} values from the experimental data. Threshold shifts accounted for by OHC damage in isolation were always less than observed threshold shifts in the AN data. Values of C_{IHC} were then chosen as a function of CF to increase the model threshold shift to match the experimental data. This approach relies on two assumptions: (1) total hearing loss at a given CF is determined by the contributions from both OHC and IHC damage and (2) reduced frequency selectivity at a given CF is solely due to OHC damage. Similar approaches have been taken with loudness models, where the total hearing loss (HL_{TOTAL} in dB) as measured by the audiogram is split into additive components associated with OHC (HL_{OHC}) and IHC (HL_{IHC}) damage (Moore and Glasberg 1997, 2004).

The physiologically based modeling predictions indicated that even for the CF region with the most severely degraded frequency selectivity, OHC damage only accounted for approximately two thirds of the threshold shift in dB (Bruce et al. 2003; Zilany and Bruce 2006). At frequencies further away, IHC damage was found to be the major factor. This finding is inconsistent with the general view that OHC damage is typically the primary factor in SNHL. However, these results are consistent with cochleograms derived from structure/function studies that have examined the status of hair cells after acoustic trauma, where the CF region of significant IHC damage is often broader than the region of OHC damage (Liberman and Dodds 1984b). These physiological results are also consistent with recent psychophysical estimates of the relative contribution of OHC (65%) and IHC damage (35%) derived from cochlear-gain estimates from a population of human listeners with mild to moderate SNHL (Plack et al. 2004). A physiologically based modeling approach similar to that used with the AN data could be developed for human patients (e.g., based on noninvasive estimates of frequency selectivity); however, this will require a better quantitative link between physiological and psychoacoustical estimates of frequency selectivity (Heinz et al. 2002; also see Sect. 7.3.1).

7.2.2 High Sound Level Effects

The fact that hearing aids amplify sounds to high levels makes it important for models of SNHL to incorporate high-level irregularities that occur in AN responses. AN fibers from normal-hearing animals can have rate-level functions with a sharp decrease in rate (notch) and/or a sharp phase transition of up to 180° at high levels, typically 80–100 dB SPL (Liberman and Kiang 1984; Ruggero et al. 1996; Wong et al. 1998). These high-level effects have been hypothesized to arise from a two-factor cancellation mechanism, in which two separate modes of BM motion act to excite the IHC and/or AN fiber (Kiang 1990). Typical terminology refers to the component-1 (C1) response as arising from the mode that dominates at sound levels below the sharp rate and/or phase transition, and the component-2 (C2) response as dominating at higher sound levels. The sharp transition is hypothesized to result from cancellation that occurs over a narrow level range for which the C1 and C2 magnitudes are similar and the phases differ by approximately 180°. The significance of these components for SNHL comes from the observation that the C1 response is vulnerable to acoustic trauma (e.g., C1 reductions correlate with damage to the tallest row of IHC stereocilia), whereas the C2 response is unaffected by complete loss of the tallest IHC stereocilia (Liberman and Kiang 1984). Similar effects on C1 and C2 occur with reduced endocochlear potential due to furosemide (Sewell 1984b).

Zilany and Bruce (2006, 2007b) extended the dynamic range over which the computational AN model can accurately represent both normal-hearing and hearing-impaired responses. They did so by combining the control-path structure from their previous models (Zhang et al. 2001; Bruce et al. 2003) with a multiple-path framework that has been successful in accounting for some of the high-level effects associated with multiple excitation modes in normal hearing (Goldstein 1995;

Meddis et al. 2001; reviewed in Chapter 2) and in impaired hearing (Schoonhoven et al. 1994). Figure 7.1 illustrates the implementation used in this model. The upper path represents the C2 path, which has its own linear broad filter and IHC transduction function. The C2 filter was set to have a bandwidth equal to the maximum C1 bandwidth. The C2 transduction function was implemented to produce responses that had a high threshold (near 90 dB SPL), very steep suprathreshold responses growth, and an inverted output relative to the C1 path. Thus, the C1/C2 transition occurs at high levels when the C1 and C2 responses have similar bandwidths, similar response amplitudes, and inverted phases. The output of the C2 path is summed within the IHC module prior to the membrane low-pass filter. The use of two separate IHC transduction functions differs from previous models that included the interaction before IHC transduction. This approach suggests that data taken from BM do not fully represent these high-level effects. This observation is significant when considering the implications of these effects for human hearing based on psychoacoustical models that are motivated primarily by BM input–output functions. The C2 path in the model is not influenced by parameters used to simulate SNHL, C_{OHC} and C_{IHC}. This implementation represents the physiological finding that C2 responses are typically unaffected by SNHL, while C1 responses can be influenced by both OHC and IHC damage (Liberman and Kiang 1984; Sewell 1984b). The response properties of the C1 path are very similar to the earlier model (Bruce et al. 2003), and thus each of the low-moderate sound level effects associated with OHC and IHC damage described above are predicted by the updated model.

Figure 7.3 shows the effects of different configurations of hair-cell damage on the rate and phase responses of model AN fibers as a function of sound level.

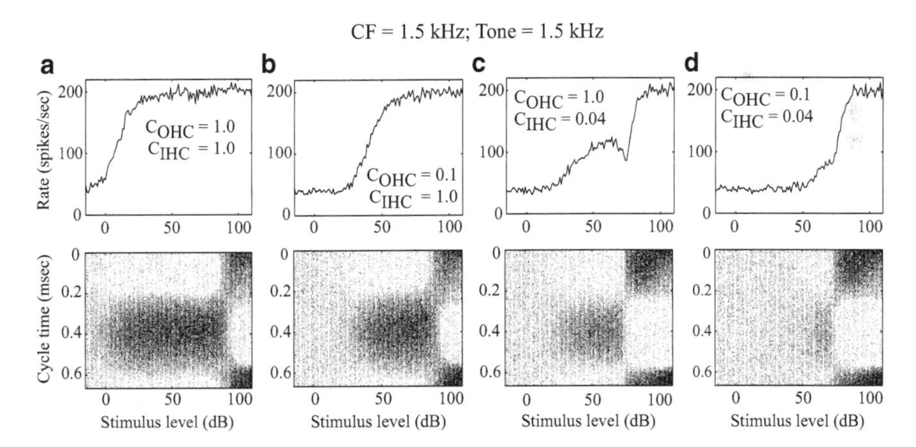

Fig. 7.3 Predicted effects of various configurations of hair-cell damage (C_{OHC} and C_{IHC}) on rate and phase responses as a function of level. Rate-level functions for a single model AN fiber responding to a characteristic-frequency (CF) tone are shown in the *top row*. Period histogram raster plots are shown as a function of level in the *bottom row*, where each *dot* represents a spike occurrence at a particular phase of the CF-tone cycle (ordinate represents one cycle). The center of the *dark region* indicates the phase at which the AN fiber phase locks to the stimulus at each level. (**a**) Normal OHC and IHC function. (**b**) OHC damage with normal IHC. (**c**) IHC damage with normal OHCs. (**d**) Mixed OHC and IHC damage (Reprinted with permission from Zilany and Bruce 2006. Copyright 2006, American Institute of Physics.)

The left panels show predicted responses when both OHC and IHC function are normal. A typical saturating rate-level function with narrow dynamic range is shown for this high-spontaneous-rate fiber. Although a high-level notch is not apparent, the C1/C2 transition at 90 dB SPL is obvious in the phase response, where the dominant phase shifts by 180° over a very narrow range of levels. With selective OHC damage (Fig. 7.3b), the only effect that is seen is an increase in fiber threshold (i.e., the rate-level function simply shifts to the right without an increase in slope). With selective IHC damage (Fig. 7.3c), a dramatic effect is observed due to the selective attenuation of the C1 response and the associated reduction in rate-level slope. Reduction of the C1 response leads to a prominent notch just prior to the steep C2 response, which starts at levels just above the sharp phase transition and is insensitive to SNHL. With significant mixed OHC and IHC damage (Fig. 7.3d), the C1 response is almost entirely eliminated, leaving the steep C2 response at high levels. Each of these types of predicted effects on AN rate-level functions were observed in experimental data collected from cats with noise-induced hearing loss, and have significant implications for theories of intensity coding after SNHL (Heinz and Young 2004; Heinz et al. 2005a). Although the phenomenon of loudness recruitment appears to correlate well with OHC damage and steeper BM growth, the neural correlates of loudness recruitment in AN responses remain unknown. Thus, physiological models that include the confounding effects of OHC and IHC damage on intensity coding are likely to be needed if modeling-based approaches to hearing-aid design are to reach their full potential (Sect. 7.3.3).

7.2.3 *Modeling the Effects of SNHL on the Neural Representation of Speech*

Before the collection of detailed AN responses to speech in cats with SNHL (Miller et al. 1997), several computational models were used to provide useful insight into effects of SNHL on neural coding of speech (Geisler 1989; Kates 1991a, b; Giguère and Woodland 1994b; Giguère and Smoorenburg 1998). Models were developed based on normal hearing responses and then modified based on properties of SNHL, primarily elevated thresholds and broadened tuning. Based on these basic effects, the models were then used to extrapolate predictions of speech responses following SNHL. The models predicted that broadened tuning would cause a degraded tonotopic representation of speech in which there was significant upward spread of the first formant (F1) response to higher CFs. This prediction correlated well with perceptual observations of enhanced upward spread of masking after SNHL in human listeners (Gagné 1988). Models were also used to predict that encoding of CV syllables was more susceptible to noise following SNHL due to broader filters. Although these models used a number of simplifying assumptions, they were still useful at the time because of the lack of detailed neurophysiological data.

Detailed neurophysiological responses to speech collected in cats with SNHL (Miller et al. 1997) made it possible to use models in a more rigorous way to provide additional insight into other factors that contribute to a degraded representation of speech. By testing their model predictions directly against physiological data from individual AN fibers, Bruce et al. (2003) demonstrated the nonintuitive result that IHC damage alone can also lead to a degradation in the tonotopic representation of speech. The shallower transduction function associated with IHC damage raises the sound level at which the IHC saturating nonlinearity leads to synchrony capture. This effect, combined with broader tuning at higher sound levels due to normal OHC function, produces a degraded tonotopic response to speech with selective IHC damage. For example, an AN fiber with a CF near F2 normally demonstrates synchrony capture to the F2 component of the vowel. Following noise-induced hearing loss, such fibers typically show a significant response to F1 in addition to F2. Model AN fibers with selective IHC damage showed similar loss of synchrony capture as those with mixed OHC and IHC damage. However, model fibers with only OHC damage did show synchrony capture at high levels due to saturation associated with normal functioning IHCs. The additional insight provided by the more detailed AN model illustrates important effects that the saturating nonlinearity associated with IHC function can have in both normal-hearing and hearing-impaired cases, sometimes even confounding the intuitive effects of OHC damage.

The addition of the C2 path and the level-dependent BF shifts in the C1 filter improved several predictions of this model for both normal and impaired hearing (Zilany and Bruce 2007b). The BF shifts with level improved predictions of the transition between F2 and F1 responses at moderate levels in normal-hearing fibers, that is, at sound levels lower than C2 threshold. Predictions of SNHL effects at very high levels were also improved by properly accounting for the substantial decrease in synchrony capture to F2. The separate transduction functions for the C1 and C2 paths were critical in accounting for the observations that all frequency components undergo the C1/C2 transition at the same overall sound level (Wong et al. 1998). This implementation also predicted that there are significant responses to non-formant components above the C2 threshold due to the extremely broadband nature of the C2 filter.

Thus, the major response properties of AN fibers in normal-hearing and hearing-impaired animals are now well accounted for by computational AN models. These models have great potential, as discussed below, to be used in applications related to understanding perceptual effects of SNHL (Sects. 7.3.1 and 7.3.2) and for designing hearing aids to maximize restoration of normal responses after SNHL (Sect. 7.3.3).

7.3 Applications of Physiologically Based Models

The focus of this chapter is on computational models of physiological effects of SNHL; however, the motivating framework for discussion has been the desire to apply these models to utilize our ever-expanding physiological knowledge to overcome auditory dysfunction in individual patients. A few key examples in several application areas are discussed here to highlight the significant insight that can be garnered from applications of physiologically based models. Although details of these applications are beyond the scope of this brief chapter and will be reviewed elsewhere, it is hoped that this brief overview will motivate further developments of computational models that can be applied to clinically relevant issues.

7.3.1 *Predicting Psychoacoustical Consequences of SNHL*

A major limitation in applying physiological knowledge to overcome auditory dysfunction in individual patients comes from the difficulty in quantitatively linking physiological and perceptual effects of SNHL. Parsimonious psychoacoustical models can be quite useful in demonstrating the perceptual significance of individual physiological properties associated with SNHL. The most successful example is the temporal window model, which has been used to demonstrate that reduced BM compression can account for the effects of SNHL in a range of temporal resolution tasks, without the need to assume a degradation in central temporal processing (reviewed by Moore and Oxenham 1998; Bacon and Oxenham 2004). However, the big-picture framework considered here is likely to require use of complex computational models that include the numerous physiological response properties associated with both OHC and IHC dysfunction (Sect. 7.2). A long-standing statistical decision theory approach to relating physiology to perception (Siebert 1970; Colburn 1973; reviewed by Delgutte 1996), was recently extended to computational models (Heinz 2000; Heinz et al. 2001a, 2002). Performance predictions based on rate and/or temporal information can be made using any model that predicts the time-varying discharge rate waveform ($r(t)$, Fig. 7.1) for individual AN fibers. This statistical framework has great potential to directly predict psychoacoustical consequences of physiological changes associated with SNHL, as well as to predict the effectiveness of hearing aids in restoring normal perception (Sect. 7.3.3); however, this approach has been used in relatively few studies to date (e.g., Huettel and Collins 2003, 2004).

One complex issue for which computational AN models are particularly useful is evaluating the perceptual relevance of spatiotemporal response patterns (i.e., relative temporal coding across AN fibers with nearby CFs, which is particularly difficult to study experimentally). Phase responses of the population of cochlear filters create robust spatiotemporal cues in CFs near stimulus features (e.g., resolved harmonics, vowel formants). These cues are encoded over a wide dynamic range because they are not limited by discharge-rate saturation. Thus, spatiotemporal cues have been hypothesized in modeling (and some experimental) studies to be perceptually relevant in a number of areas, including speech coding (Shamma 1985; Deng and Geisler 1987; Heinz 2007), pitch perception (Loeb et al. 1983; Cedolin and Delgutte 2007), loudness coding (Carney 1994; Heinz et al. 2001b), tone detection in noise (Carney et al. 2002), and binaural localization based on interaural time differences (Shamma et al. 1989; Joris et al. 2006). Because OHC damage degrades cochlear tuning and the associated phase response, spatiotemporal patterns underlying these hypotheses are predicted to be affected by SNHL. If spatiotemporal cues were perceptually relevant, this would suggest that the ability of hearing aids (and potentially cochlear implants) to restore normal perception after SNHL could be improved by considering the effects of signal-processing strategies on the phase spectrum, in addition to the magnitude spectrum. Model-based approaches to restoring phase cues have recently begun to be explored, although with limited success to date (Shi et al. 2006; Calandruccio et al. 2007). Approaches to restoring temporal response properties are particularly intriguing given recent evidence suggesting that listeners with SNHL have an inability to use temporal fine-structure cues that correlates with their difficulty in understanding speech in temporal varying background noises (e.g., Lorenzi et al. 2006). Given the lack of compelling evidence that within-fiber phase locking is affected by SNHL (Harrison and Evans 1979; Woolf et al. 1981; Miller et al. 1997), it is possible that this perceptual impairment may result from degraded fine-structure coding across CFs instead, that is, degraded spatiotemporal patterns.

7.3.2 Predicting Effects of SNHL on Speech Intelligibility

The most significant complaint of listeners with SNHL is their difficulty in understanding speech, particularly in noisy situations. Well-established modeling frameworks exist for predicting effects of degradations in speech communication channels on overall speech intelligibility, for example, the articulation index (AI) and speech transmission index (STI) (reviewed by Assmann and Summerfield 2004). Several recent computational modeling studies were motivated by the desire to relate physiological effects of SNHL to overall speech intelligibility in terms of metrics analogous to the AI or STI. Bondy et al. (2004b) developed a neural articulation index (NAI) that was derived based on instantaneous neural discharge rate waveforms from a computational AN model. The most significant benefit of neural metrics such as these is the potential to evaluate effects of SNHL and hearing-aid processing on speech intelligibility based on physiological effects (Bondy et al. 2004a). Zilany and Bruce (2007a) demonstrated that the effect of sound level on speech intelligibility, for normal-hearing and aided and unaided hearing-impaired listeners, can be accounted for using a physiologically based model.

A limitation in AI-based techniques, including neural metrics analogous to AI or STI, is that they only predict overall performance and are not able to examine specific confusion patterns. To avoid this limitation, Giguère et al. (1997) used responses from computational AN models for both normal-hearing and complete OHC damage as input to a neural-network recognition stage to predict effects of SNHL on phonemic confusion matrices derived from an extensive speech corpus. Overall, performance of the impaired model was worse than normal, but the confusion patterns for the impaired model were very similar to those of the normal model in terms of phonemic categories, consistent with reports from the literature for flat hearing losses. The class of phonemes most affected by flat OHC loss was nasals, while the class least affected was vowels. The physiological AN model was used to argue that this result was consistent with OHC loss having a greater effect on frequency selectivity for the softer nasals than for the higher-level vowels. This computational framework has great potential to evaluate effects of different physiological types of SNHL and hearing-aid algorithms on specific phonemic confusions in quiet and in noise.

7.3.3 Hearing-Aid Development and Fitting

In the last 15 years, advances in the complexity and accuracy of models of the normal and impaired auditory systems (Sect. 7.2; also see Chapter 2) have dramatically improved their potential for use in the quantitative design and fitting of hearing aids (and cochlear implants, see Chapter 9). Long before the accuracy of sensory models allowed such potential to be realized, a general theoretical framework was described for the use of mathematical models in the development of sensory prostheses (Biondi and Schmid 1972; Biondi 1978). The underlying principle of this theoretical framework is that an "optimal" prosthesis, which modifies the input to the impaired system, can be designed through adjustments to minimize the difference between the aided-impaired and normal model responses to identical stimuli. Similar quantitative goals for restoring normal responses have been proposed more recently in experimental frameworks for both neurophysiological (Miller et al. 1999; Sachs et al. 2002) and psychoacoustical responses (Edwards 2002). However, the iterative nature of the minimization problem required to quantitatively optimize hearing-aid design limits the practicality of experimental approaches. If the responses of both normal and impaired systems can be accurately predicted, then models provide a powerful approach to rigorously and quantitatively solve the optimization problem. Currently, hearing aids are fit primarily based on the audiogram, which measures the elevation in hearing threshold as a function of frequency. Although a number of fitting algorithms are based on underlying theoretical bases, for example, related to loudness or AI-based speech intelligibility (Dillon 2001), very little direct physiological knowledge has been applied to fitting algorithms. Computational models are likely to be the most efficient way to apply our vast (but underutilized) knowledge from hearing science to the development and optimization of hearing-aid signal processing algorithms (Edwards 2004, 2007; Levitt 2004). In fact, recent advances in modeling have allowed useful insight to be gained from the application of this optimization framework at various levels of the auditory system, e.g., BM responses (Giguère and Smoorenburg 1998), excitation patterns (Baer and Moore 1997), AN responses (Kates 1993; Bondy et al. 2004a), central onset responses (Chen et al. 2005), and psychophysical loudness judgments (Leijon 1990; Launer and Moore 2003).

Useful insight has been garnered from the derivation of optimal gain-frequency profiles using both simple and complex models of the AN. Kates (1993) used a simple AN model, which included the salient effects of OHC damage (i.e., loss of sensitivity, compression, suppression, and frequency selectivity), to derive the gain profile that minimized the mean squared error between the populations of aided-impaired and normal-hearing responses. The minimization framework highlighted that the optimal goal was to compensate for the difference in gain provided by the normal and impaired models responding to the particular stimulus of interest, rather than simply to correct for the threshold shift. The optimal gain at a given frequency depended not only on the hearing loss configuration and stimulus energy at that frequency, but also on the stimulus energy across a fairly wide frequency range due to the effects of suppression. These predictions are characteristically different than standard multiband compression algorithms that depend primarily on the stimulus energy near the frequency of interest. A more complex AN model was used in an optimal framework to suggest the need to apply extra attenuation to the first formant (F1) relative to F2 and F3 (Kates 1997), which is similar to the contrast-enhanced frequency shaping (CEFS) algorithm developed to reduce excess upward spread of excitation from F1 in neurophysiological AN responses in hearing-impaired cats (Miller et al. 1999). Normal and impaired AN models were used to show that the benefits provided by CEFS in improving the tonotopic representation of F1 and F2 were not reduced with the addition of a multiband compression algorithm (Bruce 2004), in contrast to most previous spectral enhancement strategies (Franck et al. 1999). A learning-based approach to hearing-aid design was used within the optimization framework to derive time-varying gains at different frequencies (Bondy et al. 2004a). Figure 7.4 illustrates a "neurocompensator" that fits its weights to an individual listener by using a training sequence to minimize the neural distance between the aided-impaired and

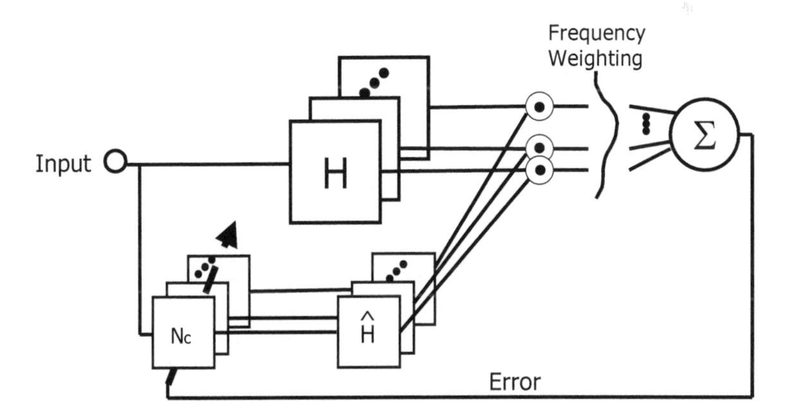

Fig. 7.4 Schematic diagram of a trainable hearing-aid algorithm that uses physiologically based auditory-nerve models. The neurocompensator (Nc) pre-processes sound for the hearing-impaired model (Ĥ). A learning rule adjusts parameters of the Nc block to minimize the neural distance (error) between spike trains from the normal-hearing model (H) and the aided-impaired model. A frequency weighting function is used to combine information across best frequencies (Reprinted with permission from Bondy et al. 2004a. Copyright 2004, Elsevier.)

normal-hearing model responses. Consistent with previous modeling work (Kates 1993, 1997), optimal gains were found to depend on both the stimulus and the degree and configuration of SNHL. Recent modeling work has also suggested that optimal gains depend on which aspect of neural responses (e.g., average-rate or temporal) are being optimized (Bruce et al. 2007). Optimal gains for restoring average-rate and temporal information in model AN responses to speech were found to differ, with rate-based optimal gains being roughly 20 dB higher than optimal temporal-based gains. Thus, physiologically based models have provided useful insight into hearing-aid fitting algorithms; however, their application remains limited by the lack of knowledge as to which aspects of the neural response are most important to restore.

7.3.4 Physiological Limitations in Restoring Normal Cochlear Responses with Amplification

Computational models have been used to demonstrate fundamental limitations imposed by the cochlear amplifier on the ability of hearing aids to restore normal BM or AN response patterns after SNHL (Kates 1991a; Giguère and Smoorenburg 1998). Giguère and Smoorenburg (1998) provided an intuitive description and modeling demonstration (Fig. 7.5) of this issue in terms of normal OHC function, which depends on both cochlear place and stimulus frequency. OHC function can

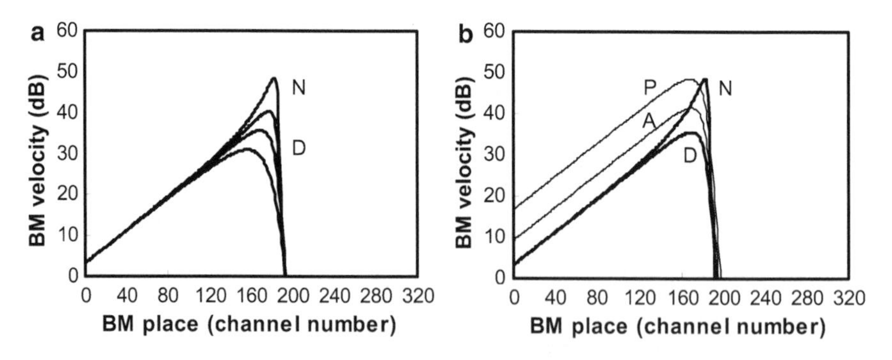

Fig. 7.5 Model demonstration of the physiological limitations in restoring normal basilar-membrane (BM) response patterns with hearing aids. (**a**) BM response patterns across the cochlea (base: channel 0; apex: channel 320) to a 1-kHz tone. A normal-hearing (N) model is compared to three damaged (D) models corresponding to 25, 50, and 100% OHC damage. (**b**) Two amplification strategies illustrate the fundamental difficulty in restoring the damaged model response (50% OHC damage) to normal. A scheme to restore the BM peak (P) response suffered from an undesirable upward shift and spread of activity towards the base. A scheme to restore the BM average (A) response reduced the undesirable upward spread of activity, but failed to restore the loss of cochlear gain near the 1-kHz place (Reprinted with permission from Giguère and Smoorenburg 1998. Copyright 1998, World Scientific.)

be represented by a two-dimensional (frequency and place) transfer function, which depends on relative frequency (f/CF), rather than absolute frequency. The fundamental limitation in restoring BM responses occurs because acoustic hearing aids only have access to one of the two dimensions (stimulus frequency), and thus cannot perfectly overcome the changes in the two-dimensional BM transfer function that occur with OHC damage. For example, because compression occurs only near CF, the normal-hearing response to a 1-kHz tone is compressive at the 1-kHz place and linear at the 2-kHz CF. In contrast, the 1-kHz tone response is linear at both CFs in the case of complete OHC loss. A hearing aid could perfectly restore the BM response to the 1-kHz tone at the 1-kHz place by providing amplification at low levels equal to the normal OHC gain and reducing this gain as sound level increases. However, this gain compensation would create a response to the 1-kHz tone that was too large at the 2-kHz place, resulting in an unwanted upward (toward the base) spread of excitation (Fig. 7.5b). Thus, even with complete knowledge of the damaged BM response, a perfect hearing aid is fundamentally limited by the physiological properties of the cochlear amplifier in that it can at best restore BM responses at one BM place. Giguère and Smoorenburg (1998) suggested (also see Kates 1991a) that although hearing aids can be very useful in restoring audibility and overcoming the effects of loudness recruitment, the potential benefit for improving speech intelligibility is limited and that focus should be spent on techniques for improving signal-to-noise ratios in various situations.

However, it is important to consider that the transmission-line model used by Giguère and Smoorenburg (1998) is most appropriate for high CFs and does not include the fact that the compressive region is much less frequency specific at low CFs than at high CFs, a property that appears true both in animals (Robles and Ruggero 2001) and in humans (Lopez-Poveda et al. 2003). Although the lack of compression frequency specificity is a property that limits the ability of some psychoacoustic methods to estimate BM compression at low CFs (Lopez-Poveda et al. 2003), it may in fact be a fortuitous property that could be taken advantage of in hearing-aid strategies that seek to restore normal BM compression. Further research with physiologically based models and complex stimuli is needed to explore the implications of differences between apical and basal cochlear mechanics and neural coding with respect to SNHL and hearing aids. These differences may be particularly significant with respect to speech, for which low-frequency information is particularly important.

7.3.5 Development of Treatment Strategies for Tinnitus

Tinnitus, the sometimes debilitating condition in which sound is perceived without an acoustic stimulus, is often associated with SNHL in humans (reviewed by Bauer and Brozoski 2008). Because the physiological correlates of tinnitus remain unknown, treatment options are limited in their effectiveness. A computational model was recently developed to explore the relation between tinnitus and SNHL by specifying explicit models for different physiological types of SNHL, including OHC loss, IHC loss, and stereocilia damage (Schaette and Kempter 2006). A simple model of homeostatic plasticity, a physiologically plausible mechanism hypothesized to play a role in tinnitus, was shown to produce central hyperactivity depending on the exact type and spatial distribution of SNHL. The computational framework was used to suggest that prolonged stimulation with a low-level spectrally designed noise could lead to prolonged reductions in hyperactivity even after the noise was turned off. This tinnitus treatment strategy differs from more common stimulation with broadband noise, which can reduce tinnitus via masking, but only during presentation of the noise. Particularly intriguing, as discussed by Schaette and Kempter (2006), is the observation that hearing aids, as well as noise devices and cochlear implants, can reduce the percept of tinnitus. A unified computational framework of peripheral and central effects may lead to a better understanding of how to optimally fit hearing aids to not only provide better speech perception, but also to reduce tinnitus.

7.4 Summary and Future Directions

Significant progress has been made in the last 15–20 years in physiologically based computational models of the peripheral effects of SNHL. These models now include the major physiological details likely to be perceptually relevant, including numerous properties associated with reduced cochlear-amplifier function at low-moderate sound levels as well as high-level AN response irregularities that are particularly relevant for hearing-aid applications. A significant feature of recent physiological models is the quantitative inclusion of the differential and combined effects of OHC and IHC damage, which are hypothesized to contribute to large across-subject variability typically observed in psychoacoustic performance (e.g., patients with similar audiograms but very different speech recognition). Whereas it is difficult in experimental approaches to estimate or control the degree of OHC/IHC damage in individual subjects, computational models now provide great potential for predicting the complicated physiological effects of combined OHC/IHC damage, as well as their perceptual significance. However, application of these models to individual patients will require improvements in estimating quantitatively and noninvasively the degree of OHC/IHC damage in humans (e.g., Lopez-Poveda et al. 2009). Clever psychoacoustic and otoacoustic techniques to estimate cochlear nonlinearity continue to be improved (Bacon and Oxenham 2004; Johnson et al. 2007); however, further work is needed to translate quantitatively and reliably results from human subjects to specific physiological model parameters (e.g., C_{OHC} and C_{IHC}). This will likely require implementation of physiological models into a quantitative framework to predict psychoacoustical performance on the same tasks used to estimate cochlear function (e.g., Lopez-Poveda and Meddis 2001; Heinz et al. 2002).

The conceptual framework in which the total degree of hearing loss can be accounted for by the combined contributions of OHC and IHC damage has also emerged from psychoacoustic models of impaired loudness perception (Moore and Glasberg 1997, 2004). Although the links between loudness models and the physiological effects of SNHL on intensity coding remain unsettled (Heinz et al. 2005a; Cai et al. 2009), loudness models have been applied to hearing-aid fitting procedures much more extensively than physiological models, for example, based on loudness restoration (Moore 2000) or loudness equalization (Moore et al. 1999). The applications of both psychoacoustic and physiological models have typically assumed that it is sufficient to model the degree of OHC and IHC dysfunction, independent of the cause (e.g., noise-induced or age-related hearing loss). This assumption is clearly a useful first step, but ultimately its validity needs to be tested through advances in modeling the physiological bases for variability in psychoacoustical performance across hearing-impaired subjects.

Although physiological models are now being applied to predicting perceptual consequences of SNHL (Sects. 7.3.1 and 7.3.2) and to fitting hearing aids (Sect. 7.3.3), these approaches remain limited by not knowing the relevant aspects of the neural code (e.g., which aspects of neural responses should be optimized in hearing-aid fitting, Bruce et al. 2007). At the time Biondi proposed the optimization framework for sensory prostheses, he suggested that although the easiest place to apply this approach was at the periphery (because peripheral models were most accurate), the most appropriate place was at the perceptual level given the ultimate goal of sensory prostheses (Biondi and Schmid 1972; Biondi 1978). Today, there is much greater potential for quantitatively linking physiological and perceptual effects of SNHL. A signal-detection theory framework that incorporates computational models has great potential to predict the psychoacoustical consequences of physiological changes associated with SNHL (Heinz 2000), and thus to ultimately use psychoacoustic tasks as a basis for evaluating the effectiveness of hearing aids (Edwards 2002). The ability to predict speech intelligibility (Sect. 7.3.2) from neural responses also provides much promise (Zilany and Bruce 2007a); however, it is likely that approaches that investigate performance in terms of specific confusion patterns as well as average performance will be needed to explain the physiological basis for across-subject variability (Giguère et al. 1997).

No matter which domain the hearing-aid optimization framework is implemented in (physiological or perceptual), it will ultimately be most effective to optimize in real time. Digital hearing aids provide the potential for online optimization that would allow amplification algorithms to factor in the specific sounds being heard in addition to the estimated physiological configuration of hair-cell damage. One example of this potential is the online loudness-restoration algorithm implemented by Launer and Moore (2003). An online loudness model was used to predict normal and impaired specific-loudness patterns, which were then matched online in the digital hearing aid by adjusting channel gains. The development of efficient digital versions of physiological models (Chapter 10) provides promise that such online implementations can be extended to theoretical goals based on predicted physiological or psycho-physiological responses.

Acknowledgments Preparation of this chapter was partially supported by a grant from the National Institute on Deafness and Other Communication Disorders (R03-DC007348). Thanks are expressed to Kimberly Chamberlain for her assistance with manuscript preparation.

References

Assmann P, Summerfield Q (2004) The perception of speech under adverse conditions. In: Greenberg S, Ainsworth WA, Popper AN, Fay RR (eds), Speech Processing in the Auditory System. New York: Springer, pp. 231–308.

Bacon SP, Oxenham AJ (2004) Psychophysical manifestations of compression: hearing-impaired listeners. In: Bacon SP, Fay RR, Popper AN (eds), Compression: From Cochlea to Cochlear Implants. New York: Springer, pp. 107–152.

Baer T, Moore BCJ (1997) Evaluation of a scheme to compensate for reduced frequency selectivity in hearing-impaired subjects. In: Jesteadt W (ed), Modeling Sensorineural Hearing Loss. Mahwah, NJ: Erlbaum, pp. 329–341.

Bauer CA, Brozoski TJ (2008) Tinnitus: theories, mechanisms and treatments. In: Schacht J, Popper AN, Fay RR (eds), Auditory Trauma, Protection and Repair. New York: Springer, pp. 101–130.

Biondi E (1978) Auditory processing of speech and its implications with respect to prosthetic rehabilitation. The bioengineering viewpoint. Audiology 17:43–50.

Biondi E, Schmid R (1972) Mathematical models and prostheses for sense organs. In: Mohler RR, Ruberti A (eds), Theory and Applications of Variable Structure Systems. London: Academic, pp. 183–211.

Bondy J, Becker S, Bruce IC, Trainor L, Haykin S (2004a) A novel signal-processing strategy for hearing-aid design: neurocompensation. Signal Process 84:1239–1253.

Bondy J, Bruce IC, Becker S, Haykin S (2004b) Predicting speech intelligibility from a population of neurons. In: Thrun S, Saul L, Scholkopf B (eds), NIPS 2003 Conference Proceedings: Advances in Neural Information Processing Systems, Vol. 16. Cambridge, MA: MIT Press, pp. 1409–1416.

Bruce IC (2004) Physiological assessment of contrast-enhancing frequency shaping and multiband compression in hearing aids. Physiol Meas 25:945–956.

Bruce IC, Zilany MSA (2007) Computational modelling of the cat auditory periphery: recent developments and future directions. In: Proceedings of 19th International Congress on Acoustics, Madrid, Spain, pp. PPA-07-004-IP: 001–006.

Bruce IC, Sachs MB, Young ED (2003) An auditory-periphery model of the effects of acoustic trauma on auditory nerve responses. J Acoust Soc Am 113:369–388.

Bruce IC, Dinath F, Zeyl TJ (2007) Insights into optimal phonemic compression from a computational model of the auditory periphery. In: Dau T, Buchholz J, Harte JM, Christiansen TU (eds), Auditory Signal Processing in Hearing-Impaired Listeners, International Symposium on Audiological and Auditory Research (ISAAR). Denmark: Danavox Jubilee Foundation, pp. 73–81.

Cai S, Ma WL, Young ED (2009) Encoding intensity in ventral cochlear nucleus following acoustic trauma: implications for loudness recruitment. J Assoc Res Otolaryngol 10:5–22.

Calandruccio L, Doherty KA, Carney LH, Kikkeri HN (2007) Perception of temporally processed speech by listeners with hearing impairment. Ear Hear 28:512–523.

Carney LH (1993) A model for the responses of low-frequency auditory-nerve fibers in cat. J Acoust Soc Am 93:401–417.

Carney LH (1994) Spatiotemporal encoding of sound level: models for normal encoding and recruitment of loudness. Hear Res 76:31–44.

Carney LH, McDuffy MJ, Shekhter I (1999) Frequency glides in the impulse responses of auditory-nerve fibers. J Acoust Soc Am 105:2384–2391.

Carney LH, Heinz MG, Evilsizer ME, Gilkey RH, Colburn HS (2002) Auditory phase opponency: a temporal model for masked detection at low frequencies. Acust Acta Acust 88:334–347.

Cedolin L, Delgutte B (2007) Spatio-temporal representation of the pitch of complex tones in the auditory nerve. In: Kollmeier B, Klump G, Hohmann V, Langemann U, M. Mauermann, Uppenkamp S, Verhey J (eds), Hearing: From Sensory Processing to Perception. Berlin: Springer, pp. 61–70.

Chen Z, Becker S, Bondy J, Bruce IC, Haykin S (2005) A novel model-based hearing compensation design using a gradient-free optimization method. Neural Comput 17:2648–2671.

Colburn HS (1973) Theory of binaural interaction based on auditory-nerve data. I. General strategy and preliminary results on interaural discrimination. J Acoust Soc Am 54:1458–1470.

Dallos P, Harris D (1978) Properties of auditory nerve responses in absence of outer hair cells. J Neurophysiol 41:365–383.

Delgutte B (1996) Physiological models for basic auditory percepts. In: Hawkins HL, McMullen TA, Popper AN, Fay RR (eds), Auditory Computation. New York: Springer, pp. 157–220.

Deng L, Geisler CD (1987) A composite auditory model for processing speech sounds. J Acoust Soc Am 82:2001–2012.

Dillon H (2001) Hearing Aids. New York: Thieme.

Edwards B (2002) Signal processing, hearing aid design, and the psychoacoustic Turing test. IEEE Proc Int Conf Acoust Speech Signal Proc 4:3996–3999.

Edwards B (2004) Hearing aids and hearing impairment. In: Greenberg S, Ainsworth WA, Popper AN, Fay RR (eds), Speech Processing in the Auditory System. New York: Springer, pp. 339–421.

Edwards B (2007) The future of hearing aid technology. Trends Amplif 11:31–45.

Franck BA, van Kreveld-Bos CS, Dreschler WA, Verschuure H (1999) Evaluation of spectral enhancement in hearing aids, combined with phonemic compression. J Acoust Soc Am 106:1452–1464.

Gagné JP (1988) Excess masking among listeners with a sensorineural hearing loss. J Acoust Soc Am 83:2311–2321.

Geisler CD (1989) The responses of models of "high-spontaneous" auditory-nerve fibers in a damaged cochlea to speech syllables in noise. J Acoust Soc Am 86:2192–2205.

Giguère C, Smoorenburg GF (1998) Computational modeling of outer hair cells damage: Implications for hearing aid and signal processing. In: Dau T, Hohmann V, Kollmeier B (eds), Psychophysics, Physiology and Models of Hearing. Singapore: World Scientific, pp. 155–164.

Giguère C, Woodland PC (1994a) A computational model of the auditory periphery for speech and hearing research. I. Ascending path. J Acoust Soc Am 95:331–342.

Giguère C, Woodland PC (1994b) A computational model of the auditory periphery for speech and hearing research .II. Descending paths. J Acoust Soc Am 95:343–349.

Giguère C, Bosman AJ, Smoorenburg GF (1997) Automatic speech recognition experiments with a model of normal and impaired peripheral hearing. Acust Acta Acust 83:1065–1076.

Goldstein JL (1995) Relations among compression, suppression, and combination tones in mechanical responses of the basilar membrane: data and MBPNL model. Hear Res 89:52–68.

Harrison RV, Evans EF (1979) Some aspects of temporal coding by single cochlear fibres from regions of cochlear hair cell degeneration in the guinea pig. Arch Otorhinolaryngol 224:71–78.

Heinz MG (2000) Quantifying the effects of the cochlear amplifier on temporal and average-rate information in the auditory nerve. PhD dissertation, Massachusetts Institute of Technology, Cambridge, MA.

Heinz MG (2007) Spatiotemporal encoding of vowels in noise studied with the responses of individual auditory nerve fibers. In: Kollmeier B, Klump G, Hohmann V, Langemann U, M. Mauermann, Uppenkamp S, Verhey J (eds), Hearing: From Sensory Processing to Perception. Berlin: Springer, pp. 107–115.

Heinz MG, Young ED (2004) Response growth with sound level in auditory-nerve fibers after noise-induced hearing loss. J Neurophysiol 91:784–795.

Heinz MG, Colburn HS, Carney LH (2001a) Evaluating auditory performance limits: I. One-parameter discrimination using a computational model for the auditory nerve. Neural Comput 13:2273–2316.
Heinz MG, Colburn HS, Carney LH (2001b) Rate and timing cues associated with the cochlear amplifier: level discrimination based on monaural cross-frequency coincidence detection. J Acoust Soc Am 110:2065–2084.
Heinz MG, Zhang X, Bruce IC, Carney LH (2001c) Auditory-nerve model for predicting performance limits of normal and impaired listeners. Acoust Res Lett Online 2:91–96.
Heinz MG, Colburn HS, Carney LH (2002) Quantifying the implications of nonlinear cochlear tuning for auditory-filter estimates. J Acoust Soc Am 111:996–1011.
Heinz MG, Issa JB, Young ED (2005a) Auditory-nerve rate responses are inconsistent with common hypotheses for the neural correlates of loudness recruitment. J Assoc Res Otolaryngol 6:91–105.
Heinz MG, Scepanovic D, Issa JB, Sachs MB, Young ED (2005b) Normal and impaired level encoding: effects of noise-induced hearing loss on auditory-nerve responses. In: Pressnitzer D, de Cheveigné A, McAdams S, Collet L (eds), Auditory Signal Processing: Physiology, Psychoacoustics and Models. New York: Springer, pp. 40–49.
Huettel LG, Collins LM (2003) A theoretical comparison of information transmission in the peripheral auditory system: normal and impaired frequency discrimination. Speech Commun 39:5–21.
Huettel LG, Collins LM (2004) A theoretical analysis of normal- and impaired-hearing intensity discrimination. IEEE Trans Speech Audio Process 12:323–333.
Johnson TA, Gorga MP, Neely ST, Oxenham AJ, Shera CA (2007) Relationships between otoacoustic and psychophysical measures of cochlear function. In: Manley GA, Fay RR, Popper AN (eds), Active Processes and Otoacoustic Emissions in Hearing. New York: Springer, pp. 395–420.
Joris PX, Van de Sande B, Louage DH, van der Heijden M (2006) Binaural and cochlear disparities. Proc Natl Acad Sci U S A 103:12917–12922.
Kates JM (1991a) Modeling normal and impaired hearing – implications for hearing-aid design. Ear Hear 12:S162–S176.
Kates JM (1991b) A time-domain digital cochlear model. IEEE Trans Signal Process 39:2573–2592.
Kates JM (1993) Toward a theory of optimal hearing-aid processing. J Rehabil Res Dev 30:39–48.
Kates JM (1997) Using a cochlear model to develop adaptive hearing-aid processing. In: Jesteadt W (ed) Modeling Sensorineural Hearing Loss. Mahwah, NJ: Erlbaum, pp. 79–92.
Kiang NYS (1990) Curious oddments of auditory-nerve studies. Hear Res 49:1–16.
Kiang NYS, Moxon EC (1974) Tails of tuning curves of auditory-nerve fibers. J Acoust Soc Am 55:620–630.
Launer S, Moore BCJ (2003) Use of a loudness model for hearing aid fitting. V. On-line gain control in a digital hearing aid. Int J Audiol 42:262–273.
Leijon A (1990) Hearing aid gain for loudness-density normalization in cochlear hearing losses with impaired frequency resolution. Ear Hear 12:242–250.
Levitt H (2004) Compression Amplification. In: Bacon SP, Popper AN, Fay RR (eds), Compression: From Cochlea to Cochlear Implants. New York: Springer, pp. 153–183.
Liberman MC (1984) Single-neuron labeling and chronic cochlear pathology. I. Threshold shift and characteristic-frequency shift. Hear Res 16:33–41.
Liberman MC, Dodds LW (1984a) Single-neuron labeling and chronic cochlear pathology. II. Stereocilia damage and alterations of spontaneous discharge rates. Hear Res 16:43–53.
Liberman MC, Dodds LW (1984b) Single-neuron labeling and chronic cochlear pathology. III. Stereocilia damage and alterations of threshold tuning curves. Hear Res 16:55–74.
Liberman MC, Kiang NYS (1984) Single-neuron labeling and chronic cochlear pathology. IV. Stereocilia damage and alterations in rate- and phase-level functions. Hear Res 16:75–90.
Liberman MC, Gao J, He DZ, Wu X, Jia S, Zuo J (2002) Prestin is required for electromotility of the outer hair cell and for the cochlear amplifier. Nature 419:300–304.

Loeb GE, White MW, Merzenich MM (1983) Spatial cross-correlation – a proposed mechanism for acoustic pitch perception. Biol Cybern 47:149–163.

Lopez-Poveda EA (2005) Spectral processing by the peripheral auditory system: facts and models. Int Rev Neurobiol 70:7–48.

Lopez-Poveda EA, Meddis R (2001) A human nonlinear cochlear filterbank. J Acoust Soc Am 110:3107–3118.

Lopez-Poveda EA, Plack CJ, Meddis R (2003) Cochlear nonlinearity between 500 and 8000 Hz in listeners with normal hearing. J Acoust Soc Am 113:951–960.

Lopez-Poveda EA, Johannesen PT, Merchán MA (2009) Estimation of the degree of inner and outer hair cell dysfunction from distortion product otoacoustic emission input/output functions. Audiol Med 7:22–28.

Lorenzi C, Gilbert G, Carn H, Garnier S, Moore BCJ (2006) Speech perception problems of the hearing impaired reflect inability to use temporal fine structure. Proc Natl Acad Sci U S A 103:18866–18869.

Meddis R, O'Mard LP, Lopez-Poveda EA (2001) A computational algorithm for computing nonlinear auditory frequency selectivity. J Acoust Soc Am 109:2852–2861.

Miller RL, Schilling JR, Franck KR, Young ED (1997) Effects of acoustic trauma on the representation of the vowel /ε/ in cat auditory nerve fibers. J Acoust Soc Am 101:3602–3616.

Miller RL, Calhoun BM, Young ED (1999) Contrast enhancement improves the representation of /ε/-like vowels in the hearing-impaired auditory nerve. J Acoust Soc Am 106:2693–2708.

Moore BCJ (1995) Perceptual Consequences of Cochlear Damage. New York: Oxford University Press.

Moore BCJ (2000) Use of a loudness model for hearing aid fitting. IV. Fitting hearing aids with multi-channel compression so as to restore 'normal' loudness for speech at different levels. Br J Audiol 34:165–177.

Moore BCJ (2004) Dead regions in the cochlea: conceptual foundations, diagnosis, and clinical applications. Ear Hear 25:98–116.

Moore BCJ, Glasberg BR (1997) A model of loudness perception applied to cochlear hearing loss. Aud Neurosci 3:289–311.

Moore BCJ, Glasberg BR (2004) A revised model of loudness perception applied to cochlear hearing loss. Hear Res 188:70–88.

Moore BCJ, Oxenham AJ (1998) Psychoacoustic consequences of compression in the peripheral auditory system. Psychol Rev 105:108–124.

Moore BCJ, Glasberg BR, Stone MA (1999) Use of a loudness model for hearing aid fitting: III. A general method for deriving initial fittings for hearing aids with multi-channel compression. Br J Audiol 33:241–258.

Oxenham AJ, Bacon SP (2003) Cochlear compression: perceptual measures and implications for normal and impaired hearing. Ear Hear 24:352–366.

Patuzzi R (1996) Cochlear micromechanics and macromechanics. In: Dallos P, Popper AN, Fay RR (eds), The Cochlea. New York: Springer, pp. 186–257.

Plack CJ, Drga V, Lopez-Poveda EA (2004) Inferred basilar-membrane response functions for listeners with mild to moderate sensorineural hearing loss. J Acoust Soc Am 115:1684–1695.

Robles L, Ruggero MA (2001) Mechanics of the mammalian cochlea. Physiol Rev 81:1305–1352.

Ruggero MA (1992) Physiology and coding of sound in the auditory nerve. In: Popper AN, Fay RR (eds), The Mammalian Auditory Pathway: Neurophysiology. New York: Springer, pp. 34–93.

Ruggero MA, Rich NC (1991) Furosemide alters organ of Corti mechanics: evidence for feedback of outer hair cells upon the basilar membrane. J Neurosci 11:1057–1067.

Ruggero MA, Rich NC, Shivapuja BG, Temchin AN (1996) Auditory-nerve responses to low-frequency tones: intensity dependence. Aud Neurosci 2:159–185.

Ruggero MA, Rich NC, Recio A, Narayan SS, Robles L (1997) Basilar-membrane responses to tones at the base of the chinchilla cochlea. J Acoust Soc Am 101:2151–2163.

Sachs MB, Kiang NY (1968) Two-tone inhibition in auditory-nerve fibers. J Acoust Soc Am 43:1120–1128.
Sachs MB, Bruce IC, Miller RL, Young ED (2002) Biological basis of hearing-aid design. Ann Biomed Eng 30:157–168.
Schaette R, Kempter R (2006) Development of tinnitus-related neuronal hyperactivity through homeostatic plasticity after hearing loss: a computational model. Eur J Neurosci 23:3124–3138.
Schmiedt RA, Zwislocki JJ, Hamernik RP (1980) Effects of hair cell lesions on responses of cochlear nerve fibers. I. Lesions, tuning curves, two-tone inhibition, and responses to trapezoidal-wave patterns. J Neurophysiol 43:1367–1389.
Schmiedt RA, Lang H, Okamura HO, Schulte BA (2002) Effects of furosemide applied chronically to the round window: a model of metabolic presbyacusis. J Neurosci 22:9643–9650.
Schoonhoven R, Keijzer J, Versnel H, Prijs VF (1994) A dual filter model describing single-fiber responses to clicks in the normal and noise-damaged cochlea. J Acoust Soc Am 95:2104–2121.
Sewell WF (1984a) The effects of furosemide on the endocochlear potential and auditory-nerve fiber tuning curves in cats. Hear Res 14:305–314.
Sewell WF (1984b) Furosemide selectively reduces one component in rate-level functions from auditory-nerve fibers. Hear Res 15:69–72.
Shamma SA (1985) Speech processing in the auditory system. I: The representation of speech sounds in the responses of the auditory nerve. J Acoust Soc Am 78:1612–1621.
Shamma SA, Shen NM, Gopalaswamy P (1989) Stereausis: binaural processing without neural delays. J Acoust Soc Am 86:989–1006.
Shi LF, Carney LH, Doherty KA (2006) Correction of the peripheral spatiotemporal response pattern: a potential new signal-processing strategy. J Speech Lang Hear Res 49:848–855.
Siebert WM (1970) Frequency discrimination in auditory system – place or periodicity mechanisms? Proc IEEE 58:723–730.
Tan Q, Carney LH (2003) A phenomenological model for the responses of auditory-nerve fibers. II. Nonlinear tuning with a frequency glide. J Acoust Soc Am 114:2007–2020.
Wang J, Powers NL, Hofstetter P, Trautwein P, Ding D, Salvi R (1997) Effects of selective inner hair cell loss on auditory nerve fiber threshold, tuning and spontaneous and driven discharge rate. Hear Res 107:67–82.
Wong JC, Miller RL, Calhoun BM, Sachs MB, Young ED (1998) Effects of high sound levels on responses to the vowel /ε/ in cat auditory nerve. Hear Res 123:61–77.
Woolf NK, Ryan AF, Bone RC (1981) Neural phase-locking properties in the absence of cochlear outer hair cells. Hear Res 4:335–346.
Zeng FG, Kong YY, Michalewski HJ, Starr A (2005) Perceptual consequences of disrupted auditory nerve activity. J Neurophysiol 93:3050–3063.
Zhang X, Heinz MG, Bruce IC, Carney LH (2001) A phenomenological model for the responses of auditory-nerve fibers: I. Nonlinear tuning with compression and suppression. J Acoust Soc Am 109:648–670.
Zheng J, Shen W, He DZ, Long KB, Madison LD, Dallos P (2000) Prestin is the motor protein of cochlear outer hair cells. Nature 405:149–155.
Zilany MSA, Bruce IC (2006) Modeling auditory-nerve responses for high sound pressure levels in the normal and impaired auditory periphery. J Acoust Soc Am 120:1446–1466.
Zilany MSA, Bruce IC (2007a) Predictions of speech intelligibility with a model of the normal and impaired auditory-periphery. In: Proceedings of 3rd International IEEE EMBS Conference on Neural Engineering. Piscataway, NJ: IEEE, pp. 481–485.
Zilany MSA, Bruce IC (2007b) Representation of the vowel /ε/ in normal and impaired auditory nerve fibers: model predictions of responses in cats. J Acoust Soc Am 122:402–417.

Chapter 8
Physiological Models of Auditory Scene Analysis

Guy J. Brown

8.1 Introduction

Human listeners are remarkably adept at perceiving speech and other sounds in unfavorable acoustic environments. Typically, the sound source of interest is contaminated by other acoustic sources, and listeners are therefore faced with the problem of unscrambling the mixture of sounds that arrives at their ears. Nonetheless, human listeners can segregate one voice from a mixture of many voices at a cocktail party, or follow a single melodic line in a performance of orchestral music. Much as the visual system must combine information about edges, colors and textures in order to identify perceptual wholes (e.g., a face or a table), so the auditory system must solve an analogous *auditory scene analysis* (ASA) problem in order to recover a perceptual description of a single sound source (Bregman 1990). Understanding how the ASA problem is solved at the physiological level is one of the greatest challenges of hearing science, and is one that lies at the core of the "systems" approach of this book.

The emerging field of computational auditory scene analysis (CASA) aims to develop machine systems that mimic the ability of human listeners to perceptually segregate acoustic mixtures (Wang and Brown 2006). However, most CASA systems are motivated by engineering applications (e.g., robust automatic speech recognition in noise), and take inspiration from psychophysical and physiological accounts of human ASA without slavishly adhering to them. The review in this chapter is therefore necessarily selective, and concerns only those computational models of ASA that are based on physiologically plausible mechanisms.

The next section briefly reviews the psychology of ASA, and then Sect. 8.3 considers the likely physiological basis of ASA in general terms, and its relationship with the wider issue of feature binding in the brain. Computer models of specific ASA

G.J. Brown (✉)
Speech and Hearing Research Group, Department of Computer Science,
University of Sheffield, Sheffield S1 4DP, UK
e-mail: g.brown@dcs.shef.ac.uk

R. Meddis et al. (eds.), *Computational Models of the Auditory System*,
Springer Handbook of Auditory Research 35, DOI 10.1007/978-1-4419-5934-8_8,

phenomena are reviewed in Sect. 8.4, which considers the perceptual grouping of simple tone sequences and speech sounds, binaural models and the role of auditory attention. Section 8.5 focuses on approaches for segregating speech from other sounds. The chapter concludes with a summary and identifies key areas for future research.

8.2 Auditory Scene Analysis

Conceptually, ASA may be regarded as a two-stage process in which *segmentation* is followed by *grouping* (Bregman 1990). In the segmentation stage, the auditory periphery decomposes the acoustic signal into its constituent frequency components. Subsequently, components that are likely to have arisen from the same environmental source are grouped to form a mental representation of the source, which is called a *stream*. Streams are formed both by grouping acoustic events that are separated in time (*sequential grouping*) and by grouping frequency components that overlap in time (*simultaneous grouping*). If the auditory scene is viewed as a time-frequency spectrogram, then sequential and simultaneous grouping correspond to horizontal and vertical organization, respectively.

One of the earliest laboratory demonstrations of sequential grouping in ASA was the "auditory streaming" phenomenon associated with sequences of pure tones (van Noorden 1975), which is illustrated in Fig. 8.1. The listener is played a

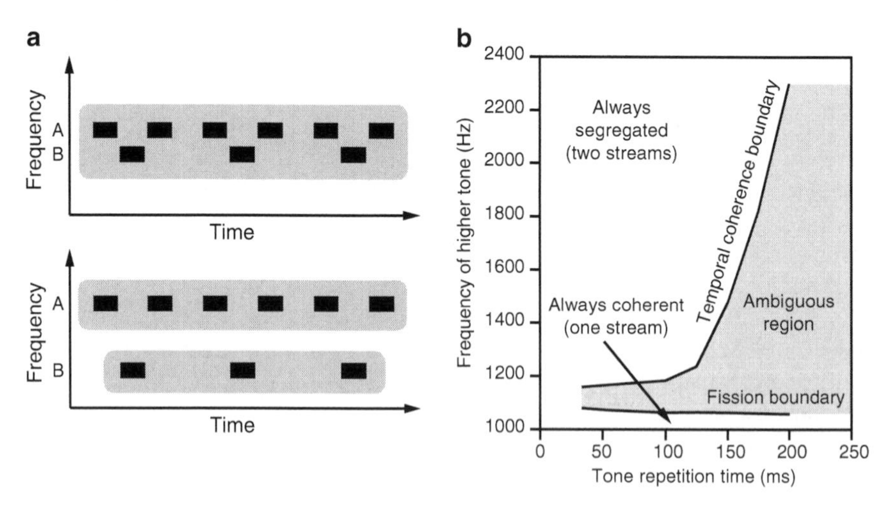

Fig. 8.1 An example of sequential integration (horizontal organization). (**a**) Schematic illustration of the perceptual grouping of tone sequences; streams are indicated by *gray rectangles*. When successive tones are close in frequency, the sequence binds into a single perceptual stream (*top*). When the frequency separation is large, high-frequency and low-frequency tones segregate into separate streams (*bottom*). (**b**) Data from McAdams and Bregman (1979) showing the dependence of auditory streaming on tone-repetition time and frequency separation

continuously repeating sequence of tones of the form ABA-ABA-..., where A and B denote tones of different frequency. If the time between tone onsets (the tone repetition time, TRT) is greater than about 150 ms, then the perceptual organization of the tones depends upon the frequency difference between them. When the frequency difference is small, listeners perceive a single stream in which the high and low tones alternate in a galloping rhythm. However, a large frequency difference causes this rhythm to be abolished, and instead the tones segregate into a rapid stream of high tones (A-A-A-) and a slower stream of low tones (B---B---). For intermediate frequency separations, listeners can switch between coherent (one stream) and segregated (two stream) organizations depending on their attentional focus. In addition, when two streams are heard they separate into figure and ground, such that the attended stream is perceived as louder than the unattended stream.

The phenomenon of auditory streaming can be understood in terms of *primitive* grouping processes, which are innate and exploit the intrinsic structure of environmental sounds (Bregman 1990). These can be characterized as heuristics that closely resemble those invoked by the Gestalt psychologists to explain the perception of visual patterns (Palmer 1999). Referring to Fig. 8.1, it is apparent that consecutive tones tend to be grouped into the same stream if they are close in time and frequency; this is analogous to the Gestalt principle of *proximity*. Other cues that favor sequential integration include smooth changes in fundamental frequency (F0), spatial location, spectrum and intensity. For example, Darwin and Bethell-Fox (1977) have shown that a smoothly varying F0 contour is needed to bind together a sequence of speech syllables; listeners interpret abrupt discontinuities in F0 as the onset of a new voice.

Primitive grouping processes are also implicated in simultaneous organization. A much-studied laboratory demonstration of this is the perceptual segregation of concurrent vowels ("double vowels"). Listeners are presented with a mixture of two (usually synthetic) vowel sounds that start and stop together, and are asked to identify both constituents of the mixture. When the vowels have the same F0, listeners are able to identify the vowels correctly on approximately 50% of trials. However, identification performance improves markedly when a small difference in F0 (ΔF0) is introduced between the two vowels, and performance improves with increasing ΔF0 up to a difference of four semitones (Fig. 8.2a).

One explanation for the perceptual segregation of double vowels is that simultaneous grouping mechanisms exploit harmonicity, as shown in Fig. 8.2b, c (other factors may also be at play, as discussed in Sect. 8.4.2). When the vowels have the same F0, their harmonics overlap and cannot be distinguished based on primitive grouping cues. However, when a ΔF0 is introduced, the harmonics of each vowel are interleaved and could be separated by a mechanism that groups regularly spaced frequency components. Such a process may be viewed as a "harmonic sieve" that has holes at integer multiples of a particular F0 (Scheffers 1983). Harmonics that are aligned with the holes in the sieve "fall through" and contribute to the vowel percept, whereas other frequency components are blocked.

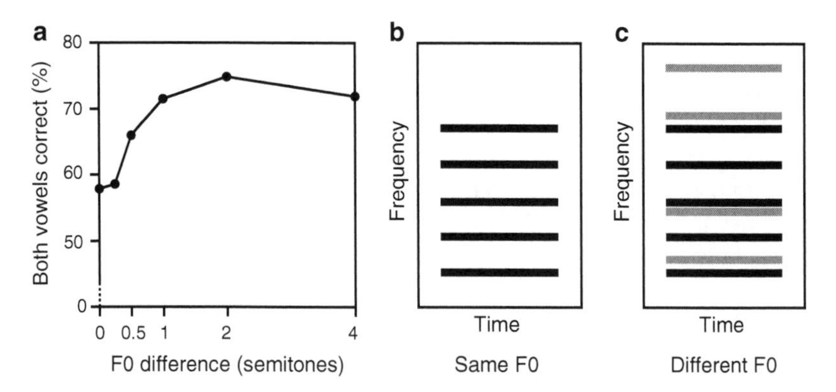

Fig. 8.2 An example of simultaneous grouping (vertical organization). (**a**) Data from the perceptual experiment by Assmann and Summerfield (1990), showing that identification of two concurrent vowels improves as the difference in F0 between them is increased. Chance performance level in this task was 6.7%. (**b**) Two vowel sounds that start and stop synchronously and have the same F0 cannot be segregated by primitive grouping cues. The figure shows a schematic spectrogram, in which a single harmonic series is present. (**c**) When a difference in F0 is introduced between the vowels, the two harmonic series are interleaved and can be segregated

Simultaneous grouping is also promoted by correlated changes in amplitude and frequency, as suggested by a Gestalt principle of *common fate*; listeners tend to perceptually group acoustic components that change proportionally and synchronously. For example, frequency components that appear at the same time (common onset) and disappear at the same time (common offset) tend to be perceptually fused, because they are likely to have originated from the same acoustic event. An *old-plus-new heuristic* may also contribute to the perceptual segregation of sounds that differ in their onset times (Bregman 1990). This heuristic states that the auditory system prefers to perceive the current sound as a continuation of a previous one; but if there is evidence against doing so, then the difference between the current sound and the previous one forms the basis for a new group. For instance, listeners are better able to segregate double vowels with the same F0 if the onset of one of the vowels is delayed relative to the other.

An interesting feature of Fig. 8.2a is that listener's performance in the double vowel identification task is above chance even when the two vowels are equally loud, have the same F0, and start and stop together. This suggests that even in the absence of primitive grouping cues, listeners are able to use their knowledge of speech sounds to identify the constituents of the vowel mixture. Bregman (1990) refers to the application of prior knowledge in auditory organization as *schema-driven grouping*. In terms of information-processing theory, primitive grouping may be regarded as a bottom-up (data-driven) process whereas schema-driven grouping is top-down (hypothesis-driven). A schema stores knowledge about familiar acoustic patterns (e.g., speech, music) and is able to select the acoustic features that it requires from the auditory scene. A further example of schema-based organization is the phenomenon of perceptual restoration (auditory induction), in which listeners perceptually restore parts of an acoustic signal that have been replaced with noise bursts (Warren 1970). In the case of speech, the phonemes restored depend on the surrounding context; in other words, language-specific schemas appear to select acoustic evidence from the signal in such a way that the restored section fits meaningfully into the utterance.

8.3 The Neural Basis of Auditory Scene Analysis

Bregman's account of ASA offers a coherent view of perceptual organization in hearing, and can be regarded as an auditory equivalent of Marr's computational theory of vision (Marr 1982). As such, ASA is a high-level explanation of the information processing problems that listeners must solve in order to perceptually segregate acoustic mixtures, and it makes little reference to the underlying physiological mechanisms. Gestalt grouping principles are most easily visualized in terms of schematic spectrograms, as shown in Figs. 8.1 and 8.2, but it is unlikely that the auditory system reasons with symbolic representations of sound in the same way.

8.3.1 Auditory Scene Analysis and the Binding Problem

What, then, is the physiological substrate of ASA? The initial frequency analysis of sound performed in the cochlea is usually modeled as a bank of bandpass filters with overlapping pass-bands, such that each filter channel models the frequency response associated with a particular position along the cochlear partition (Chapter 2). The decomposition of sound into frequency channels can be regarded as analogous to the segmentation stage of ASA. Furthermore, this model suggests that auditory figure–ground segregation could be achieved by a process of *channel selection*. At a conceptual level, channel selection involves a "gain control" on each frequency channel: figure–ground segregation is achieved either by increasing the gain on channels that correspond to the foreground stream, or by decreasing the gain on channels that correspond to the background stream. This general idea has formed the basis for a number of computational models of ASA (e.g., Lyon 1983; Beauvois and Meddis 1991; Brown and Cooke 1994).

The notion of channel selection requires some refinement before it can be justified in physiological terms. First, a "channel" is itself an abstraction; frequency-specific nerve fibres are arranged in an orderly (tonotopic) fashion throughout the auditory pathway, but their best frequencies vary continuously. It is therefore necessary to explain how a "gain control" might be implemented at the level of individual auditory neurons. Second, information about the characteristics of a sound source is encoded at many different levels of the auditory pathway (Chapters 3 and 6). A key question, then, is how the percept of a coherent auditory stream arises from neural activity that is distributed across many different regions of the auditory system.

This issue is really a specific case of what is known in neuroscience as the *binding problem*: how is information encoded in different areas of the brain bound together into a coherent whole? The binding problem is more general than ASA because it is inherently multimodal. However, the two are inextricably linked because auditory perception cannot be considered in isolation; for example, it is well known that visual cues can influence auditory sound localization (Wallach 1940) and speech perception (McGurk and MacDonald 1976).

8.3.2 Solutions to the Binding Problem

Solutions to the binding problem fall into two main categories; those based on *hierarchical coding* and *temporal correlation*. The hierarchical coding scheme posits a hierarchy of increasingly specialized cells that represent ever more specialized features of an object. Following this reasoning, one might expect cells at the top of the hierarchy to represent complete objects, such as a face; for this reason, hierarchical coding is also known as the "grandmother cell" or "cardinal cell" theory (Barlow 1972). Support for hierarchical coding has come from numerous physiological studies, such as the investigation of visual cortex by Hubel and Weisel (1962). Similarly, hierarchical organization is apparent in the auditory cortex (Chapter 5).

Nonetheless, it seems unlikely that the binding problem can be solved by hierarchical coding alone. A practical objection to hierarchical coding is that it apparently requires a prohibitive number of neurons in order to represent every possible combination of features that might be encountered. An alternative solution, suggested by Milner (1974) and substantially developed by von der Malsburg (1981), is based on the temporal correlation of neural responses. The temporal correlation theory proposes that neurons which code features of the same object are bound together by synchronization of their temporal fine structure; this allows features to be bound in a dynamic and flexible way. Von der Malsburg and Schneider (Von der Malsburg and Schneider, 1986) implemented this idea in a computational model of auditory grouping, which was able to segregate two (synthetic) auditory events based on their onset times. Their model employed neurons with oscillatory firing behavior, and introduced the idea of using the phase of neural oscillations to encode feature binding. In this scheme, features belonging to the same object are encoded by neural oscillators whose responses are synchronized (i.e., phase locked with zero phase lag), and are desynchronized from oscillators that encode different objects. The notion of oscillatory correlation as a solution to the binding problem has been further developed by Wang and co-workers (e.g., Wang and Terman 1995), and is supported by physiological studies that report coherent oscillations in the cortex.

8.3.3 An Illustrative Computer Model

To illustrate these ideas, a neural network of the form described by Wang and Terman (1995) is now described. The building block of the network is a neural oscillator, which is modeled as a feedback loop between an excitatory unit x and an inhibitory unit y, as shown in Fig. 8.3a. Formally, the dynamics of x and y are given by

$$\frac{dx}{dt} = 3x - x^3 + 2 - y + \rho + I + S \tag{8.1a}$$

$$\frac{dy}{dt} = \varepsilon(\gamma[1 + \tanh(x/\beta)] - y) \tag{8.1b}$$

where I is the external input to the oscillator; S represents input from other oscillators in the network; ρ is a Gaussian noise term; and ε, γ, and β are parameters. The parameter ε is chosen to be small ($\varepsilon \ll 1$).

It is instructive to consider the behavior of a single oscillator in the *phase plane*, as shown in Fig. 8.3b. Here, x and y are plotted on orthogonal axes so that the solution of (8.1) corresponds to a trajectory in the xy plane. The nullclines of the system, obtained by setting $dx/dt=0$ and $dy/dt=0$, convey important information about its behavior. The x-nullcline is a cubic curve, whereas the y-nullcline is a sigmoid. When $I \le 0$, the nullclines intersect at a stable fixed point on the left branch of the cubic, and the oscillator is inactive. However if $I>0$, the two nullclines intersect at a point on the middle branch of the cubic curve, and a stable oscillation (limit cycle) is obtained. This consists of periodic alternation between high values of x (the "active phase") and low values of x (the "silent phase"). Transitions between the active and silent phases occur rapidly, but within each phase the oscillator shows nearly steady-state behavior. The oscillator therefore

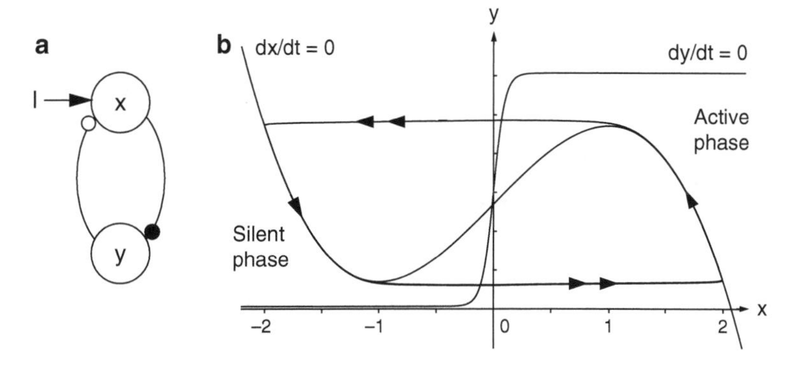

Fig. 8.3 (**a**) Schematic diagram of the Terman–Wang neural oscillator. The *small filled circle* indicates an excitatory connection; the *small open circle* indicates an inhibitory connection. (**b**) Phase plane behavior of the Terman–Wang oscillator for the case $I>0$, which leads to a periodic orbit (*thick line*) (Adapted from Wang and Brown 1999. Copyright © IEEE.)

exhibits two time scales, which is a characteristic of so-called *relaxation oscillators* (van der Pol 1926). When x is plotted over time, as shown in the lower panels of Fig. 8.4, it can be seen that the trace resembles a periodic sequence of action potentials. Indeed, (8.1) may be interpreted as a model of a single neuron in which x corresponds to the membrane potential and y corresponds to the activation state of ion channels. It is therefore closely related to the Fitzhugh–Nagumo equations for action potential generation (Fitzhugh 1961; Nagumo et al. 1962). Alternatively, (8.1) may be viewed as a mean-field approximation to the activity of a group of interconnected excitatory and inhibitory neurons.

The network properties of the Terman–Wang oscillator are illustrated in Fig. 8.4, which shows a small *locally excitatory globally inhibitory network* (LEGION) consisting of four oscillators, $x_1 \ldots x_4$, and a "global inhibitor," z. Oscillators i and j are linked by local excitatory connections which have a corresponding weight w_{ij}. The global inhibitor z receives excitation from every oscillator, and in turn it inhibits each oscillator. The dynamics are arranged such that when one or more oscillators move to their active phase, the global inhibitor is activated and it inhibits all of the oscillators in the network (see Wang and Terman 1995 for details). In the absence of local excitatory connections, the global inhibitor prevents more than

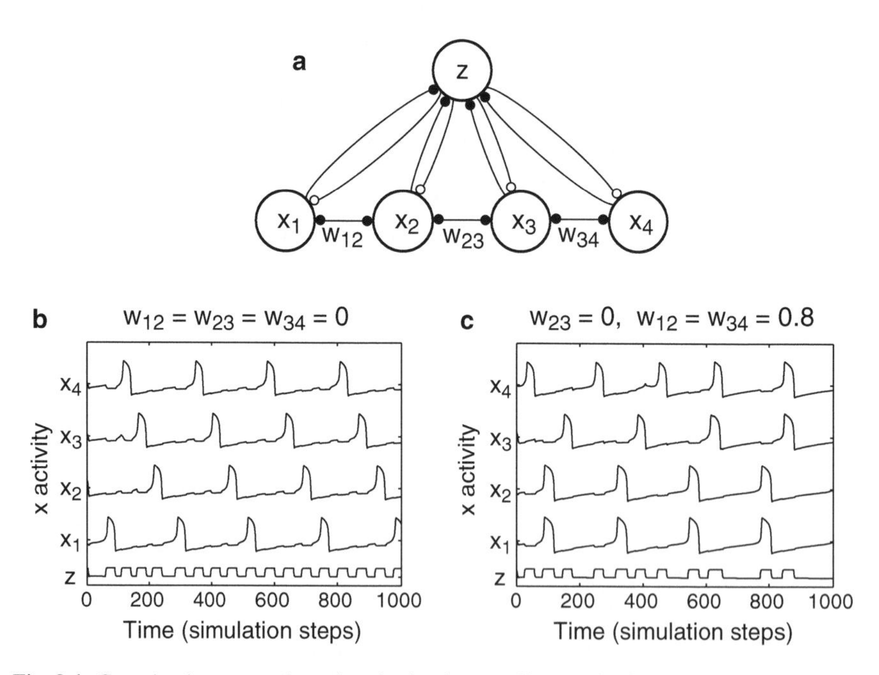

Fig. 8.4 Grouping by temporal synchronization in a small network of Terman–Wang oscillators. (**a**) Network topology, showing connections between the oscillators x_i and global inhibitor z. *Small open circles* indicate inhibitory connections; *small filled circles* indicate excitatory connections. (**b**) In the absence of local excitatory connections (all w_{ij} set to 0) all oscillators are desychronized. (**c**) Local excitatory connections between x_1 and x_2, and between x_3 and x_4, cause two groups of synchronized oscillators to be formed

one oscillator from "jumping" to the active phase at the same time; as a result, each oscillator is desynchronized from the others (also, the Gaussian noise ρ in (8.1) serves to desynchronize oscillators that might otherwise become synchronized by chance). This behavior is shown in Fig. 8.4b. However, local excitatory connections can mitigate the effects of global inhibition and allow groups of synchronized oscillators to form; the global inhibitor then serves to desynchronize one group from another. This is shown in Fig. 8.4c, where oscillators x_1 and x_2 have an excitatory connection, as do oscillators x_3 and x_4. After a couple of cycles, two groups of synchronized oscillators form, but the groups are desynchronized from one another. Figure 8.4 therefore demonstrates how, at the neural level, groups of features might be bound together based on the temporal synchronization of their responses. Computational models of auditory grouping based on this concept are reviewed in Sects. 8.4 and 8.5.

8.3.4 Physiological Studies of Auditory Scene Analysis

Relatively few physiological studies have directly addressed the neural basis of ASA. However, physiological correlates of ASA have been identified in event-related brain potentials (ERPs) by Alain et al. (2002) using a harmonic mistuning paradigm. Listeners were presented with a harmonic complex in which one component was mistuned so that it was no longer an integer multiple of the F0, and the ERP was simultaneously recorded. When the harmonic was substantially mistuned, listeners reported hearing two sounds rather than one, and a biphasic negative–positive potential appeared in the ERP shortly after the onset of the stimulus. The negative part of this wave (referred to as the object-related negativity, ORN) was present when listeners were actively attending to stimulus, and during passive listening. On the other hand, the later positive part of the wave was apparent only during active listening. Alain et al. suggest that the ORN is correlated with primitive ASA and the late positive wave indicates top-down processing; their findings can therefore be seen as supporting the distinction made by Bregman (1990) between top-down and bottom-up mechanisms. Subsequent work by McDonald and Alain (2005) has shown that the ORN is also correlated with grouping by spatial location.

The role of attention in sequential grouping has also been studied using ERP recordings. Alain and Izenburg (2003) show that sequential grouping processes are more affected by auditory attentional load than simultaneous grouping processes, as indexed by the mismatch negativity (MMN) wave of the ERP. Sussman et al. (1999) also show that the MMN is correlated with sequential grouping. They presented alternating tone sequences to listeners and found that the MMN did not occur when the TRT was long (i.e., when listeners heard a single stream). However, when the TRT was short, listeners heard two perceptual streams and a MMN was apparent in the ERP recording. Since the MMN is regarded as an index of preattentive acoustic processing, Sussman et al. interpret their findings as evidence that auditory streaming is a preattentive phenomenon. However, the role of attention in auditory streaming remains a controversial issue, as further discussed in Sect. 8.4.4.

Physiological evidence for the role of temporal correlation in feature binding has come from electroencephalogram (EEG) studies that report coherent oscillations in the "gamma" range of 30–60 Hz (see also Chapter 5). Originally reported in the visual system (e.g., Gray et al. 1989), gamma oscillations have also been identified throughout the auditory cortex and thalamus. Using magnetic field tomography (MFT), Ribary et al. (1991) find that synchronized 40 Hz thalamocortical oscillations accompany auditory processing and occur independently of stimulus parameters. Barth and MacDonald (1996) present evidence that gamma oscillations are generated within the auditory cortex but modulated by the acoustic thalamus, and that they serve to synchronize the activity of primary and secondary cortical areas. Brosch et al. (2002) used a multielectrode system to study gamma oscillations in the primary and caudomedial auditory cortex of the macaque monkey. They find that oscillations are correlated with the match between the frequency of a tonal stimulus and the preferred frequency of cortical units. Furthermore, synchronization of gamma oscillations at different recording sites was stronger when the preferred frequencies of the two recorded units matched. The authors conclude that gamma oscillations may provide a means of binding the responses of auditory neurons according to the similarity of their receptive fields. Physiological studies in the monkey also suggest that synchronized neural oscillations underlie the integration of auditory cortical responses with visual inputs (Ghazanfar et al. 2008).

8.4 Computer Models of Auditory Grouping Phenomena

Physiologically motivated computer models of ASA usually focus on specific laboratory demonstrations of auditory grouping. The next two sections review models that concern the auditory streaming of tone sequences and double vowel segregation, respectively. The role of binaural cues and attentional factors in physiological models of ASA are then considered.

8.4.1 Streaming of Alternating Tone Sequences

Perceptual streaming of tone sequences was introduced in Sect. 8.2, and a number of computational models of this phenomenon are now considered. Such models face two challenges. The first is to explain the dependence of the streaming percept on the frequency separation of the tones and the tone repetition time (TRT). Van Noorden (1975) identified two perceptual boundaries that are related to these parameters, as shown in Fig. 8.1b. Above the *temporal coherence boundary*, the A and B tones always segregate into different perceptual streams. Below the *fission boundary*, the sequence of tones is always heard as a coherent whole. Between these two boundaries, there is an *ambiguous region* such that listeners may hear streaming or coherence; spontaneous shifts between streaming and coherence may occur, and listeners can switch between the two percepts by means of conscious attentional effort. A second challenge for computational models is to explain the time course of the streaming percept. Alternating tone sequences are initially heard as coherent, and streaming builds up over several cycles of the stimulus at a rate that depends on the TRT (Anstis and Saida 1985).

As noted in Sect. 8.2, streaming of tone sequences can be explained in terms of primitive sequential integration processes that group successive tones according to their proximity in time and frequency (Bregman 1990). However, this explanation is phrased in terms of heuristics that operate on a symbolic pattern: what is the corresponding mechanism at the physiological level? Beauvois and Meddis (1991, 1996) describe a physiological model of auditory streaming that is based on two key ideas. First, streaming is presumed to arise from selective accentuation of peripheral frequency channels (this is a specific form of the channel selection theory introduced in Sect. 8.3.1). Second, they propose that spontaneous shifts between foreground and background organization are underlain by the stochastic nature of auditory nerve firing. Auditory grouping is therefore assumed to arise from simple circuits at low (subcortical) levels of the auditory pathway.

The first stage in the Beauvois and Meddis model is a simulation of auditory peripheral processing. This consists of a number of parallel frequency channels, each of which is modeled by an ERB filter and a model of neuromechanical transduction by inner hair cells. The output of each peripheral channel is then routed via two pathways, as shown in Fig. 8.5. The *temporal-integration path* smoothes auditory nerve firing patterns by passing them through a leaky integrator with a

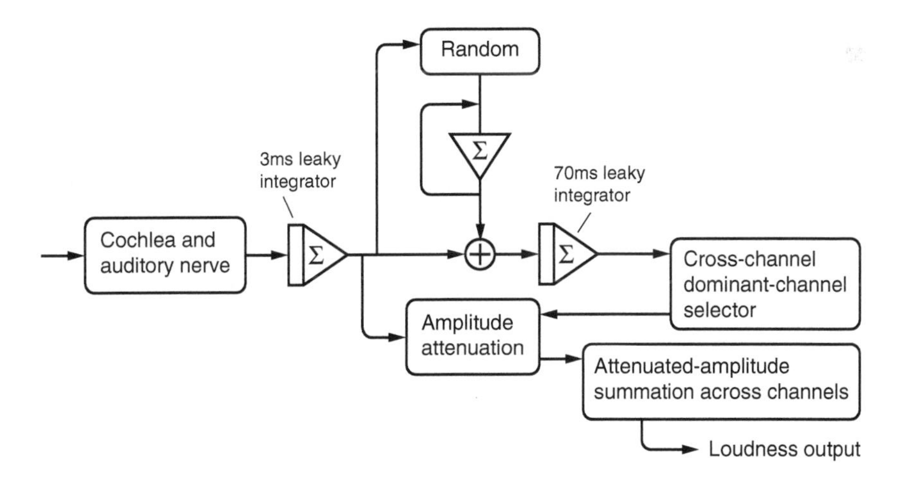

Fig. 8.5 Schematic diagram showing one channel of the Beauvois and Meddis model of auditory streaming (Adapted from Beauvois and Meddis 1996, with permission. Copyright 1996, Acoustical Society of America.)

short time constant (3 ms). The output of the temporal-integration path is then fed to an *excitation-level path*, which adds a cumulative random bias and performs further smoothing with a longer time constant (70 ms). The behavior of the model can be illustrated by considering three peripheral channels, with best frequencies corresponding to the low-frequency tone, the high-frequency tone, and midway between the two. At each time step in the simulation, the outputs of the excitation-level paths for the three channels are compared, and the one with the highest activity is defined as the "dominant" (foreground) channel. The activity in the temporal-integration path is then attenuated by a factor of 0.5, for all channels except the dominant channel (i.e., the nondominant channels are relegated to the "background"). Stream segregation is assessed based on the relative amplitude levels of the high and low tones in the output of the model. If the two tones elicit a comparable level of activity then perceptual coherence is reported; otherwise streaming is assumed to occur.

As a result of the cumulative random bias, each frequency channel exhibits a different random walk. This causes the dominant channel to fluctuate between the high and low tones, causing spontaneous changes in the foreground and background. The random bias also underlies the ability of the model to simulate the buildup of streaming over time. Initially, the high and low channels have a similar level of activity (coherence), but over time the random bias accumulates and causes an appreciable difference in level (streaming). The model is also sensitive to the TRT, because the random bias is made proportional to the amount of activity in each channel. When the temporal gaps between successive tones are long, the random bias tends to dissipate and hence slow rates of presentation yield perceptual coherence. At faster presentation rates the random bias accumulates rapidly, leading to a substantial difference in the activity of the high and low channels, and hence a streaming percept. Overall, the model provides a close fit to the fission boundary and temporal coherence boundary reported by Van Noorden (1975).

The Beauvois and Meddis model focuses specifically on two-tone streaming, and hence considers competitive interactions between a limited number of frequency channels. A more general model has been proposed by McCabe and Denham (1997), in which competition occurs between tonotopically arranged arrays of "foreground" and "background" neurons. The two arrays are reciprocally connected by inhibitory inputs. As a result, activity at a particular frequency in one array suppresses activity at the same frequency in the other array. The strength of inhibition is determined by the frequency proximity between inputs to the network, and each unit also receives a recurrent inhibitory input that is related to the reciprocal of its own activity. The latter form of inhibition serves to suppress the response of the network to differences in subsequent incoming stimuli, and may therefore be regarded as a neural implementation of Bregman's "old plus new" heuristic (see Sect. 8.2). Like the Beauvois and Meddis model, McCabe and Denham's scheme is able to account for the influence of frequency proximity and TRT on the streaming of tone sequences, and reproduces the buildup of streaming over time. In addition, the interacting neural arrays in their model allow it to simulate auditory grouping phenomena in which background organization influences foreground organization. For example, it correctly predicts the effect of unattended "captor tones" on listeners' ability to judge the order of successive high and low frequency tones, as in the study of Bregman and Rudnicky (1975). The attentional factors involved in this experiment, and computer models that can account for them, are further discussed in Sect. 8.4.4.

Both the Beauvois and Meddis (1991, 1996) and McCabe and Denham (1997) models are channel selection schemes in which the "gain" on each channel determines the perceptual foreground and background. As noted in Sect. 8.3.2, temporal correlation of neural responses offers a more general means of implementing channel selection, and models of auditory streaming based on this idea have been proposed by Wang (1996), Baird (1997) and Brown and Cooke (1998). Wang's model is based on a LEGION neural architecture similar to that shown in Fig. 8.4, but the array of neural oscillators is two-dimensional with axes corresponding to time and frequency. It is assumed that the time axis is obtained via a system of neural delay lines. Each oscillator receives excitatory input from its neighbors, with connection strengths set according to a Gaussian function in time and frequency. As a result, the tendency of an oscillator to synchronize with another diminishes with increasing distance between them in the neural array. This arrangement implements grouping by temporal and frequency proximity in Wang's model. Oscillators that receive an input and are close in time and/or frequency tend to synchronize, due to the influence of local excitatory connections. Oscillators that are distant in time and/or frequency do not receive strong mutual excitation, and are segregated into different streams due to the influence of a global inhibitor. It should be noted that Wang's model is somewhat abstract, because input is provided to the oscillator network in the form of an idealized binary pattern rather than a sampled acoustic signal. Nonetheless, it demonstrates how streaming of tone sequences can arise in an oscillator network, and accounts for other grouping phenomena such as the grouping of frequency-modulated tones and competition between competing alternative organizations.

Brown and Cooke's (1998) oscillator-based model of auditory streaming is similar in concept to that of Wang, but differs in a number of important respects. First, their model takes a sampled acoustic signal as input, and includes a model of peripheral auditory processing prior to the oscillator network. Second, grouping is signaled by a one-dimensional array of neural oscillators (one per frequency channel) rather than a two-dimensional time/frequency grid. Each channel has an associated onset detector, and a simple learning rule reduces the coupling between oscillators whose onset detectors do not exhibit the same level of activity at the same time. Initially, all oscillators receive strong mutual excitation and are therefore synchronized, indicating a default state of perceptual coherence (Bregman 1990). When a stimulus is presented, coupling is reduced at a rate that depends on the frequency and temporal separation between successive tones. This is because differences in onset level are greater for tones that are distant in frequency (due to the bandpass nature of auditory filters), and coupling declines at a rate that is proportional to the number of tone onsets that occur. Oscillators representing the high- and low-frequency tones therefore tend to desynchronize (segregate into different streams) when the frequency separation between the tones is large and the TRT is short.

Brown and Cooke's model also reproduces the buildup of streaming over time, because coupling between oscillators diminishes over a time course of several seconds. In this respect it has an advantage over Wang's model; LEGION networks require no more than *N* oscillator cycles to segregate *N* objects in the input (Terman and Wang 1995), and hence segregation proceeds very rapidly in Wang's model rather than building up in a TRT-dependent fashion. On the other hand, Brown and Cooke's model has a number of limitations compared to Wang's approach. Most significantly, there is no time axis in Brown and Cooke's model, so it cannot express different grouping relationships at different times within the same frequency channel. Wang's two-dimensional time-frequency grid of oscillators is a more flexible representation in this regard.

Further extensions of Wang's model have been proposed by Norris (2003) and Chang (2004). Norris introduces a metric for measuring the synchrony between oscillators in Wang's network, and uses this to quantify its match with human performance. He reports that the temporal coherence boundary and fission boundary for the model have plausible shapes, but do not agree closely with human data. Norris also extends Wang's model to include grouping by onset synchrony, using a scheme in which onsets are detected by a parallel process and passed to the oscillator network as additional inputs. Oscillators that receive inputs from the onset detectors at approximately the same time become more tightly coupled. Norris shows that the modified model is sensitive to the onsets of acoustic events, but it fails to reproduce listener's behavior when onset cues are placed in competition with frequency and temporal proximity. Chang (2004) also describes a modified version of Wang's model, in which coupling weights vary according to the TRT. His model provides a good match with the fission boundary and temporal coherence boundary for human listeners. Chang also introduces a random element into the activity of the global inhibitor in Wang's model, thus enabling it to exhibit spontaneous shifts between streaming and perceptual coherence.

The channel selection schemes reviewed in the preceding text fall into two classes: those in which grouping is signaled via a spatial distribution of activity in one or more neural arrays (e.g., Beauvois and Meddis), and those in which grouping is represented by a temporal code (e.g., Wang). However, these two approaches need not be mutually exclusive. An interesting middle ground is suggested by Todd (1996), who describes a model of auditory streaming based on a spatial representation of the temporal pattern of the acoustic stimulus. Within his model, acoustic events are grouped if they have a highly correlated amplitude modulation (AM) spectrum, whereas those with uncorrelated AM spectra are segregated. Underlying his model is a neural representation of the AM spectrum, which is presumed to be computed by a cortical map of periodicity-sensitive cells. Although his model is functional, Todd identifies specific cortical circuits that might compute and correlate AM spectra. Although he does not present a quantitative evaluation of his model, it agrees qualitatively with the effects of frequency proximity and TRT on the streaming of tone sequences.

8.4.2 Perceptual Segregation of Double Vowels

The perceptual segregation of double vowels was introduced in Sect. 8.2 as a well-studied laboratory demonstration of simultaneous auditory organization. Recall that when two concurrent vowels have the same F0, the ability of listeners to identify them is mediocre (but above chance). Identification performance improves substantially when a difference in F0 (ΔF0) is introduced between the vowels, as shown in Fig. 8.2a.

This phenomenon has proven to be a popular topic for computational modeling studies, most of which assume that the effect is underlain by a mechanism that groups harmonically related acoustic components (recall Fig. 8.2c). The model of Meddis and Hewitt (1992) has been particularly successful at modeling the perceptual data. It assumes that simulated auditory nerve firing patterns, derived from a peripheral model based on the gammatone filterbank and Meddis hair cell model (Meddis and Hewitt 1991), are subjected to a running autocorrelation of the form:

$$r(t,f,\tau) = \sum_{k=1}^{\infty} x(t-T,f)\, x(t-T-\tau,f)\, \exp(-T / \Omega) \qquad (8.2a)$$

$$T = kdt \qquad (8.2b)$$

Here, $x(t,f)$ is the probability of firing in peripheral channel f at time t, τ is the autocorrelation lag, and dt is the sample period. The summation over time is limited by a time constant Ω, which is typically chosen to be in the range 10–25 ms. The resulting representation is termed the *correlogram*, and has been used in a number of other computational models of auditory pitch analysis and F0-based sound segregation (e.g., Slaney and Lyon 1990; Meddis and Hewitt 1991; Brown and Cooke 1994). As shown in Fig. 8.6, the correlogram of a periodic sound shows a characteristic spine at the fundamental period. This pitch-related structure can be emphasized by pooling the channels of the correlogram to give a *summary correlogram*:

$$s(t,\tau) = \sum_{f=1}^{N} r(t,f,\tau) \qquad (8.3)$$

where N is the number of peripheral channels. Meddis and Hewitt divide the summary correlogram into two regions; activity at relatively long lags (4.5 ms or greater) is regarded as the *pitch region*, since peaks in this region indicate the presence of a fundamental period. For lags of less than 4.5 ms, activity is related to the high-frequency structure of the sound and is therefore regarded as the *timbre region*. In the Meddis and Hewitt model, the pitch region is used to segregate the vowels and the timbre region is used to identify them. First, the largest peak in the summary correlogram is detected, and it is assumed that this corresponds to the fundamental period of one of the two vowels. The channels of the correlogram are then partitioned into two disjoint sets. Channels that have a peak at the detected fundamental period are allocated to the first vowel, and the remaining channels are

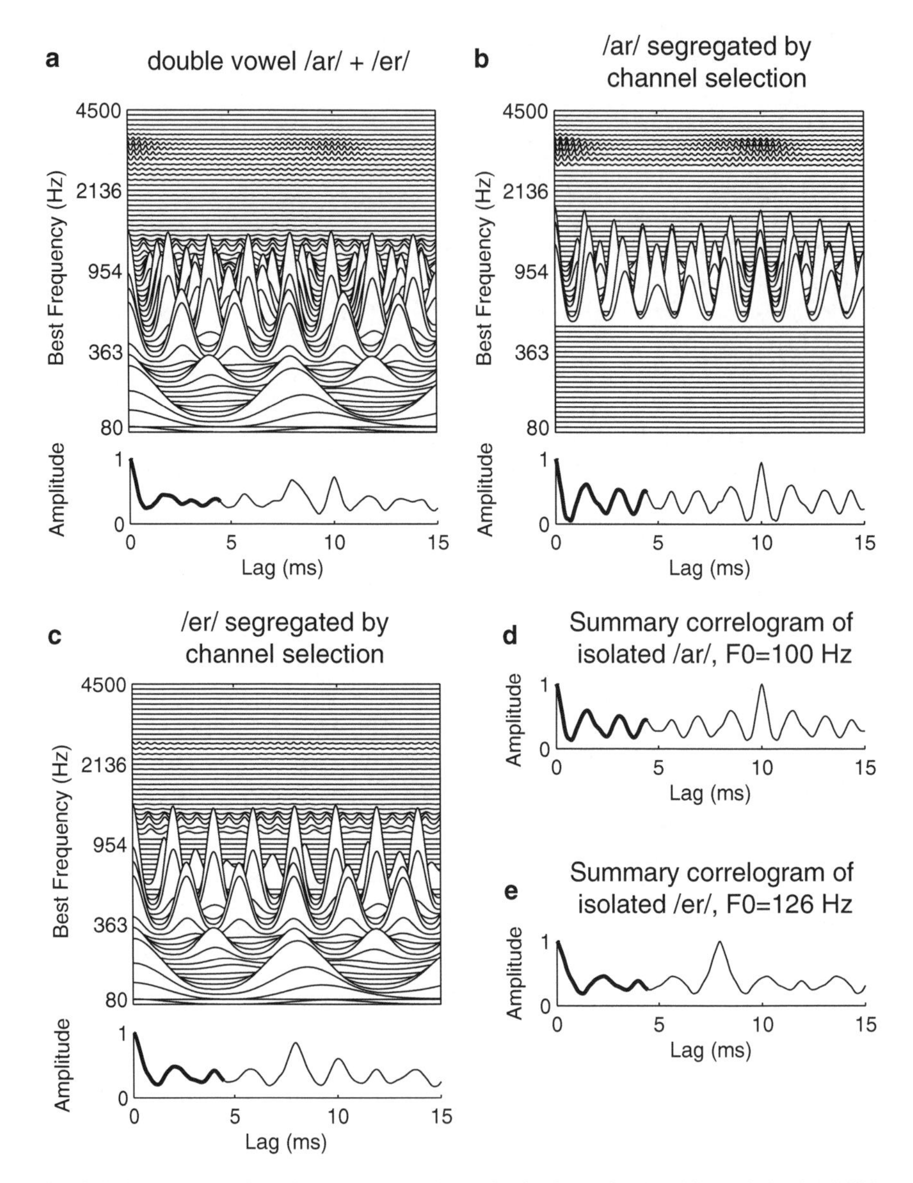

Fig. 8.6 Segregation of double vowels by channel selection, after Meddis and Hewitt (1992). (**a**) Correlogram (*upper panel*) and summary correlogram (*lower panel*) for a mixture of two vowels, /ar/ (F0 = 100 Hz) and /er/ (F0 = 126 Hz). The "timbre region" of the summary correlogram is shown as a *thick line*. (**b**) Channels that have a peak at a lag of 10 ms correspond to the /ar/ vowel. (**c**) The remaining channels correspond to the /er/ vowel. (**d**) Summary correlogram of the vowel /ar/, presented alone. (**e**) Summary correlogram of the vowel /er/, presented alone. Note that the timbre regions of the segregated vowels (**b**, **c**) bear a close resemblance to those of the uncontaminated vowels (**d**, **e**)

allocated to the second vowel. A template-matching procedure is then used to identify the vowels, in which the Euclidean distance is computed between the timbre regions of the segregated vowels and "template" timbre regions derived from isolated vowels. The template with the smallest distance indicates the vowel identity.

This model accounts for the role of F0 in double vowel segregation as follows. When there is no ΔF0 between the two vowels, a single peak occurs in the summary correlogram and the vowel identity must be judged from the timbre region of the mixture, giving mediocre performance. When a ΔF0 is introduced, the pitch periods of the two vowels become separated in the summary correlogram, allowing the channels of the correlogram to be partitioned into two sets. Estimation of the pitch period of one of the vowels becomes more accurate as the ΔF0 increases, and hence the match between the timbre regions of the segregated vowels and their respective templates also improves with increasing ΔF0. However, the pitch periods of the two vowels are resolved as distinct peaks in the summary correlogram by a ΔF0 of two semitones, so identification performance plateaus with further increases in ΔF0.

The Meddis and Hewitt channel selection scheme is illustrated in Fig. 8.6 for a mixture of two synthetic vowels, /ar/ and /er/. The vowels have F0s of 100 and 126 Hz respectively, giving a ΔF0 of four semitones. The largest peak in the summary correlogram of the double vowel (Fig. 8.6a) occurs at 10 ms, which corresponds to the period of /ar/. Figure 8.6b shows the set of correlogram channels that have a peak at this 10-ms period; other channels have been set to zero ("turned off"). Note that the timbre region of the corresponding summary correlogram bears a very close resemblance to that of an isolated /ar/ vowel, as shown in Fig. 8.6d. In Fig. 8.6c, channels have been selected that do *not* have a peak at a period of 10 ms, and therefore constitute the vowel /er/. Again, the timbre region of the segregated /er/ vowel is an excellent match to that of the isolated /er/ vowel, shown in Fig. 8.6e. Also note that the summary correlogram of the segregated /er/ vowel contains a prominent peak at the period of its F0 (about 7.9 ms).

This approach to modeling the segregation and identification of double vowels has a number of interesting features. A key issue with channel selection is that the spectral representation of each vowel contains "holes" in regions where channels have been omitted. One solution to this problem is to recognize the incomplete spectral pattern using missing data techniques (Cooke et al. 2001; see also Sect. 8.5). Meddis and Hewitt use an alternative approach, in which a temporal representation of the vowel spectrum (the timbre region) is used for vowel identification; even when the spectrum contains "holes," the timbre region does not. A second notable feature of their model is that it only estimates the pitch period of one of the two vowels. Models of double vowel segregation that attempt to identify both F0s have proven to be less successful in reproducing the trend of increasing vowel identification performance with increasing ΔF0 (e.g., Assmann and Summerfield 1990).

How might Meddis and Hewitt's model be implemented in physiological terms? The model extracts F0 information by a running autocorrelation, which could be implemented by a systematic arrangement of neural delay lines and coincidence detectors. However, there is no evidence for such a structure in the auditory brainstem. Nonetheless, Meddis and O'Mard (2006) have shown that periodicity detection can be implemented by a detailed physiological model in which the delay-and-multiply operation of autocorrelation is replaced by a cascade of ventral cochlear nucleus (VCN) chopper units and inferior colliculus (IC) cells. In this scheme, pitch analysis arises from the tendency of VCN chopper cells to synchronize with stimuli that match their intrinsic firing rate. It is also necessary to explain how channel selection is implemented at the physiological level. Meddis and Hewitt (1992) propose that their model could be underlain by neural circuits in which frequency-selective channels inhibit each other if they are dominated by different periodicities. For example, inhibitory mechanisms could suppress the response to the "foreground" vowel in order to get an unobstructed view of the "background" vowel, an approach that is similar in concept to that of McCabe and Denham (1997; see Sect. 8.4.1). Alternatively, Brown and Wang (1997) have suggested that inhibitory circuits could serve to desynchronize the neural activity associated with the two vowels. They describe a computational model in which Meddis and Hewitt's scheme is implemented within a neural oscillator architecture, similar to that shown in Fig. 8.4. In their model, periodicity analysis is performed via a correlogram, but each frequency channel has an associated neural oscillator that encodes grouping with other channels. Oscillators are desynchronized by a global inhibitor if their corresponding frequency channels are not dominated by the same F0, so that each vowel is represented by a synchronized block of oscillators.

A novel feature of Brown and Wang's neural oscillator model of double vowel segregation is that it allows a single channel to be allocated to *both* vowels. In Meddis and Hewitt's model, a *principle of exclusive allocation* is enforced (Bregman 1990), which ensures that each frequency channel is assigned to either one vowel or the other. However, a single channel may be excited by both vowels in cases where harmonics of the two F0s overlap. For example, in Fig. 8.6 the fourth harmonic of 126 Hz and fifth harmonic of 100 Hz are both close to 500 Hz, so the energy in the 500 Hz channel might be shared between the two vowels. Brown and Wang's model cannot apportion the energy within a single channel, but it allows an oscillator associated with a shared channel to synchronize with two groups by means of a slow inhibition mechanism. However, this causes all of the energy in a shared channel to be allocated to both vowels, and the shared-allocation version of Brown and Wang's model does not match listeners' performance as closely as one in which exclusive allocation is used.

The neural cancellation model of de Cheveigné (1997) has been more successful in modelling processes that partition the energy within a single frequency channel. de Cheveigné et al. (1997) note that the Meddis and Hewitt scheme fails to model human performance in double vowel experiments when the amplitude of one vowel is 10 or 20 dB less than the other. For certain vowel pairs, this discrepancy in amplitude causes *all* channels to be dominated by the more intense vowel. The Meddis and Hewitt model therefore allocates all channels to the dominant vowel, and predicts no improvement in vowel identification performance when a ΔF0 is introduced; however, a strong effect of ΔF0 was found in the perceptual experiment.

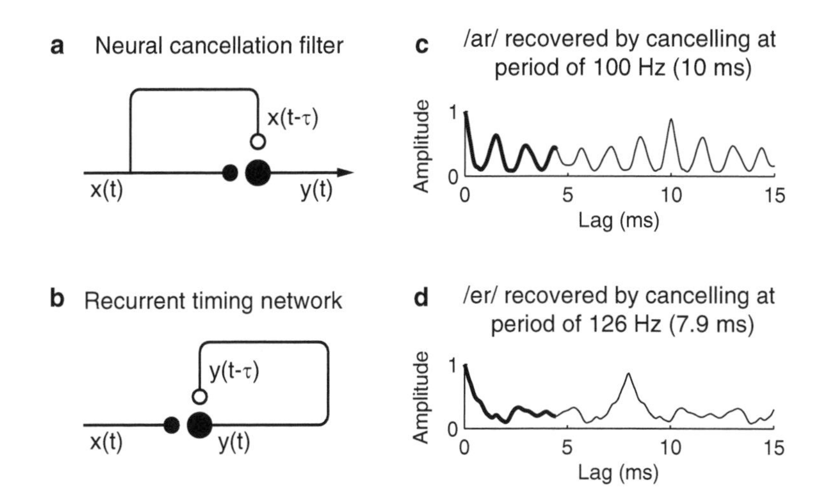

Fig. 8.7 (**a**, **b**) de Cheveigné's neural cancellation filter (in which inhibition arises from the delayed input) and Cariani's recurrent timing network (in which inhibition arises from the delayed output). *Small closed circles* indicate an excitatory synapse, and *small open circles* indicate an inhibitory synapse. (**c**, **d**) Segregation of double vowels using a neural cancellation filter. Tuning the delay t to the period of one of the vowels cancels it, allowing the other vowel to be recovered

de Cheveigné's model is based on a time-domain cancellation process, which can be regarded as a comb filter of the form

$$y(t) = \max[0, x(t) - x(t - \tau)] \tag{8.4}$$

where $x(t)$ is the probability of an input spike at time t, $y(t)$ is the output spike probability and τ is the time lag before inhibition occurs (note that $y(t)$ is always non-negative due to the max operator). Equation (8.4) can be realized as the simple neural circuit shown in Fig. 8.7a, in which excitatory and delayed inhibitory inputs converge on the same unit.

Vowel segregation using the cancellation filter is illustrated in Fig. 8.7c, d for the mixture of /ar/ and /er/ used in Fig. 8.6. The vowel mixture was processed by a model of the auditory periphery, and the output from each channel was passed through a cancellation filter with the delay t tuned to the period of one of the vowels. Summary correlograms were then computed from the filtered output to allow comparison with Fig. 8.6. In panel C the delay was set to 7.9 ms; this suppresses the /er/ vowel, allowing the /ar/ to be recovered from the mixture. The timbre region of the segregated /ar/ closely matches the timbre region of the isolated vowel, shown in Fig. 8.6d. Similarly, the /er/ vowel can be recovered by tuning the cancellation filter to the period of /ar/ (10 ms), as shown in Fig. 8.7d.

A similar approach based on *neural timing networks* has been advocated by Cariani (2001, 2003). However, it differs in that the inhibitory input arises from a recurrent neural circuit; the output of each processing unit is fed back to its input, as shown in Fig. 8.7b. This scheme involves an array of coincidence detectors that receive the same direct input, but also have a recurrent input that is tuned to a particular delay. The activity of each unit is updated according to the rule

$$y(t) = y(t-\tau) + \beta[x(t) - y(t-\tau)] \tag{8.5}$$

where $x(t)$ is the probability of a direct input spike, t is the delay associated with the recurrent input, and the constant β determines the rate at which the output adapts to the input. Periodic patterns in the input are amplified by recurrent circuits with a matching delay, so that vowels with different F0s are extracted at different points on the neural array. Cariani (2003) shows that a recurrent timing network can recover two vowels from their mixture, and that the quality of separation improves with increasing ΔF0. A notable feature of this approach is that it does not require explicit information about the F0s of the two vowels; however, it remains to be seen how well a neural timing network could segregate speech with a continuously varying F0, as opposed to the static vowels used in Cariani's simulations.

Numerous other models of double vowel segregation have been proposed, such as those based on amplitude modulation (AM) maps (Berthommier and Meyer 1997). The majority of these models assume that harmonic grouping underlies the improvement in vowel identification performance with increasing ΔF0. An exception is a study by Culling and Darwin (1994), who suggest that listeners segregate double vowels with a small ΔF0 by exploiting the beating between unresolved harmonic components. Their reasoning is that beating gives rise to fluctuations in the spectral envelope of the vowel mixture, so that at particular time instants it may closely resemble one vowel or the other. Culling and Darwin describe a computer model that matches short-term spectral estimates of the vowel mixture, and show that it provides a good match to human performance for small ΔF0s.

8.4.3 Localization and Segregation of Spatially Separated Sound Sources

It is well established that human listeners are able to exploit a difference in spatial location between sound sources in order to perceptually segregate them. For example, Spieth et al. (1954) report that the ability of listeners to recognize two concurrent utterances improves when the spatial separation between them is increased. This finding has motivated a number of binaural auditory models that address the role of spatial cues in ASA, although few of these make strong links with physiological mechanisms.

As noted by Jennings and Colburn (Chapter 4), interaural time differences (ITDs) and interaural intensity differences (IIDs) are the two primary cues for coding the spatial location of sounds. Most binaural models of ASA emphasize the role of ITDs, which are extracted by a spatial array of ITD-sensitive units as suggested by Jeffress (1948). In practice, this is achieved by cross-correlating the simulated auditory nerve response in left- and right-ear channels with the same best frequency, as in the influential model of Colburn (1973). By analogy with (8.2a), a running cross-correlogram can be formed by computing

$$c(t,f,\tau)=\sum_{k=1}^{\infty}x(t-T,f)\,y(t-T-\tau,f)\exp(-T/\Omega) \qquad (8.6)$$

where $x(t,f)$ and $y(t,f)$ represent the simulated auditory nerve response for the left and right ears respectively, t is the cross-correlation lag (interaural delay), and the other parameters are defined as in (8.2).

Lyon (1983) was the first to apply the cross-correlogram to the segregation of spatially distinct sounds. He considered the segregation and dereverberation of a mixture of two sounds, in which ITD estimates from a cross-correlogram were used to derive time-varying gains for each frequency channel (again, this may be regarded as a channel-selection approach). Lyon noted that in many instances the ITD within a particular time-frequency region did not correspond to that of either of the two sound sources, and suggested summing the cross-correlogram over all frequency channels to give a pooled function in which the ITD of each sound source was indicated by a prominent peak. Note that the same operation was used to compute a summary correlogram for pitch analysis, as given in (8.3). Lyon did not evaluate the summary cross-correlogram as a means for multisource sound localization in any systematic way, but the idea has been adopted in many subsequent systems. For example, Bodden (1993) describes a "cocktail party processor" that is based on the binaural auditory models of Lindemann (1986) and Gaik (1993). In his system, the reliability of multisource localization is improved by computing the running cross-correlogram with a relatively long time constant (100 ms), and each channel of the cross-correlogram is weighted before forming a sum. The weight associated with each channel depends upon the reliability of localization information obtained from it, which is assessed during a learning stage. Bodden's model performs a mapping from ITD to azimuth, giving a summary binaural pattern that is indexed by azimuth and time. Peaks within this binaural pattern are regarded as candidate source locations, and a Wiener filter is used to extract the sound source that corresponds to each time-varying azimuth track.

Bodden reports reasonable performance for his model, but it fails when two sources are close in azimuth and in the presence of reverberation. When two sound sources interact within a single cross-correlogram channel, the position of the peak in the cross-correlation function depends on the relative level of the two sounds as well as their ITDs, leading to error in the estimated azimuths (see Roman et al. 2003 for an analysis). To address this problem, several methods have been proposed for sharpening the cross-correlogram along the delay axis. Stern and Trahiotis (1992) propose a "straightness weighting" which emphasizes the modes of the cross-correlogram that occur at the same lag in different frequency regions. The "skeleton cross-correlogram" (Palomäki et al. 2004; Roman et al. 2003) adopts a similar approach, in which broadly tuned peaks in the cross-correlogram are replaced with narrower Gaussian functions; such an operation could be implemented by lateral inhibition between adjacent ITD-sensitive neurons in a physiological model. This idea is somewhat related to the use of contralateral inhibition in Lindemann's (1986) binaural auditory model. Also relevant is the "stencil filter" of Liu et al. (2000), which uses the theoretical cross-correlation pattern associated with a particular source azimuth as a template for matching against observed cross-correlograms of sound mixtures.

A number of workers have attempted to improve the performance of binaural models for multisource sound localization and segregation by incorporating other acoustic cues. Braasch (2002) describes an "interaural cross-correlation difference" model which exploits a difference in onset time between a target sound and background noise. If the target and background sounds are uncorrelated, then the cross-correlogram of the background can be estimated during periods when the target is silent; this is then subtracted from the cross-correlogram of the mixture in order to recover the target, when both the target and background noise are present. This approach is most effective when the background sound is stationary noise, but is less suited to nonstationary interference such as a competing voice. Kollmeier and Koch (1994) also describe an approach that integrates additional cues into a binaural model for speech segregation. Their system extracts low-frequency (below 400 Hz) envelope modulation from each peripheral frequency channel, giving a two-dimensional representation of AM frequency against best frequency. Their approach is motivated by physiological findings of similar neural maps in the inferior colliculus of the cat (Langner and Schreiner 1988). The second stage of their model weights the AM spectrum by a function derived from binaural localization cues, so that speech-related energy arising from a particular location is passed, and energy from other locations is suppressed.

Although most of the models discussed above use ITD as a cue for segregating concurrent sounds, it should be noted that human listeners might be unable to do so. Culling and Summerfield (1995) show that listeners are unable to perceptually segregate mixtures of vowel-like bandpass noise sounds when ITD is the only grouping cue available. Instead, they propose that the target sound is unmasked by a process that works independently in each frequency band, via an interaural cancellation mechanism similar to the equalization-cancellation (EC) model of Durlach (1963). Further evidence to support this idea has come from a perceptual study by Edmonds and Culling (2005), who demonstrate that listeners can exploit a difference in ITD between speech and noise sources, but the ITD of each source need not be consistent across frequency. Again, this suggests a within-band unmasking mechanism rather than one in which acoustic components are grouped by common ITD. Brown and Palomäki (2005) implement this idea in a computer model that combines correlogram-based F0 analysis with a cross-correlation based EC mechanism, and acts as a front-end to an automatic speech recognizer. Their model gives a qualitative match with the human data obtained by Edmonds and Culling (2005).

8.4.4 Modeling Attention in Auditory Scene Analysis

The computational models discussed so far acknowledge the role of attention in ASA, but do not model it explicitly. For example, the streaming model of McCabe and Denham (1997) has an "attentional input" that influences the activity of the foreground neural array, but the mechanism by which it achieves this is not fully developed. Similarly, the model of Beauvois and Meddis (1991, 1996) allows the foreground and background streams to change spontaneously due to a cumulative random bias, but does not allow for conscious shifts in attention between the high and low tones.

Wang (1996) notes that the global inhibitor in his streaming model bears a structural and functional resemblance to the thalamic reticular complex, which has been implicated in selective attention (Crick 1984). This suggests that selective attention could be achieved by selective synchronisation of neural oscillators, an idea that has also been developed by Baird (1997) and Wrigley and Brown (2004).

Baird's model is motivated by a psychophysical experiment reported by Jones et al. (1981), which suggests that rhythmic expectancy plays a role in sequential grouping. Jones et al. asked listeners to judge the order of presentation of two target tones with different frequencies, A and B, which were embedded in a sequence of flanking tones F and captor tones C (i.e., C–C–C–FABF–C–C–C where dashes indicate silent intervals). An earlier experiment by Bregman and Rudnicky (1975) using the same stimulus showed an effect of grouping by frequency proximity. Listeners found it easier to determine the order of the target tones when they were distant in frequency from the flanking tones, presumably because the target tones were grouped into a separate stream rather than embedded in a stream FABF. However, Jones et al. noted that the target tones are also distinguished from the F and C tones based on their rhythmic pattern. They found that when the sequence was played isochronously (i.e., C–C–C–F–A–B–F–C–C–C) listeners found it harder to determine the order of the target tones. This suggests that grouping by frequency proximity only partially accounts for the ability of listeners to do the task; in order for the A and B tones to be effectively segregated, they must differ in rhythmic pattern from the F and C tones.

This finding led Baird (1997) to propose a physiological model of auditory streaming in which rhythmic expectancy plays a key role. In his model, neural oscillators with periodic firing patterns synchronize with periodicities in the input signal (in the case of a tone sequence, the dominant periodicity corresponds to the TRT). As a result, rhythmic expectancies are generated that focus attention on future points in time. If auditory events occur at these anticipated times, their corresponding neural activity patterns are amplified and pulled into synchrony with oscillators that model the attentional set of the listener (the so-called "attentional stream"). Likewise, new streams are formed if there are discrepancies between rhythmic expectations and the times at which acoustic events actually occur.

This model explains the findings of Jones et al. (1981) as follows. In the case of Bregman and Rudnicky's original stimulus (C–C–C–FABF–C–C–C), there is a discrepancy between the rhythm of the target tones and that established by the preceding C and F tones; hence the A and B tones are synchronized with the attentional stream and are segregated from the sequence, allowing their order to be judged. However, the isochronous sequence (C–C–C–F–A–B–F–C–C–C) tends to form a single attentional stream because differences in frequency proximity are insufficient to overcome sequential grouping due to the common rhythmic pattern of the tones. Baird also claims some justification for his model at the physiological level; he suggests that the discrepancies between predicted and actual neural activity patterns may underlie the mismatch negativity response seen in ERP recordings (see Sect. 8.3.4).

Attention has been characterized as having two components: an involuntary bottom-up (*exogenous*) process and a voluntary top-down (*endogenous*) process (Spence and Driver 1994). Baird's model primarily addresses exogenous attention, since rhythmic expectancy in his model is data-driven; however, it also provides for endogenous attention by allowing the attentional stream to be voluntarily synchronized with oscillators that exhibit a particular periodicity.

In fact, the role of endogenous attention in auditory streaming has proven controversial. Bregman and Rudnicky (1975) suggest that endogenous attention is not required for stream formation; they note that even though subjects were instructed to attend to the A and B tones in the C–C–C–FABF–C–C–C sequence, this did not prevent the C tones (which were presumably unattended) from forming a stream with the F tones. However, a flaw in this argument has been noted by Carlyon et al. (2001); they point out that there was no competing task to draw listener's attention away from the C and F tones, and hence it is possible that listeners attended to them, despite being given instructions to focus on the A and B tones (see also Sussman et al. 2007). To clarify this issue, Carlyon et al. designed an experiment that examined the effect of a competing task on the buildup of auditory streaming. They presented a tone sequence to one ear, and during the first half of this sequence a competing task (a sequence of noise bursts) was presented to the opposite ear. When listeners were instructed to attend to the tone sequence only, the buildup of streaming occurred over a time course that was consistent with previous studies (Anstis and Saida 1985). However, when listeners were instructed to do the competing task before switching their attention to the tone sequence, the buildup of streaming was delayed.

The experiment of Carlyon et al. suggests a role for endogenous attention in auditory streaming, and has motivated a computational modeling study by Wrigley and Brown (2004). Their model consists of three stages, as shown schematically in Fig. 8.8. In the first stage, the acoustic input to each ear is processed by a model of the auditory periphery to yield simulated auditory nerve firing patterns, which are then sharpened by lateral inhibition. The second stage identifies periodicities in the auditory nerve representation by means of a correlogram. Periodicity information is combined across both ears in order to give an estimate of the dominant F0 present. In the third stage of the model, auditory grouping is encoded by an array of neural oscillators (one per frequency channel). As in the system of Wang and Brown (1999), a cross-channel correlation mechanism is used to identify prominent frequency components ('segments') in the correlogram, which are represented by synchronized blocks of oscillators. Furthermore, long-range excitatory connections cause oscillators to synchronize if their corresponding frequency channels are dominated by the same F0.

The key component of the attentional mechanism in this model is an *attentional leaky integrator* (ALI), which sums and smoothes the output from the neural oscillator array. The output of the ALI is an attentional stream, similar in concept to that

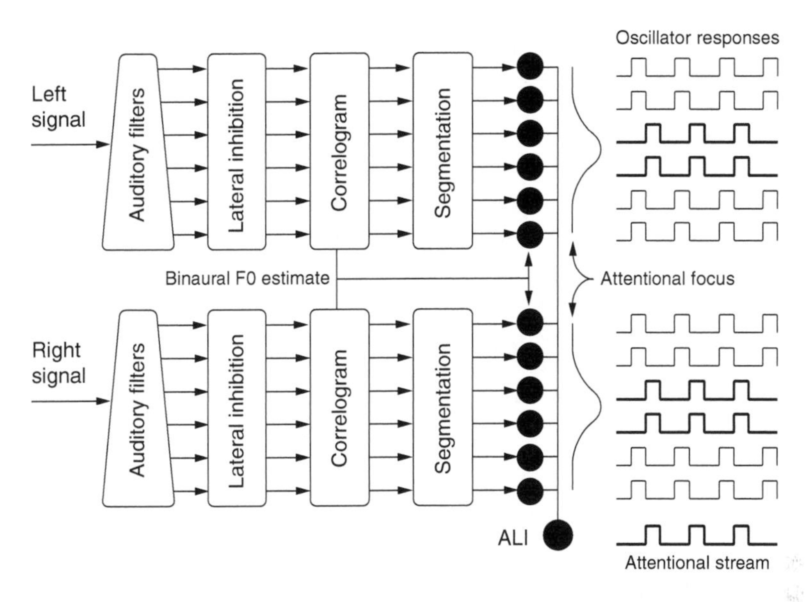

Fig. 8.8 Schematic diagram of Wrigley and Brown's neural oscillator model of auditory attention. Oscillators are pulled into synchrony with an attentional leaky integrator (ALI) according to grouping by F0 and the influence of a frequency-dependent bias ("attentional focus"). Oscillator responses that are synchronized with the output of the ALI ("attentional stream") are shown in *bold*

of Baird's model, which is synchronized with oscillator groups that are in the attentional foreground. The connection weights between the neural oscillator network and the ALI are modulated by endogenous processes, including "conscious" preference. The latter is modeled by a Gaussian distribution ('attentional focus') across frequency. Initially, there are strong connection weights between all oscillators and the ALI, indicating that all frequency channels are grouped by default (see also Bregman 1990). As time proceeds, connection weights adapt towards the shape of the attentional focus, so that the selected source is in the attentional foreground. However, sufficiently intense sounds are able to cause (exogenous) redirection of attention by overcoming the influence of the attentional focus. Wrigley and Brown demonstrate that their model shows the same behavior as listeners in Carlyon's experiment; in addition, it is able to explain other auditory grouping phenomena, such as the segregation of a mistuned harmonic from a complex tone, grouping by onset synchrony, and the sequential capture of harmonics from a complex sound.

8.5 Modeling the Separation of Speech from Other Sounds

As noted above, most physiologically motivated models of auditory scene analysis have focused on phenomena involving relatively simple laboratory stimuli, such as tone sequences and double vowels. A key issue is whether these models can be extended to process complex real-world sounds, such as speech. The segregation of speech from other sounds is a key problem in computational hearing, as it has a practical application in robust automatic speech recognition devices (e.g., Barker 2007). However, relatively few modeling approaches to the speech segregation problem have adopted a physiologically motivated approach. An early study by Brown and Cooke (1994) describes a system for speech segregation in which acoustic features are derived from "auditory maps"; these are motivated by physiological evidence for auditory neurons that are tuned to periodicity, amplitude and frequency modulation, onsets and offsets. However, their system essentially takes an engineering approach, and does not make strong reference to physiological mechanisms beyond the feature extraction stage.

Wang and Brown (1999) describe a neural oscillator-based model of speech segregation that is perhaps the most complete "physiological" model of ASA to date. Their model is effectively a fusion of two previous systems: Wang's (1996) neural oscillator model of streaming (see Sect. 8.4.1) and Brown and Cooke's (1994) speech segregation system. By combining the physiologically motivated feature extraction stage of Brown and Cooke's model with Wang's neural oscillator network, they propose a complete architecture that is both biologically plausible and able to operate on real-world acoustic signals.

The first stage of Wang and Brown's (1999) system passes a digitally sampled input signal (a mixture of speech and noise) through a computational model of the auditory periphery. Two "mid-level" representations are then computed from the simulated auditory nerve response. Specifically, a correlogram and summary correlogram are computed at 20-ms intervals, and adjacent channels of the correlogram are cross-correlated to identify frequency regions that are excited by the same harmonic or formant. The latter is used to determine local excitatory connections in the third stage of the model, which is a two-dimensional time-frequency network of neural oscillators arranged into two layers. The structure of this network mirrors the two conceptual stages of ASA described in Sect. 8.2; the first ("segmentation") layer represents salient acoustic events, whereas the second ("grouping") layer organizes those events into auditory streams.

Wang and Brown's network is shown in Fig. 8.9 and its behavior is now considered in detail. In the segmentation layer, oscillators receive an external input if the energy in their associated frequency channel exceeds a threshold value. In addition, each oscillator receives a weighted input from its four neighbours in the time-frequency grid. Weights along the time axis are uniformly set to one, whereas the weight between oscillators associated with adjacent frequency channels is set to one if their cross-correlation is high, and zero otherwise. As a result, segments form in the first layer of the network that correspond to contiguous regions of strong acoustic energy, such as harmonics and formants of speech. Oscillators that encode the same segment are synchronized due to the influence of local excitatory connections, whereas a global inhibitor desynchronizes oscillators that represent different segments. In the second layer of the model, long-range excitatory connections link oscillators at the same point on the time axis, but in different frequency regions. The strength of these connections is determined by F0 information from the correlogram. At each time instant, oscillators receive mutual excitatory connections if their corresponding correlogram channels both agree with the dominant F0, or if

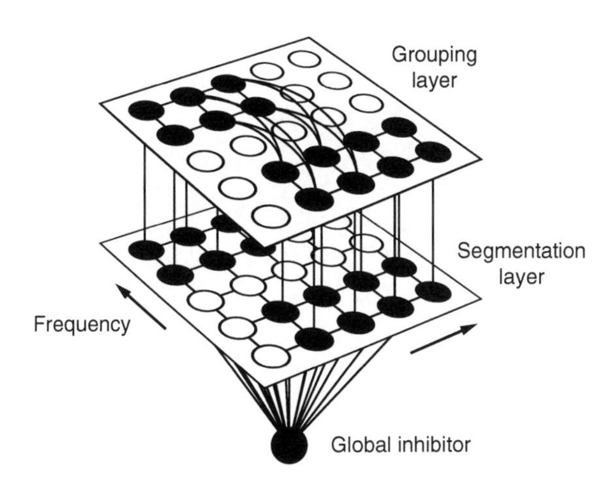

Fig. 8.9 A two-layer neural network for auditory scene analysis proposed by Wang and Brown (1999) (Copyright © IEEE.)

they both disagree with the dominant F0; otherwise they receive mutual inhibitory connections. The dynamics of the grouping layer are arranged so that oscillators corresponding to the longest segment in the segmentation layer jump to the active phase first (recall Fig. 8.3b). This segment then recruits others in the grouping layer, so long as they have consistent F0 information in the majority of overlapping time frames. As a result, oscillators in the grouping layer rapidly synchronize to form two blocks of oscillators (streams), which represent sources in the acoustic input with different F0s.

This behavior is illustrated in Fig. 8.10 for a mixture of speech and a ringing telephone. The figure shows a snapshot of activity in the grouping layer of Wang and Brown's network taken at two different times, where white pixels indicate active oscillators and black pixels indicate inactive oscillators. The pattern of oscillator activity can be interpreted as a binary time-frequency mask, which indicates the time-frequency regions that contribute to a particular stream. By weighting the output of each peripheral channel by the mask and summing over frequency, a time-domain signal for the desired source can be resynthesized from the acoustic mixture (see also Brown and Cooke 1994). Alternatively, the mask can provide an automatic speech recognition system with information about reliable and unreliable time-frequency regions, which can be used to improve robustness during speech decoding (Cooke et al. 2001). A similar approach has been described by Rouat et al. (2007), who also use a network of Terman–Wang oscillators but employ low-frequency envelope modulation to group related time-frequency regions.

Although some degree of neurobiological plausibility can be claimed for Wang and Brown's model, it omits a factor that is likely to play a major role in speech segregation; top-down (schema-driven) application of learned knowledge. Recall from Fig. 8.2a that human listeners are able to segregate double vowels with a performance above chance level even when there is no ΔF0 between the vowels; this

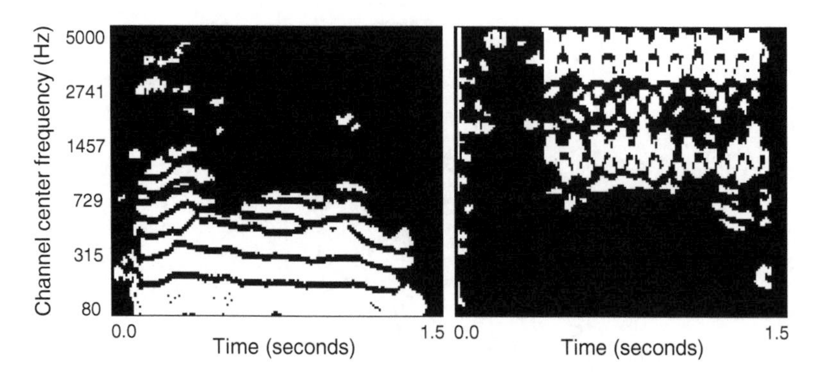

Fig. 8.10 Illustration of auditory grouping in the neural oscillator model of Wang and Brown (1999). Two snapshots of activity in the oscillator network are shown, in which *white pixels* indicate active oscillators and *black pixels* indicate inactive oscillators. The input to the network was a mixture of speech and a telephone ring, and the two sources have been segregated into two synchronized groups of oscillators (streams). The *left panel* shows oscillators corresponding to the speech source, the *right panel* shows oscillators corresponding to the telephone ring (Adapted from Wang and Brown 1999 with permission. Copyright © IEEE.)

suggests a role for schema-driven grouping, since no primitive grouping cues are available. Numerous other studies have supported the idea of schema-based integration, such as those concerning the 'duplex' perception of speech sounds in which schema-based cues for grouping appear to overcome primitive cues for segregation (see Chapter 4 of Bregman (1990) for a review).

Only a few studies have addressed schema-driven grouping within a physiologically plausible model. Sagi et al. (2001) describe a neural model based on the "cortronic" architecture of Hecht-Nielsen (1998), which focuses on the problem of understanding a single target talker in the presence of other talkers. The cortronic architecture is a hierarchical associative network that learns the structural relationships between sounds and linguistic units (sentences and words). In the input layer of their network, neurons that respond most strongly to a short-time spectral analysis of the input signal are selected, forming a sparse binary code (so-called "sound tokens"). Subsequent stages of the network recall sound tokens that were learned during training, and it is shown that words can be recovered in the presence of interfering sounds. However, the model makes a number of restrictive assumptions; in particular, it is assumed that the input contains a fixed number of sound tokens and that these are processed by the same number of sound-processing regions in the network. It is unclear how well such an approach would generalize to real-world signals.

Neural oscillations may also play a role in memory, and hence in schema-driven grouping. It has been suggested that long-term synaptic potentiation can be facilitated by the convergence of synchronous firing activity on postsynaptic cells (Basar et al. 2000), and hence temporal synchronization of synaptic inputs may be implicated in neural plasticity (Singer 1993). If neural oscillations play a role in both feature binding and memory, then they may act as a common code that links top-down and bottom-up sensory information processing. This idea has been implemented in a computational model of speech segregation by Liu et al. (1994). Their model consists of a three-layer oscillator network, in which each layer (labeled A, B, and C) is identified with an area of the auditory cortex. The input to the network is typically a vowel in noise. Initially, the vowel formants are identified as spectral peaks in the acoustic input, and each formant is encoded as a group of synchronized oscillations in the A ("feature extracting") layer. The B ("feature linking") layer acts as a simple associative memory, which receives its input from the A layer and encodes the formant frequencies of particular vowels via hard-wired connections. The vowel category is determined by the C ("categorizing") layer, which assesses the degree of synchronization in each formant region. The authors demonstrate that top-down and bottom-up interactions between the A and B layers are necessary for robust vowel recognition in the presence of babble noise. However, their model has a number of limitations. Specifically, it can only process static spectral patterns, and does not include primitive grouping mechanisms. Additionally, it is unclear whether the model can separate two sounds if both are encoded in its B layer (e.g., a double vowel). Resonance between ascending and descending neural pathways is also central to the ARTSTREAM model of Grossberg et al. (2004), although their model addresses pitch-based auditory grouping of relatively simple sounds, rather than speech segregation.

A more general approach to schema-driven grouping has been proposed by Elhilali (2004). Inspired by a model of cortical processing, her system attempts to reconcile the observed acoustic input with an internal world representation. This is achieved by a combination of unsupervised learning and source estimation based on Kalman filtering. The model is able to reproduce simple auditory grouping phenomena, such as the streaming of tone sequences, sequential capture of tones and segregation of crossing time-frequency trajectories. In addition, it separates mixtures of pre-learned speech signals very effectively, and can partially segregate speech mixtures without prior knowledge of the constituent signals. The model also displays behavior consistent with the "old plus new" heuristic (Elhilali and Shamma 2008).

Finally, recent computational modeling studies suggest a role for the medial olivocochlear (MOC) efferent feedback system in noise-robust speech recognition. Previous studies have indicated that temporal information must be used by the auditory system in order to robustly encode speech sounds in noise (e.g., Miller and Sachs 1983). However, Ghitza (2007) shows that a rate/place representation of auditory nerve activity is able to accurately predict the ability of human listeners to discriminate diphones in background noise, so long as efferent feedback is included in the model.

8.6 Summary

Physiological models of auditory scene analysis are still in their infancy. Much remains to be discovered about the coding of auditory grouping at the neural level, and computer models are likely to play an important role in future research by providing a link between physiological and psychophysical findings. Channel selection appears to be a plausible physiological mechanism for source segregation, but the relationship between this approach and the wider issue of the binding problem requires further study. Solutions to the binding problem based on temporal correlation may be important in this regard, as they are compatible with channel selection and also constitute a general mechanism for feature binding.

One way in which computer models can contribute to our understanding is by making predictions that can be tested through physiological or psychophysical experimentation. For example, de Cheveigné's studies of double vowel segregation (1997) demonstrate a synergy between psychophysical experiments and computer modeling. In addition, his cancellation model has been evaluated on auditory nerve firing patterns obtained from physiological recordings, which represents an intriguing convergence of computational and physiological approaches (de Cheveigné 1993). Similarly, the neural oscillator models of Wang (1996) and Wrigley and Brown (2004) make specific predictions regarding the role of neural oscillations in feature binding and attentional processing, which can be tested by physiological experimentation.

Neural time delays are required by many of the computer models described here, but direct physiological implementation of such circuits appear to be of doubtful plausibility. Although neural time delays are apparently generated within the auditory midbrain where they are believed to underlie the detection of ITDs, there is no strong physiological support for the neural 'time axis' required by the models of Wang (1996) and Wang and Brown (1999). Similarly, the F0-based segregation models of Meddis and Hewitt (1992), de Cheveigné (1997) and Cariani (2001, 2003) require neural delays that are as long as the longest fundamental period encountered (which is likely to be at least 20 ms). However, such schemes may not be implemented in such a literal way at the neural level. As noted in Sect. 8.4.2, behavior akin to autocorrelation may arise from the intrinsic membrane properties of auditory neurons, rather than a systematic arrangement of neural delay lines.

The integration of top-down and bottom-up processing in ASA is an important topic for future work, as is the role of attention. There is no consensus from psychophysical and physiological studies as to whether stream segregation precedes attention or requires it, and this is reflected in computer models that take opposing views (e.g., Beauvois and Meddis 1991, 1996; Wrigley and Brown 2004). In addition, most computer models of ASA assume that the listener and sound sources are static. In natural environments, sound sources move over time and the listener is active; as a result, factors such as head movement and dynamic tracking of spatial location need to be accounted for in more sophisticated models.

Finally, physiological models of ASA may in turn lead to useful engineering systems for sound separation and speech enhancement. Although there is no requirement that a practical system should be based on a physiological account, ASA approaches based on neural models are attractive because they are comprised of simple, parallel and distributed components and are therefore well suited to hardware implementation (see also Chapter 10). For example, Cosp and Madrenas (2003) describe a very large-scale integration (VLSI) implementation of Wang and Terman's LEGION network, comprising a grid of 16×16 neural oscillators (see also Girau and Torres-Huitzil 2007). Similar devices built on a larger scale could provide the basis for hardware implementation of physiological ASA models, and their compact size and low power consumption could make them suitable for application in hearing aids.

References

Alain C, Izenburg A (2003) Effects of attentional load on auditory scene analysis. J Cogn Neurosci 15:1063–1073.

Alain C, Schuler BM, McDonald KL (2002) Neural activity associated with distinguishing concurrent auditory objects. J Acoust Soc Am 111:990–995.

Anstis S, Saida S (1985) Adaptation to auditory streaming of frequency-modulated tones. J Exp Psychol Hum Percept Perform 11:257–271.

Assmann PF, Summerfield Q (1990) Modeling the perception of concurrent vowels: vowels with different fundamental frequencies. J Acoust Soc Am 88:680–697.

Baird B (1997) Synchronized auditory and cognitive 40 Hz attentional streams, and the impact of rhythmic expectation on auditory scene analysis. In: Jordan M, Kearns M, Solla S (eds), Neural Information Processing Systems, Vol. 10. Cambridge, MA: MIT Press, pp 3–10.

Barker J (2007) Robust automatic speech recognition. In: Wang DL, Brown GJ, Computational Auditory Scene Analysis: Principles, Algorithms and Applications. Piscataway, NJ: IEEE Press/Wiley Interscience.

Barlow HB (1972) Single units and cognition: a neuron doctrine for perceptual psychology. Perception 1:371–394.

Barth DS, MacDonald KD (1996) Thalamic modulation of high-frequency oscillating potentials in auditory cortex. Nature 383:78–81.

Basar E, Basar-Eroglu C, Karakas S, Schurmann M (2000) Brain oscillations in perception and memory. Int J Psychophysiol 35:95–124.

Beauvois MW, Meddis R (1991) A computer model of auditory stream segregation. Q J Exp Psychol 43A:517–541.

Beauvois MW, Meddis R (1996) Computer simulation of auditory stream segregation in alternating tone sequences. J Acoust Soc Am 99:2270–2280.

Berthommier F, Meyer G (1997) Improving amplitude modulation maps for F0-dependent segregation of harmonic sounds. In: Proceedings of EUROSPEECH, Rhodes, Greece, September 22–25, pp 2483–2486.

Bodden M (1993) Modeling human sound-source localization and the cocktail party effect. Acta Acust 1:43–55.

Braasch J (2002) Localization in the presence of a distractor and reverberation in the frontal horizontal plane: II. Model algorithms. Acta Acust/Acustica 88:956–969.

Bregman AS (1990) Auditory Scene Analysis. Cambridge, MA: MIT Press.

Bregman AS, Rudnicky AI (1975) Auditory segregation: stream or streams? J Exp Psychol 1:263–267.

Brosch M, Budinger E, Scheich H (2002) Stimulus-related gamma oscillations in primate auditory cortex. J Neurophysiol 87:2715–2725.

Brown GJ, Cooke MP (1994) Computational auditory scene analysis. Comput Speech Lang 8:297–336.

Brown GJ, Cooke MP (1998) Temporal synchronization in a neural oscillator model of primitive auditory stream segregation. In: Rosenthal DF, Okuno HG (eds), Computational Auditory Scene Analysis. Mahwah, NJ: Lawrence Erlbaum, pp 87–101.

Brown GJ, Palomäki KJ (2005) A computational model of the speech reception threshold for laterally separated speech and noise. In: Proceedings of Interspeech, Lisbon, September 4–8, pp 1753–1756.

Brown GJ, Wang DL (1997) Modelling the perceptual segregation of concurrent vowels with a network of neural oscillators. Neural Netw 10:1547–1558.
Cariani P (2001) Neural timing nets. Neural Netw 14:737–753.
Cariani P (2003) Recurrent timing nets for auditory scene analysis. Proc IJCNN-2003, Portland, OR, July 20–24.
Carlyon RP, Cusack R, Foxton JM, Robertson IH (2001) Effects of attention and unilateral neglect on auditory stream segregation. J Exp Psychol Hum Percept Perform 27:115–127.
Chang P (2004) Exploration of Behavioural, Physiological and Computational Approaches to Auditory Scene Analysis. MSc Thesis, The Ohio State University, Department of Computer Science and Engineering.
Colburn HS (1973) Theory of binaural interaction based on auditory-nerve data. I. General strategy and preliminary results on interaural discrimination. J Acoust Soc Am 54:1458–1470.
Cooke M, Green P, Josifovski L, Vizinho A (2001) Robust automatic speech recognition with missing and unreliable acoustic data. Speech Commun 34:267–285.
Cosp J, Madrenas J (2003) Scene segmentation using neuromorphic oscillatory networks. IEEE Trans Neural Netw 14:1278–1296.
Crick F (1984) Function of the thalamic reticular complex: the searchlight hypothesis. Proc Natl Acad Sci U S A 81:4586–4590.
Culling JF, Darwin CJ (1994) Perceptual and computational separation of simultaneous vowels: cues arising from low-frequency beating. J Acoust Soc Am 95:1559–1569.
Culling JF, Summerfield Q (1995) Perceptual separation of concurrent speech sounds: Absence of across-frequency grouping by common interaural delay. J Acoust Soc Am 98:785–797.
Darwin CJ, Bethell-Fox CE (1977) Pitch continuity and speech source attribution. J Exp Psychol Hum Percept Perform 3:665–672.
de Cheveigné A (1993) Separation of concurrent harmonic sounds: fundamental frequency estimation and a time-domain cancellation model of auditory processing. J Acoust Soc Am 93:3271–3290.
de Cheveigné A (1997) Concurrent vowel identification. III. A neural model of harmonic interference cancellation. J Acoust Soc Am 101:2857–2865.
de Cheveigné A, Kawahara H, Tsuzaki M, Aikawa K (1997) Concurrent vowel identification. I. Effects of relative amplitude and F0 difference. J Acoust Soc Am 101:2839–2847.
Durlach NI (1963) Equalization and cancellation theory of binaural masking level differences. J Acoust Soc Am 35:1206–1218.
Edmonds BA, Culling JF (2005) The spatial unmasking of speech: evidence for within-channel processing of interaural time delay. J Acoust Soc Am 117:3069–3078.
Elhilali M (2004) Neural Basis and Computational Strategies for Auditory Processing. PhD Thesis, Department of Electrical and Computer Engineering, University of Maryland.
Elhilali M, Shamma SA (2008) A cocktail party with a cortical twist: how cortical mechanisms contribute to sound segregation. J Acoust Soc Am 124:3751–3771.
Fitzhugh R (1961) Impulses and physiological states in models of nerve membrane. Biophys J 1:445–466.
Gaik W (1993) Combined evaluation of interaural time and intensity differences: Psychoacoustic results and computer modeling. J Acoust Soc Am 94:98–110.
Ghazanfar AA, Chandrasekaran C, Logothetis NK (2008) Interactions between the superior temporal sulcus and auditory cortex mediate dynamic face/voice integration in rhesus monkeys. J Neurosci 28:4457–4469.
Ghitza O (2007) Using auditory feedback and rhythmicity for diphone discrimination of degraded speech. In: Proceedings of ICPhS, Saarbrücken, August 6–10.
Girau B, Torres-Huitzil C (2007) Massively distributed digital implementation of an integrate-and-fire LEGION network for visual scene segmentation. Neurocomputing 70:1186–1197.
Gray CM, König P, Engel AK, Singer W (1989) Oscillatory responses in cat visual cortex exhibit inter-columnar synchronization which reflects global stimulus properties. Nature 338:334–337.

Grossberg S, Govindarajan KK, Wyse L, Cohen MA (2004) ARTSTREAM: a neural network model of auditory scene analysis and source segregation. Neural Netw 17:511–536.
Hecht-Nielsen R (1998) A theory of the cerebral cortex. In: Proceedings of the International Conference on Neural Information Processing (ICONIP), Burke, VA, October 21–23. Amsterdam: IOS Press, pp 1459–1464.
Hubel DH, Weisel TN (1962) Receptive fields, binocular interaction and functional architecture in the cat's visual cortex. J Physiol 160:106–154.
Jeffress LA (1948) A place theory of sound localization. J Comp Physiol Psychol 41:35–39.
Jones MR, Kidd G, Wetzel R (1981) Evidence for rhythmic attention. J Exp Psychol: Hum Percept Perform 7:1059–1073.
Kollmeier B, Koch R (1994) Speech enhancement based on physiological and psychoacoustical models of modulation perception and binaural interaction. J Acoust Soc Am 95:1593–1602.
Langner G, Schreiner CE (1988) Periodicity coding in the inferior colliculus of the cat. I. Neuronal mechanisms. J Neurophysiol 60:1799–1822.
Lindemann W (1986) Extension of a binaural cross-correlation model by contralateral inhibition. I. Simulation of lateralization for stationary signals. J Acoust Soc Am 80:1608–1622.
Liu C, Wheeler BC, O'Brien WD, Bilger RC, Lansing CR, Feng AS (2000) Localization of multiple sound sources with two microphones. J Acoust Soc Am 108:1888–1905.
Liu F, Yamaguchi Y, Shimizu H (1994) Flexible vowel recognition by the generation of dynamic coherence in oscillator neural networks: Speaker-independent vowel recognition. Biol Cybern 71:105–114.
Lyon RF (1983) A computational model of binaural localization and separation. In: Proceedings of IEEE ICASSP, Boston, April 14–16, pp 1148–1151.
Marr D (1982) Vision. San Francisco: WH Freeman.
McAdams S, Bregman AS (1979) Hearing musical streams. Comput Music J 3:26–43.
McCabe SL, Denham MJ (1997) A model of auditory streaming. J Acoust Soc Am 101:1611–1621.
McDonald K, Alain C (2005) Contribution of harmonicity and location to auditory object formation in free field: Evidence from event-related brain potentials. J Acoust Soc Am 118:1593–1604.
McGurk H, McDonald J (1976) Hearing lips and seeing voices. Nature 264:746–748.
Meddis R, Hewitt MJ (1991) Virtual pitch and phase sensitivity of a computer model of the auditory periphery. I. Pitch identification. J Acoust Soc Am 89:2866–2882.
Meddis R, Hewitt MJ (1992) Modeling the identification of concurrent vowels with different fundamental frequencies. J Acoust Soc Am 91:233–245.
Meddis R, O'Mard LP (2006) Virtual pitch in a computational physiological model. J Acoust Soc Am 120:3861–3869.
Miller MI, Sachs MB (1983) Representation of stop consonants in the discharge patterns of auditory nerve fibers. J Acoust Soc Am 74:502–517.
Milner PM (1974) A model for visual shape recognition. Psychol Rev 81:521–535.
Nagumo J, Arimoto S, Yoshizawa S (1962) An active pulse transmission line simulating nerve axon. Proc IRE 50:2061–2070.
Norris M (2003) Assessment and Extension of Wang's Oscillatory Model of Auditory Stream Segregation. PhD Thesis, University of Queensland, School of Information Technology and Electrical Engineering.
Palmer SE (1999) Vision Science. Cambridge, MA: MIT Press.
Palomäki KJ, Brown GJ, Wang DL (2004) A binaural processor for missing data speech recognition in the presence of noise and small-room reverberation. Speech Commun 43:361–378.
Ribary U, Ioannides AA, Singh KD, Hasson R, Bolton JP, Lado F, Mogilner A, Llinás R (1991) Magnetic field tomography of coherent thalamocortical 40-Hz oscillations in humans. Proc Natl Acad Sci U S A 88:11037–11041.
Roman N, Wang DL, Brown GJ (2003) Speech segregation based on sound localization. J Acoust Soc Am 114:2236–2252.

Rouat J, Loiselle S, Pichevar R (2007) Towards neurocomputational speech and sound processing. Lect Notes Comput Sci 4391:58–77.
Sagi B, Nemat-Nasser SC, Kerr R, Hayek R, Downing C, Hecht-Nielsen R (2001) A biologically motivated solution to the cocktail party problem. Neural Comput 13:1575–1602.
Scheffers MTM (1983) Sifting Vowels. PhD Thesis, University of Gröningen.
Singer W (1993) Synchronization of cortical activity and its putative role in information processing and learning. Ann Rev Physiol 55:349–374.
Slaney M, Lyon RF (1990) A perceptual pitch detector. In: Proceedings of ICASSP-1990, Albuquerque, NM, April 3–6, pp 357–360.
Spence CJ, Driver J (1994) Covert spatial orienting in audition: Exogenous and endogenous mechanisms. J Exp Psychol Hum Percept Perform 20:555–574.
Spieth W, Curtis JF, Webster JC (1954) Responding to one of two simultaneous messages. J Acoust Soc Am 26:391–396.
Stern RM, Trahiotis C (1992) The role of consistency of interaural timing over frequency in binaural lateralization. In: Cazals Y, Horner K, Demany L (eds), Auditory Physiology and Perception. Oxford: Pergamon, pp. 547–554.
Sussman ES, Ritter W, Vaughan HG (1999) An investigation of the auditory streaming effect using event-related brain potentials. Psychophysiology 36:22–34.
Sussman ES, Horváth J, Winkler I, Orr M (2007) The role of attention in the formation of auditory streams. Percept Psychophys 69:136–152.
Terman D, Wang D (1995) Global competition and local cooperation in a network of neural oscillators. Physica D 81:148–176.
Todd NPM (1996) An auditory cortical theory of primitive auditory grouping. Network Comput Neural Syst 7:349–356.
Van der Pol B (1926) On "relaxation oscillations". Philos Mag 2:978–992.
Van Noorden LPAS (1975) Temporal Coherence in the Perception of Tone Sequences. PhD Thesis, Eindhoven University of Technology.
Von der Malsburg C (1981) The Correlation Theory of Brain Function. Internal Report No. 81-2, Max-Planck-Institut for Biophysical Chemistry, Göttingen, Germany.
Von der Malsburg C, Schneider W (1986) A neural cocktail-party processor. Biol Cybern 54:29–40.
Wallach H (1940) The role of head movements and vestibular and visual cues in sound localization. J Exp Psychol 27:339–368.
Wang DL (1996) Primitive auditory segregation based on oscillatory correlation. Cogn Sci 20:409–456.
Wang DL, Brown GJ (1999) Separation of speech from interfering sounds using oscillatory correlation. IEEE Trans Neural Netw 10:684–697.
Wang DL, Brown GJ (2006) Computational Auditory Scene Analysis: Principles, Algorithms and Applications. Piscataway, NJ: IEEE Press/Wiley Interscience.
Wang DL, Terman D (1995) Locally excitatory globally inhibitory oscillator networks. IEEE Trans Neural Netw 6:283–286.
Warren RM (1970) Perceptual restoration of missing speech sounds. Science 167:392–393.
Wrigley SN, Brown GJ (2004) A computational model of auditory selective attention. IEEE Trans Neural Netw 15:1151–1163.

Chapter 9
Use of Auditory Models in Developing Coding Strategies for Cochlear Implants

Blake S. Wilson, Enrique A. Lopez-Poveda, and Reinhold Schatzer

9.1 Introduction

Auditory models are used at least to some extent in all current designs of cochlear implant (CI) systems. For example, all current designs use a filter bank to mimic in a coarse way the filtering that occurs in the normal auditory periphery. However, the models used are relatively simple and do not include the intricacies of the normal processing or the interactions (e.g., feedback loops) among processing steps.

The purposes of this chapter are to (1) provide an overview of the current designs; (2) indicate the present levels of performance obtained with unilateral CIs; (3) mention two recent advances in implant design and performance, involving bilateral CIs and combined acoustic and electric stimulation of the auditory system for persons with some residual, low-frequency hearing; (4) list the remaining problems standing in the way of still-better implant systems; (5) describe the use of auditory models in implant design; and (6) offer some possibilities for the future. In broad terms, great advances have been made in the design and performance of implantable auditory prostheses, but much remains to be done. Some of the remaining problems and deficits may be addressed with the use of better and more sophisticated auditory models.

B.S. Wilson (✉)
Duke Hearing Center, Duke University Medical Center,
Durham, NC 27710, USA
and
Division of Otolaryngology, Head and Neck Surgery,
Department of Surgery, Duke University Medical Center,
Durham, NC 27710, USA
and
MED-EL Medical Electronics GmbH, 6020 Innsbruck, Austria
e-mail: blake.wilson@duke.edu

R. Meddis et al. (eds.), *Computational Models of the Auditory System*,
Springer Handbook of Auditory Research 35, DOI 10.1007/978-1-4419-5934-8_9,

9.2 Design of Implantable Auditory Prostheses

A detailed description of the design of implantable auditory prostheses is presented in Wilson (2004), in an earlier volume of the *Springer Handbook of Auditory Research.* Detailed descriptions of processing strategies for implant systems are provided in Wilson (2006) and in Wilson and Dorman (2009). A summary of these descriptions is presented in this chapter, to acquaint the reader with the main issues in implant design.

9.2.1 Components of Implant Systems

The essential components in a cochlear prosthesis system include (1) a microphone for sensing sound in the environment; (2) a speech processor to transform the microphone output into a set of stimuli for an implanted array of electrodes; (3) a transcutaneous link for the transmission of power and stimulus information across the skin; (4) an implanted receiver/stimulator to decode the information received from the radiofrequency signal produced by an external transmitting coil and then to generate stimuli using the instructions obtained from the decoded information; (5) a multiwire cable to connect the outputs of the receiver/stimulator to the individual electrodes; and (6) the electrode array. These components must work together as a system to support excellent performance, and a weakness in a component can degrade performance significantly.

9.2.2 Electrical Stimulation of the Auditory Nerve

The principal cause of hearing loss is damage to or complete destruction of the sensory hair cells. In the deaf or deafened cochlea, the hair cells are largely or completely absent, severing the connections (both afferent and efferent) between the peripheral and central auditory systems. The function of a cochlear prosthesis is to bypass the missing or damaged hair cells by stimulating directly the surviving neurons in the auditory nerve, to reinstate afferent input to the central system.

In some cases, the auditory nerve may be grossly compromised, severed, or missing, such as in some congenital causes of deafness, some types of basal skull fractures, and removals of tumors from the surface of or within the auditory nerve, which usually take the nerve with the resected tumor. In these (fortunately rare) cases, structures central to the auditory nerve must be stimulated to restore function. Sites that have been used include (1) the surface of the dorsal cochlear nucleus (DCN; e.g., Otto et al. 2002); (2) the surface of the DCN combined with intranucleus stimulation using penetrating electrodes in conjunction with the surface electrodes (McCreery 2008); and (3) the central nucleus of the inferior colliculus, using an array of electrodes on a penetrating shank or "carrier" (Lim et al. 2008). The number of patients who have received implants at these locations in the central auditory system is slightly higher than 500, whereas the number of patients who have received CIs to date (March 2009) exceeds 160,000. In the remainder of this chapter, discussion is restricted to CIs. Obviously, use of models in the design of implants for structures central to the auditory nerve also would need to include models of central – as well as peripheral – processing. (For now, the periphery is challenging enough, and consideration of central processing is beyond the scope of the chapter.)

In the deaf cochlea, and without the normal stimulation provided by the hair cells, the peripheral parts of the neurons – between the cell bodies in the spiral ganglion and the terminals within the organ of Corti – undergo retrograde degeneration and eventually die (Hinojosa and Marion 1983). Fortunately, the cell bodies are far more robust. At least some usually survive, even for prolonged deafness or for virulent etiologies such as meningitis (Hinojosa and Marion 1983; Miura et al. 2002; Leake and Rebscher 2004). These cells, or more specifically the nodes of Ranvier just distal or proximal to them, are the putative sites of excitation for CIs. In some cases, though, peripheral processes may survive, and excitation may possibly occur more peripherally.

Direct stimulation of remaining elements in the auditory nerve is produced by currents delivered through electrodes placed in the scala tympani (ST). Different electrodes in the implanted array may stimulate different subpopulations of neurons. Implant systems attempt to mimic or reproduce a tonotopic pattern of stimulation by activating basally situated electrodes to indicate the presence of high-frequency sounds, and by activating electrodes at more apical locations to indicate the presence of sounds with lower frequencies.

The spatial specificity of stimulation with ST electrodes most likely depends on a variety of factors, including the geometric arrangement of the electrodes; the proximity of the electrodes to the target neural structures; and the condition of the implanted cochlea in terms of nerve survival, ossification, and fibrosis around the intracochlear electrode array. Present evidence suggests, however, that no more than four to eight independent sites are available in a speech-processor context and using current electrode designs, even for arrays with as many as 22 electrodes (Lawson et al. 1996; Fishman et al. 1997; Wilson 1997; Kiefer et al. 2000; Friesen et al. 2001; Garnham et al. 2002). Most likely, the number of independent sites is limited by substantial overlaps in the electric fields from adjacent (and more distant) electrodes (e.g., Fu and Nogaki 2004; Dorman and Spahr 2006).

9.2.3 Processing Strategies

One of the simpler and most effective approaches for representing speech and other sounds with present-day CIs is illustrated in Fig. 9.1. This is the continuous interleaved sampling (CIS) strategy (Wilson et al. 1991), which is used as the default strategy or as a processing option in all implant systems now in widespread clinical use.

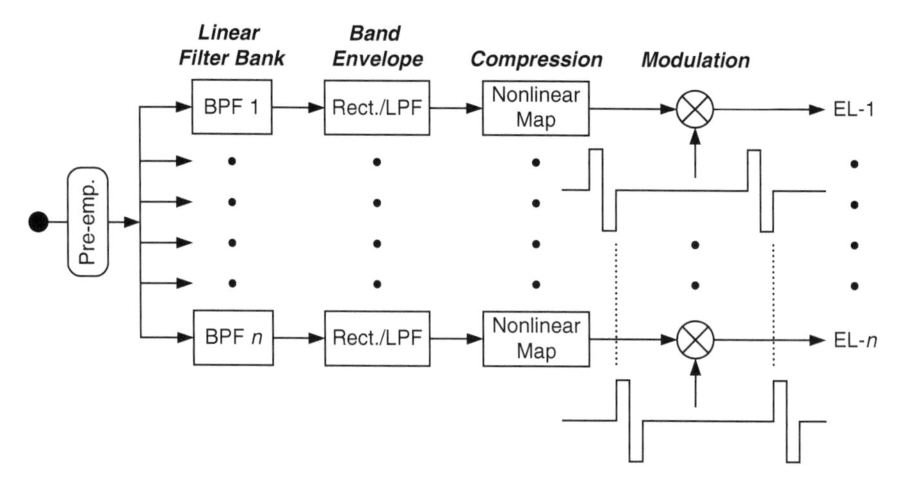

Fig. 9.1 Block diagram of the continuous interleaved sampling (CIS) strategy. The input to the strategy is indicated by the *filled circle* in the *left-most part* of the diagram. This input can be provided by a microphone or alternative sources such as an FM wireless link in a classroom. Following the input, the strategy uses a pre-emphasis filter (Pre-emp.) to attenuate strong components in speech below 1.2 kHz. This filter is followed by multiple channels of processing. Each channel includes stages of bandpass filtering (BPF), envelope detection, compression, and modulation. The envelope detectors generally use a full-wave or half-wave rectifier (Rect.) followed by a low-pass filter (LPF). A Hilbert transform or a half-wave rectifier without the LPF also may be used. Carrier waveforms for two of the modulators are shown immediately below the two corresponding multiplier blocks (*circles* with an "x" mark within them). The outputs of the multipliers are directed to intracochlear electrodes (EL-1 to EL-*n*), via a transcutaneous link or percutaneous connector (Adapted from Wilson et al. 1991 with permission of the Nature Publishing Group.)

The CIS strategy filters speech or other input sounds into bands of frequencies with a bank of bandpass filters (Fig. 9.1). Envelope variations in the different bands are represented at corresponding electrodes in the cochlea with amplitude-modulated trains of biphasic current pulses. The envelope signals extracted from the bandpass filters are compressed with a nonlinear mapping function before the modulation, to map the wide dynamic range of sound in the environment (up to ~90 dB) into the narrow dynamic range of electrically evoked hearing (~10 dB or somewhat higher). The output of each bandpass channel is directed to a single intracochlear electrode, with low-to-high channels assigned to apical-to-basal electrodes, to mimic at least the order, if not the precise locations, of frequency mapping in the normal cochlea. The pulse trains for the different channels and corresponding electrodes are interleaved in time, so that the pulses across channels and electrodes are nonsimultaneous. This eliminates a principal component of electrode interaction, which otherwise would be produced by direct vector summation of the electric fields from different (simultaneously activated) electrodes. (Other interaction components are not eliminated with the interleaving, but these other components are generally much lower in magnitude than the principal component resulting from the summation of the electric fields; see, e.g., Favre and Pelizzone 1993.) The corner or "cutoff" frequency of the lowpass filter in each envelope detector typically is set at 200 Hz or higher, so that the fundamental frequencies (F0s) of speech sounds are represented in the modulation waveforms. Pulse rates in CIS processors typically approximate or exceed 1,000 pulses/s/electrode, for an adequate "sampling" of the highest frequencies in the modulation waveforms (Busby et al. 1993; Wilson 1997; Wilson et al. 1997). CIS gets its name from the continuous sampling of the (compressed) envelope signals by rapidly presented pulses that are interleaved across electrodes. Between 4 and 22 channels (and corresponding stimulus sites) have been used in CIS implementations to date.

Other strategies also have produced outstanding results. Among these are the CIS+, *n*-of-*m*, spectral peak (SPEAK), advanced combination encoder (ACE), "HiResolution" (HiRes), HiRes with the Fidelity 120 option (HiRes 120), and fine structure processing (FSP) strategies. These strategies are described in detail in Wilson and Dorman (2009).

9.3 Status Report

Cochlear implants are among the great success stories of modern medicine. Indeed, a principal conclusion of the 1995 NIH Consensus Conference on Cochlear Implants in Adults and Children (National Institutes of Health 1995) was that "A majority of those individuals with the latest speech processors for their implants will score above 80% correct on high-context sentences, even without visual cues." This restoration of function is truly remarkable and is far greater than that achieved to date with any other type of neural prosthesis (e.g., Wilson and Dorman 2008a).

9.3.1 Performance with Present-Day Unilateral Implants

Among the strategies reviewed or mentioned in the preceding text, all but the SPEAK strategy are in current widespread use and each of these currently utilized strategies supports recognition of monosyllabic words on the order of 50% correct (using hearing alone), across populations of tested subjects (see Table 2.4 in Wilson 2006). Variability in outcomes is high, however, with some subjects achieving scores at or near 100% correct and with other subjects scoring close to zero on this most difficult test used in standard audiological practice. Standard deviations of the scores range from approximately 10% to approximately 30% for the various studies conducted to date.

Scores for other (easier) tests also are similar across the strategies and the implant systems that utilize them. However, when the tests are made even more difficult than the monosyllabic word test, differences among systems can be demonstrated. For example, if patients are tested in noise and at soft presentation levels, then systems with a large dynamic range of input processing (before the bank of bandpass filters) outperform systems with a small input dynamic range (Spahr et al. 2007). This and other findings (such as those reviewed in Section 4.3 of Wilson 2004) emphasize the importance of the details in hardware (and software) implementations of processing strategies, especially for listening under adverse conditions.

The ranges of scores and other representative findings for contemporary CIs are illustrated in Fig. 9.2, which shows scores for 55 adult users of the MED-EL COMBI 40 implant system and the CIS processing strategy. Scores for the Hochmair–Schultz–Moser (HSM) sentences are presented in the top panel, and scores for recognition of the Freiburger monosyllabic words are presented in the bottom panel. (These are standard tests in the German language.) Results for five measurement intervals are shown, ranging from 1 month to 2 years after the initial fitting of the speech processor. (Note that time is plotted along a logarithmic scale in Fig. 9.2, to reflect the approximately logarithmic spacing of the intervals.) The solid line in each panel shows the median of the individual scores and the dashed and dotted lines show the interquartile ranges. The data are a superset of those reported in Helms et al. (1997) and include scores for additional subjects at various test intervals.

As is evident from the figure, scores are broadly distributed at each test interval and for both tests. However, ceiling effects are encountered for the sentence test for many of the subjects, especially at the later test intervals. At 24 months, 46 of the 55 subjects score above 80% correct, consistent with the 1995 NIH Consensus Statement. Scores for the recognition of monosyllabic words are much more broadly distributed. For example, at the 24-month interval only nine of the 55 subjects have scores above 80% correct and the distribution of scores from about 10% correct to nearly 100% correct is almost perfectly uniform.

An interesting aspect of the results presented in Fig. 9.2 is a learning or accommodation effect, with continuous improvements in scores over the first 12 months of use. This suggests the likely importance of brain function in determining outcomes, and the reorganization (brain plasticity) that must occur to utilize such sparse inputs to the maximum extent possible.

9.3.2 Recent Advances

Two recent advances in the design and performance of CIs are (1) electrical stimulation of both ears with bilateral CIs and (2) combined electric and acoustic stimulation (EAS) of the auditory system for persons with residual hearing at low frequencies. Bilateral electrical stimulation may reinstate at least to some extent the interaural amplitude and timing difference cues that allow people with normal hearing to lateralize sounds in the horizontal plane and to selectively "hear out" a voice or other source of sound from among multiple sources at different locations. In addition, stimulation on both sides may allow users to make use of the acoustic

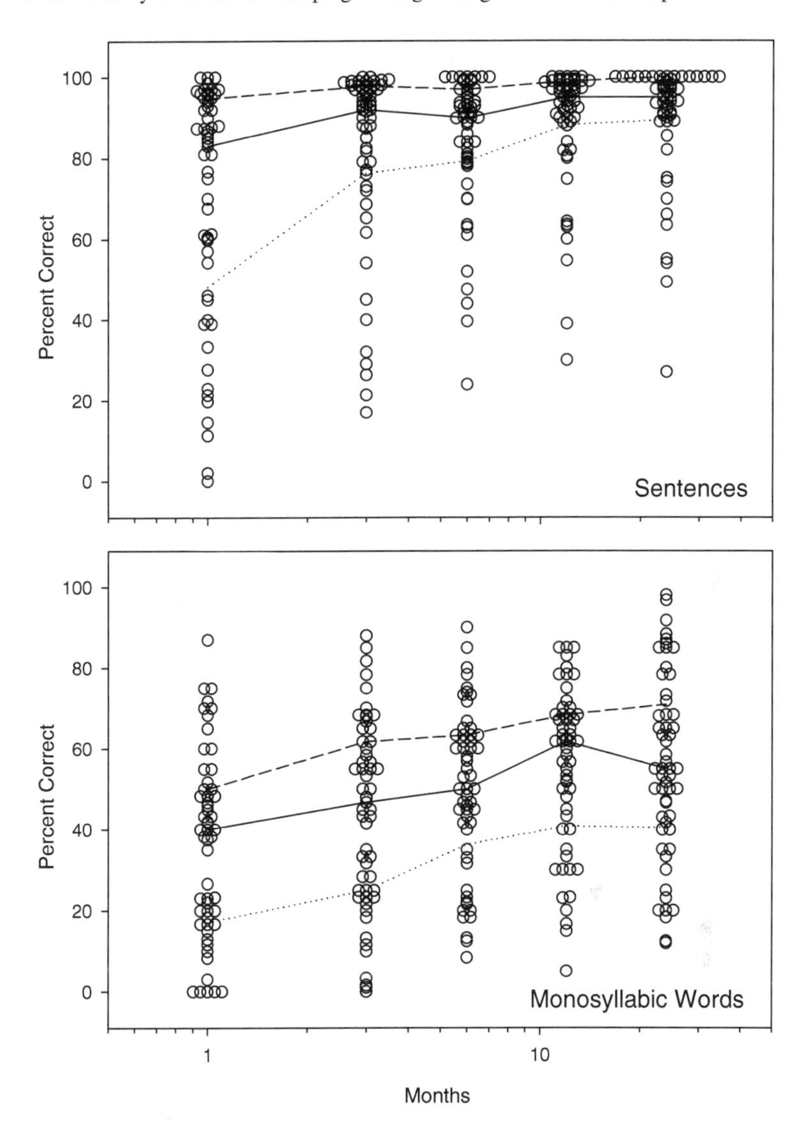

Fig. 9.2 Percentage correct scores for 55 users of the COMBI 40 implant and the continuous interleaved sampling (CIS) processing strategy. Scores for recognition of the Hochmair–Schultz–Moser (HSM) sentences are presented in the *top panel*, and scores for recognition of the Freiburger monosyllabic words are presented in the *bottom panel*. The *solid line* in each panel shows the median of the scores, and the *dashed* and *dotted lines* show the interquartile ranges. The data are an updated superset of those reported in Helms et al. (1997), kindly provided by Patrick D'Haese of MED-EL GmbH, in Innsbruck, Austria. The experimental conditions and implantation criteria are described in Helms et al. All subjects took both tests at each of the indicated intervals after initial fitting of their speech processors. Identical scores at a single test interval are displaced horizontally for clarity. Thus, for example, the horizontal "line" of scores in the *top right portion* of the *top panel* all represent scores for the 24-month test interval (From Wilson 2006. Used with permission of John Wiley & Sons.)

shadow cast by the head for sound sources off the midline. In such cases, the signal-to-noise ratio (S/N) may well be more favorable at one ear compared to the other for any one among multiple sources of sound, and users may be able to attend to the ear with the better S/N for the desired source. Combined EAS may preserve a relatively normal hearing ability at low frequencies, with excellent frequency resolution and other attributes of normal hearing, while providing a complementary representation of high frequency sounds with the CI and electrical stimulation.

Each of these relatively new approaches, bilateral electrical stimulation and combined EAS, utilizes or reinstates a part of the natural system. Two ears are better than one, and use of even a part of normal or nearly normal hearing at low frequencies can provide a highly significant advantage. Both approaches have produced large improvements in speech reception performance compared to control conditions (see review in Wilson and Dorman 2009). Detailed discussions of possible mechanisms underlying the benefits of bilateral CIs and of combined EAS are provided in Wilson and Dorman (2009); Qin and Oxenham (2006); Dorman et al. (2007); and Turner et al. (2008).

9.3.3 Remaining Problems

Present-day CIs support a high level of function for the great majority of patients, as indicated in part by sentence scores of 80% correct or higher for most patients and the ability of most patients to use the telephone. In addition, some patients achieve spectacularly high scores with present-day CIs. Indeed, their scores are in the normal ranges even for the most difficult of standard audiological tests (e.g., the top few patients in Fig. 9.2 and the patient described in Wilson and Dorman 2007). Such results are both encouraging and surprising in that the implants provide only a very crude mimicking of only some aspects of the normal physiology.

Such wonderful results might make one wonder whether any additional improvements are needed for CIs, including incorporation of more advanced or sophisticated auditory models in processing strategies for implants. Any thought of complacency is quickly dashed, however, by observing that even the high-scoring patients still do not hear as well as listeners with normal hearing, particularly in demanding situations such as speech presented in competition with noise or other talkers. In addition, and much more importantly, a wide range of outcomes persists, even with the current processing strategies and implant systems, and even with bilateral implants or combined EAS. Thus, although great progress has been made, much remains to be done.

Major remaining problems with CIs are described in detail in Wilson and Dorman (2008b). Briefly, they include:

- *A high variability in outcomes*, as just mentioned
- *Difficulty in recognizing speech in adverse situations for all patients*, but particularly those at the low end of the performance spectrum
- *Highly limited access to music and other sounds that are more complex than speech*, for all patients except those using combined EAS

- *Likely limitations imposed by impairments in auditory pathway or cortical function*, due, for example, to long periods of auditory deprivation before receiving a cochlear implant
- *Likely limitations imposed by present electrode designs and placements*, as mentioned in Sect. 9.2.2
- *An apparent disconnect between the number of discriminable sites vs. the number of effective channels with implants*, following the observation that 22 or more sites can be perceived as discriminable pitches by some patients when the sites are stimulated in isolation, but no patient tested to date has more than about eight effective channels when stimuli are rapidly sequenced across electrodes in a real-time, speech-processor context
- *A possible deficit in the representation of "fine structure" or fine-frequency information with present "envelope based" strategies such as CIS*, in which frequency variations or frequency differences of components within the passband of a bandpass channel may not be represented or represented well
- *A less-than-resolute representation of F0s for complex sounds*, which most likely requires a good representation of the first several harmonics of the F0, at the correct tonotopic sites, in addition to the F0 itself, which may be achieved by combined EAS and possibly explains at least in part the excellent results obtained with combined EAS, but would be very difficult if not impossible to achieve with electrical stimuli only
- *Little or no sound localization ability*, except for users of bilateral CIs and possibly for users of combined EAS when the electric and acoustic stimuli are presented to opposite ears (or when the acoustic stimuli are presented to both ears)
- *A high effort in listening that is required for all CI patients*, including the top performers in achieving their high scores, and especially including patients with lower levels of performance, in struggling to understand portions of speech and to "glue" those portions together into a whole using contextual cues

Possible ways to address these various problems are also outlined in Wilson and Dorman (2008b). Incorporation of more accurate and more sophisticated auditory models – beyond a crude mimicking of auditory filtering – might be helpful in addressing many of these problems, particularly those relating to perception of complex sounds and to recognizing difficult speech (e.g., monosyllabic words) or speech presented in competition with noise or other talkers. In addition, a more accurate or more natural representation that might be provided with use of better models might reduce the concentration required for implant patients to understand speech.

9.4 Use of Auditory Models in Implant Design

Auditory models may be incorporated into implant designs to provide an ever closer approximation to the signal processing that occurs in the normal cochlea and to the patterns of discharge recorded at the auditory nerve in higher mammals (e.g., cats) with normal hearing, in response to acoustic stimuli. Of course, a perfect representation at the periphery should produce scores for implant patients that are indistinguishable from those achieved by subjects with normal hearing, for patients who have a fully intact and fully functional "auditory brain." In addition, the high scores should be obtained immediately and without any learning period for such a representation and for such patients, in that the representation would be perfectly matched with what the normal brain is configured to receive.

A perfect representation is not attainable, at least with present ST electrodes, which do not provide sufficient resolution for independent control of single neurons. Instead, only gross control over largely overlapping populations of neurons is possible. In addition, the great majority of implant patients most likely suffer from at least some deficits in central processing abilities. (The only exceptions may be patients who had acquired language and who had normal hearing before becoming deaf and then received a CI very soon after deafness, or deaf children who received their implants before the age of about 3 years, when the brain is still capable of large plastic changes and for whom areas of cortex normally associated with auditory processing have not been "encroached" by other senses or functions.) For these patients, even a perfect representation at the periphery would not guarantee a good result; indeed, a simpler or sparser representation may provide a better match between what is presented and what the compromised brain can receive and process. Thus, effective applications of auditory models get down to questions of (1) stimulus control and (2) how well the brain can utilize the input from the periphery. One can imagine, for example, that an application of a sophisticated model might not produce any better performance than an application of a much simpler model, if the details of the sophisticated model cannot be represented in the stimuli. Further, even in cases for which a more faithful representation is possible, the details must be perceived and utilized to provide any additional benefit, beyond the benefits obtained with simple models.

A related consideration is parsimony in design. Engineers in the United States call this the "KISS" principle, for "keep it simple, stupid." This principle has emerged from hard-won experience and knowing that unnecessary frills often actually degrade function or performance. Most likely, this also is the case with implants. For example, complex patterns of stimulation produced with the use of sophisticated auditory models may exacerbate electrode interactions or produce other undesired (or even unanticipated) effects, whereas use of simpler models may facilitate a better understanding of a system under development and perhaps a more robust representation of the key features of the input signal. The subtleties may be lost with the use of relatively simple models, but such use has advantages and is almost certainly best when the subtleties cannot be represented or perceived.

Of course, the models used can be so simple that much of the information in the input is discarded or highly distorted, or that the resulting representation at the auditory nerve is completely unlike the normal representation and therefore cannot be interpreted by the brain, at least without extensive training. Thus, a balance must be struck, and that balance depends on how well the temporal and spatial patterns of auditory nerve discharges can be controlled by the implant, and on how well the user's brain can perceive, interpret, and utilize the patterns. The balance may be altered with (1) advances in stimulus control, for example, with new designs or placements of electrodes; (2) utilization of a part of the natural system, as with combined EAS; or (3) development and use of training procedures that may facilitate desired plastic changes in brain function and thereby enable the brain to utilize additional details in the input from the periphery.

The experience to date with CIs suggests that the balance may be too far toward the simple side in contemporary systems. This experience has shown that large improvements are possible with manipulations at the periphery – with the advent of new and better processing strategies and with combined EAS, for example. On the other hand, there is no evidence that the present models, as applied in the current processing strategies and in conjunction with ST electrodes, are too complex. Possibly, an adjustment of the present balance, toward a more faithful reproduction of processing in the normal cochlea, may be advantageous, perhaps particularly so for patients with a normal or nearly normal auditory brain.

An approach for incorporating progressively more accurate and more faithful models into implant designs is outlined in the remainder of this chapter. The approach already is showing encouraging results in tests with implant patients (e.g., Wilson et al. 2006), and the approach may be pursued further as advances in stimulus control allow representation of further subtleties in model outputs. (Such advances are likely; see, e.g., Wilson and Dorman 2008b, 2009.) The balance to be struck at each stage may be especially well informed by Einstein's famous maxim, "Make things as simple as possible, but no simpler." (This is an often-cited paraphrase of Einstein's actual statement, which was "The supreme goal of all theory is to make the irreducible basic elements as simple and as few as possible without having to surrender the adequate representation of a single datum of experience.")

9.4.1 The Target

The target for a "better mimicking" approach for CIs is the signal processing that occurs in the normal cochlea. A simplified block diagram of the processing is presented in Fig. 9.3. The processing includes (1) induction of pressure variations into the cochlear fluids by movements of the stapes or via various "bone conduction" pathways; (2) initiation of a traveling wave of displacement along the basilar membrane (BM) by the pressure variations, which always progresses from base to apex; (3) highly nonlinear filtering of the input by the BM and associated structures, including level-dependent tuning and compression, which is produced by a local feedback loop involving electromotile contractions of the outer hair cells (OHCs); (4) rectification, low-pass filtering, and further compression in the transduction of BM movements to membrane potentials at the inner hair cells (IHCs); (5) a further noninstantaneous compression and adaptation at the synapses between IHCs and adjacent type I fibers of the auditory nerve; (6) random release of chemical transmitter substance at the bases of the IHCs, into the synaptic cleft even in the absence of stimulation, which gives rise to spontaneous activity in auditory neurons and statistical independence in the discharge patterns among neurons; (7) facilitation or

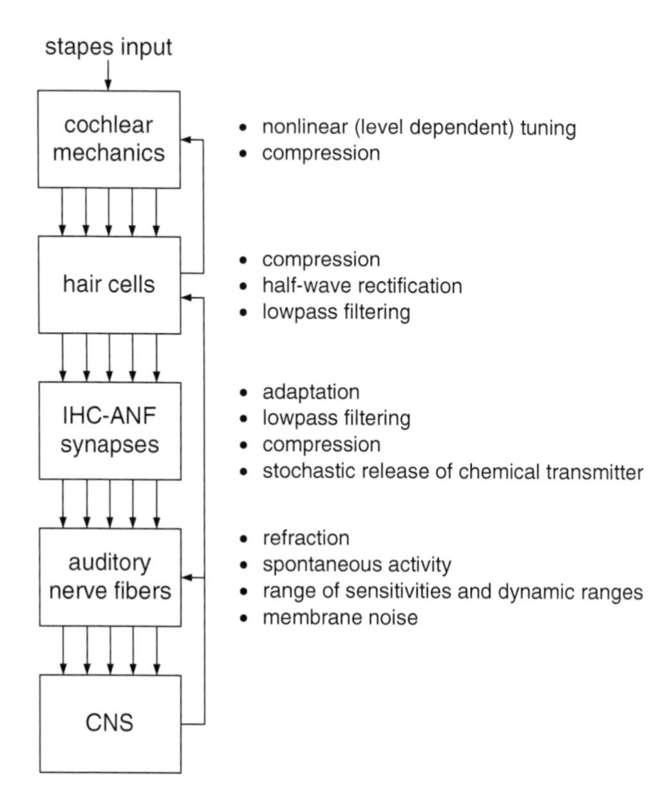

Fig. 9.3 Simplified block diagram of the normal auditory periphery. The hair cells include three rows of "outer" hair cells (OHCs) and one row of "inner" hair cells (IHCs) along the length of the organ of Corti. Electromotile actions of the OHCs alter the mechanics of the basilar membrane (BM) in a way that increases the BM's sensitivity to sound and also sharpens its spatial pattern of responses to individual frequency components in the input, particularly at low sound pressure levels. This feedback provided by the OHCs is indicated by the upper path on the right side of the diagram, ending with an arrow pointing to the box labeled "cochlear mechanics." Approximately 95% of the fibers in the auditory nerve synapse at the bases of the IHCs, and modulation of discharge activity in those fibers provides the afferent input from the auditory nerve to the brain, which is labeled "CNS" in the diagram. Efferent fibers in the nerve terminate at the bases of the OHCs and adjacent to fibers that innervate the IHCs. Efferent control may therefore alter responses of the OHCs or fibers innervating the IHCs or both. This feedback—from the brain to the cochlea—is indicated by the lower path on the right side of the diagram, ending with one arrow pointing to the box labeled "hair cells" and with another arrow pointing to the box labeled "auditory nerve fibers." The forward path, from the inward and outward movements of the stapes to discharge patterns in the auditory nerve, includes the initiation of a traveling wave of displacement along the BM; modification of the BM responses by the action of the OHCs; facilitation or inhibition of chemical transmitter release at the bases of the IHCs according to the degree and direction of the "shearing" of their cilia as a result of local displacements of the BM; and modulation of discharge activity in the auditory nerve fibers (ANFs) that innervate each IHC. The responses of the OHCs or of fibers innervating the IHCs may be modified by efferent control. The feedback path from the CNS to the stapedius muscle is not shown in the present diagram; this path controls the transmission of sound energy through the middle ear. (Adapted from Delgutte 1996. Used with permission of Springer-Verlag.)

inhibition of this "baseline" release of chemical transmitter substance into the cleft according to the depolarization or hyperpolarization of the IHC membranes; (8) excitation of neurons when the cleft contents exceed a threshold (which is different for different neurons, as described later); (9) the inability of single neurons to respond immediately after a prior response, due to refractory effects; (10) a wide distribution of spontaneous rates among the 10–20 afferent fibers that innervate each IHC; (11) a wide distribution of thresholds and dynamic ranges of those fibers, which is related to the distribution of spontaneous activities among the fibers (e.g., fibers with low average rates of spontaneous activity have high thresholds and relatively wide dynamic ranges, whereas fibers with high average rates have low thresholds and relatively narrow dynamic ranges); and (12) feedback control from the central nervous system (CNS) that can alter the response properties of the OHCs and the afferent fibers innervating the IHCs. Many additional details about the processing in normal auditory periphery are presented in Chapters 2 and 5, by Ray Meddis and Enrique Lopez-Poveda and by Bertrand Delgutte and Kenneth Hancock, respectively. Chapter 7 by Michael Heinz also indicates how the processing is affected by damage to the auditory periphery, primarily at the hair cells. Each of the chapters presents models of the processing. The models of normal processing may be incorporated into processing strategies for CIs, as indicated by the example presented in Sect. 9.4.3.

9.4.2 The Need for Better Models

Present processing strategies for CIs, such as the CIS strategy illustrated in Fig. 9.1, provide only a very crude approximation to the processing that occurs in the normal cochlea. For example, a bank of linear bandpass filters is used instead of the nonlinear and coupled filters that would model normal auditory function. Also, a single nonlinear map is used in the CIS and other strategies to produce the overall compression that the normal system achieves in multiple steps. The compression in CIS and other processors is instantaneous, whereas compression at the IHC/neuron synapse in the normal cochlea is noninstantaneous, with large adaptation effects.

Such differences between the normal processing and what present implants provide may well limit the perceptual abilities of implant patients. For example, Deng and Geisler (1987), among others, have shown that nonlinearities in BM filtering (as influenced by the action of the OHCs) greatly enhance the neural representation of speech sounds presented in competition with noise. Similarly, findings of Tchorz and Kollmeier (1999) have indicated the importance of adaptation at the IHC/neuron synapse in representing temporal events or markers in speech, especially for speech presented in noise. (Such markers are useful in the recognition and identification of stop consonants, for example.) Aspects of the normal processing are responsible for the sharp tuning, high sensitivity, wide dynamic range, and high resolution of normal hearing. Those aspects, and indeed entire steps and feedback loops, are missing in the processing used today for CIs. Incorporation of at least some of the missing parts and feedback loops, through the use of better models, may lead to improvements in performance, for example, improvements in recognizing speech presented in competition with noise or other talkers, which remains as one of the major unsolved problems with contemporary CIs.

9.4.3 A New Processor Structure

An approach for providing a closer approximation to the normal processing is suggested in Fig. 9.4. The idea is to use more accurate and more sophisticated models than have been used in the past, and further to represent the new subtleties and details in the outputs (ultimately the modulation waveforms for the stimulus pulses) by exploiting possibilities for stimulus (and neural) control to the fullest.

Comparison of Figs. 9.1 and 9.4 shows that, in the new structure (Fig. 9.4), a model of nonlinear filtering is used instead of the bank of linear filters, and a model of the IHC membrane and synapse is used instead of an envelope detector

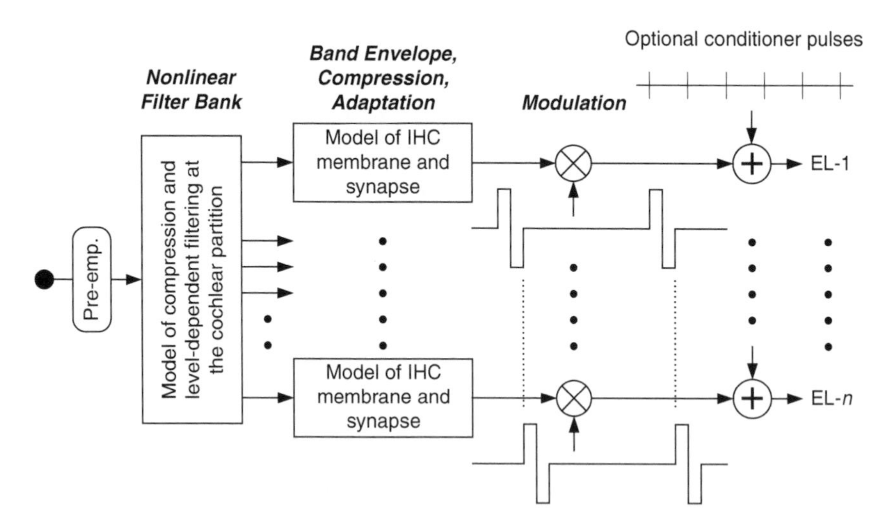

Fig. 9.4 Incorporation of more-sophisticated auditory models into a continuous interleaved sampling (CIS) structure. Models that could be used for the block labeled "Model of compression and level-dependent filtering at the cochlear partition" include those developed and described by Carney (1993), Meddis et al. (2001), Robert and Eriksson (1999), and Zhang et al. (2001). Models that could be used for the blocks labeled "Model of IHC membrane and synapse" include the one developed and described by Meddis (1986, 1988). Many additional models could be used for the various blocks and some of those models are described in Chapter 2. Abbreviations in the diagram are the same as those used in Figure 9.1. (From Wilson et al. 2003. Used with permission of the Annual Reviews.)

and nonlinear mapping function. Note that the mapping function is not needed in the new structure, because the multiple stages of compression implemented in the models should provide the overall compression required for mapping the wide dynamic range of processor inputs onto the narrow dynamic range of electrically evoked hearing. (Some scaling may be needed, but the compression functions should be approximately correct, as each of the three compression stages found in normal hearing – from the stapes input to evoked discharges in the primary auditory neurons – is included in the models.) The compression achieved in this way would be much more analogous to the way it is achieved in normal hearing.

Relatively high rates for the stimulus (carrier) pulses, or the addition of high-rate "conditioner" pulses (that are not modulated), may be used if desired, to impart spontaneous-like activity in auditory neurons and stochastic independence among neurons (Wilson et al. 1997; Rubinstein et al. 1999). This can increase the dynamic ranges of auditory neuron responses to electrical stimuli, bringing the ranges closer to those observed for normal hearing using acoustic stimuli. Stochastic independence among neurons also may be helpful in representing rapid temporal variations in the stimuli at each electrode, in that higher frequencies can be represented in the ensemble of all neural responses to a stimulus when the neurons are independent, as compared to conditions in which all (or most) neural responses to a stimulus are highly correlated (see, e.g., Parnas 1996; Wilson et al. 1997). (This may be one of the functions of spontaneous activity in normal hearing.) Representation of these higher frequencies or rapid temporal events does not guarantee perception of them, of course, but the representation may be a first necessary step in evaluating the hypotheses that the variations can in fact be perceived and that such perception leads to improvements in speech and music reception.

The approach shown in Fig. 9.4 is intended as a move in the direction of closer mimicking. It does not include feedback control from the CNS, and it does not include a way to stimulate fibers close to an electrode differentially, to mimic the distributions of thresholds and dynamic ranges of the multiple neurons innervating each IHC in the normal cochlea. In addition, the approach does not reproduce the details of the temporal and spatial patterns of responses in the auditory nerve that are found in normal hearing, for example, the precise "placements" of harmonics along the tonotopic axis that may be required for highly salient representations of the F0s of complex sounds (Oxenham et al. 2004), or the rapid increases in latencies of single-neuron responses at and near the positions of maximal displacements along the BM, which might be "read" by the central auditory system to infer the sites of the maximal displacements and therefore the frequencies of components in the input at the stapes (e.g., Loeb et al. 1983). The approach does, however, have the potential to reinstate other aspects of the normal processing that are likely to be important, including details of the filtering at the BM and associated structures, and including noninstantaneous compression and adaptation at the IHCs and their synapses. Future advances in stimulus control may allow representation of additional details.

9.4.4 Preliminary Studies by the Authors

Studies are underway in the authors' laboratories to evaluate processing strategies that follow the general structure shown in Fig. 9.4 (e.g., Wilson et al. 2006). The work is proceeding in steps, including (1) substitution of a bank of dual-resonance, nonlinear (DRNL) filters (Lopez-Poveda and Meddis 2001; Meddis et al. 2001; Chapter 2) for the bank of linear filters used in a standard CIS processor; (2) substitution of the Meddis IHC and synapse model (Meddis 1986, 1988) for the envelope detector and for some of the compression ordinarily provided by the nonlinear mapping function in a standard CIS processor; and (3) substitution of both the DRNL filters and the Meddis IHC and synapse model, with fine tuning of the interstage gains and amounts of compression at the various stages. Work thus far has focused on implementation and evaluation of processors using DRNL filters (step 1). For those processors, the envelope detectors and nonlinear mapping functions are retained, but the amount of compression provided by the mapping functions is greatly reduced because substantial compression is provided by the DRNL filters (whereas no compression is provided by the linear filters in the standard CIS structure). The DRNL filters have many parameters whose adjustment may affect performance. The starting point for the preliminary studies was to use parameter values that produced approximately uniform compressions at the most responsive frequencies (nominal "center frequencies") of the different filters. This choice departs from the highly nonuniform compression across frequencies (and filters) described in Lopez-Poveda and Meddis (2001), but corresponds to more recent findings (Lopez-Poveda et al. 2003; Williams and Bacon 2005). The DRNL filters do not model the cross-coupling between neighboring filters found in normal hearing, but they do model the principal nonlinearities of filtering at the BM/OHC complex, including compression and the variation in tuning with input level. Thus, use of DRNL filters may allow a move in the direction of closer mimicking.

In general, the frequency responses of the DRNL filters are much sharper than those of the linear bandpass filters used in standard CIS processors, at least for 6–12 channels of processing and stimulation, and at least for low to moderate input levels. Thus, if one simply substitutes DRNL filters for the standard filters without alteration, then substantial gaps will be introduced in the represented spectra of lower-level inputs to the filter bank. Such a "picket fence" effect might degrade performance, even though other aspects of DRNL processing may be beneficial.

The studies to date have included evaluation of processors using the DRNL filters with the parameter values mentioned in the preceding text, and also with parameter values that produce shallower and overlapping frequency responses, but that retain other properties of DRNL responses such as compression. This latter set of values was designed to deal with the picket-fence effect.

In addition, other ways to deal with the picket-fence effect were evaluated, including (1) assigning multiple DRNL channels (see definition of a DRNL channel later) and their outputs to single sites of stimulation, with slightly different nominal center frequencies among the channels for each site, and (2) increasing the number of discriminable sites to a high number using "virtual sites" (see later) in addition to the physical electrodes, and then assigning one DRNL channel and its output to each of the discriminable sites. In either of these cases, a high number of DRNL channels was used, with the first set of parameter values (i.e., those producing sharp responses in the frequency domain, as in normal hearing) and with overlapping responses (i.e., without the picket-fence effect).

The "*n*-to-*m*" constructs of the first approach included two variations. In one, the average of outputs from the multiple DRNL channels was calculated and then that average was used to determine the amplitude of a stimulus pulse for a particular electrode. Each DRNL channel included a DRNL filter, an envelope detector, and a compressive mapping function. Thus, the average was the average of mapped amplitudes for the number of DRNL channels assigned to the electrode. This variation was called the "avg *n*-to-*m* strategy," in which *m* was the maximum number of electrodes available (or utilized) in the implant and in which *n* was the total number of DRNL channels, an integer multiple of *m*. In the other variation, the maximum among outputs from the DRNL channels for each electrode was identified and then that maximum was used to determine the amplitude of the stimulus pulse. This second variation was called the "max *n*-to-*m* strategy." Both the avg and max strategies were designed to retain the sharp tuning of individual DRNL filters while minimizing or eliminating the picket-fence effect. However, information obviously was lost in the averaging or in the selection of the peak response, for the avg and max strategies respectively.

The second general approach used virtual sites (or "channels") as a way to increase the number of discriminable sites up to the number of DRNL filters required to produce a relatively uniform response across all represented frequencies (i.e., without the picket-fence effect). As described in detail elsewhere (e.g., Wilson et al. 1994; Bonham and Litvak 2008), virtual sites are formed by delivering currents simultaneously to pairs of adjacent electrodes. The ratio of the currents most likely "shifts" or "steers" the neural excitation field between the two electrodes. Such shifts may be perceived as shifts in pitch, and indeed many discriminable pitches can generally be produced between adjacent electrodes using different ratios for the currents, and further these pitches can be distinct from the pitches produced with stimulation of either electrode in the pair alone. Thus, combinations of virtual sites with stimulation of the single electrodes in the implant can produce a high number of discriminable pitches. For example, Koch et al. (2007) found an average of 93 discriminable pitches for a large population of subjects using virtual sites in addition to the single-electrode sites with either of two versions of the Advanced Bionics Corp. electrode array, both of which include 16 physical intracochlear electrodes spaced approximately 1 mm apart.

Results for these various preliminary studies are presented in Schatzer et al. (2003) and in Wilson et al. (2006). The studies described in Schatzer et al. included seven subjects and evaluations of various implementations of processors using DRNL filters, including (1) one-to-one assignments of DRNL channel outputs to electrodes, using either of the two parameter sets; (2) avg *n*-to-*m* processors, using the first of the parameter sets and typically using values of 22 and 11 for n and m, respectively; and (3) max n-to-m processors, also using the first of the parameter sets and the same choices for n and m. The studies described in Wilson et al. also included an additional subject tested with processors that combined DRNL filtering with virtual-site stimulation. For that subject, 21 DRNL channels were mapped onto six intracochlear electrodes and three virtual sites between each pair of adjacent electrodes, for a total of 21 sites and corresponding (distinct) pitches. The first seven subjects used MED-EL GmbH implants with their 12 intracochlear electrodes, and the eighth subject used an Ineraid (Richards Medical Corp.) implant with its six intracochlear electrodes. Additional details about the processor implementations and the subjects are presented in the references just cited.

In broad terms, the results have been encouraging. Processors that included DRNL filters and that used the n-to-m strategies supported speech reception scores that were significantly better than the scores obtained with either of the processors that also included DRNL filters but used one-to-one mappings of DRNL channel outputs to electrodes. In addition, the scores obtained with processors using the n-to-m strategies were immediately on a par with the scores obtained with control CIS processors using m channels and associated stimulus sites, which each of the subjects had used in their daily lives, usually for years. This result was encouraging because such initial equivalence can be followed by much better performance with the new approach, once subjects gain some experience with the new approach (e.g., Tyler et al. 1986).

In addition, and for the one tested subject, the processor using DRNL filters in combination with virtual-site stimulation supported significantly better performance than the control CIS processor, especially for speech reception in noise. Among other things, this finding suggests that a possibly-higher level of stimulus control provided with the use of virtual-site stimulation allowed details in the DRNL channel outputs to be represented in a way that they could be perceived. (In particular, the use of virtual sites may have improved at least somewhat the precision in the cochlear place representation of the different DRNL channels, and may have allowed the use of enough DRNL channels to provide a nearly uniform response across all frequencies spanned by the filters, without discarding or distorting information as in the n-to-m strategies. Similar results might have been obtained with an electrode array having a high number of electrodes and a high number of discriminable sites for some patients, for example, the Cochlear Ltd. array with its 22 intracochlear electrodes, perhaps obviating the apparent need for the virtual sites. The postulated increase in stimulus control using either of these approaches – virtual sites or an electrode array with many contacts – would not necessarily produce an increase in the number of effective channels with implants; however, the postulated increase may have allowed a better representation of the DRNL channel outputs and a relatively good way to deal with the picket-fence effect.)

Following these encouraging results with the first eight subjects, two additional subjects have been studied. One of the subjects uses the MED-EL PULSAR implant, which includes 12 intracochlear electrodes as in the COMBI 40+ implant but also allows for the simultaneous stimulation of adjacent electrodes and therefore the construction of virtual sites between the electrodes. Processors using 21 DRNL channels mapped to 21 sites (11 electrodes plus ten virtual sites) have been tested with this subject and compared with control CIS processors using the same number of channels and sites (also using virtual sites). The second subject was implanted with an experimental version of the Cochlear Ltd. device, which uses a percutaneous connector rather than the standard transcutaneous link, for direct electrical access to the implanted electrode array from the outside. (The direct access eliminates restrictions in specifying stimuli for the implanted electrodes.) This implant device includes 22 intracochlear electrodes, as noted earlier, which allowed a one-to-one mapping of the outputs from 21 DRNL channels to 21 of the electrodes, thereby eliminating the need to construct virtual sites. (However, all 21 sites for either of these subjects may not have been discriminable from one another; this remains to be evaluated.) Such processors and the one-to-one mapping were compared with control CIS processors, also using 21 processing channels (with linear bandpass filters) whose outputs were represented at 21 corresponding electrodes.

As with the prior subjects, each of these two additional subjects had used CIS in their daily lives all day and for every day for at least 1 year preceding the acute tests with the alternative strategies. This may have put the alternative strategies at a disadvantage, as noted earlier for the prior subjects.

Results from the tests with the additional subjects have been generally consistent with the prior results, that is, processors using the DRNL filters produced speech reception scores that were immediately on a par with the control CIS processors. Further studies obviously are needed, but the findings thus far have been encouraging.

9.4.5 Work in Other Laboratories

The initial work outlined in Sect. 9.4.4 has inspired two other studies, also to evaluate the general structure shown in Fig. 9.4. Kim et al. (2007) compared processors using DRNL filters with a processor using linear bandpass filters, and Nogueira et al. (2007) compared the ACE processing strategy with otherwise identical strategies that used either nonlinear filters, an implementation of the Meddis IHC/synapse model, or both. (The ACE strategy is described in detail in Wilson and Dorman 2009.) In the study by Kim et al., eight subjects with normal hearing listened to acoustic simulations of the processors, and in the study by Nogueira et al., different groups of cochlear implant patients served as subjects for each of the three comparisons, with three subjects in each group. (Only one of the subjects participated in more than one of the comparisons.) The nonlinear filters used in the latter study were not DRNL filters but instead were based on a model of peripheral auditory filtering and masking first described by Zwicker (1985), and then refined and elaborated by Zwicker and Peisl (1990) and by Baumgarte (1999). The structure developed by Baumgarte was used. These filters included longitudinal coupling of neighboring sections, in addition to the nonlinear effects of sharper tuning at low input levels and of compression. The inputs to the processors used in the study by Kim et al. included vowels or vowels presented in competition with noise at the S/N of +5 dB. The inputs used in the study by Nogueira et al. included subsets of the HSM sentences, presented in competition with noise at the S/N of +15 dB. The task of the subjects in the Kim et al. study was to identify the vowels, under the various processing conditions, and the task of the subjects in the Nogueira et al. study was to report the words they heard in the sentences, also for the conditions used for each of the three groups of subjects.

The results from the study by Kim et al. demonstrated significantly higher scores for identification of vowels with the use of the DRNL filters as compared to the linear filters, especially for the vowels presented in competition with noise. Spectrograms of the simulation outputs also indicated an improved representation of the formant frequencies for the vowels, again especially for the noise condition, for the DRNL processing compared to the linear filter bank, consistent with the earlier findings of Deng and Geisler (1987), who compared simulated representations at the level of the auditory nerve using nonlinear vs. linear filters to model processing peripheral to the nerve.

The results from the study by Nogueira et al. did not show an improvement with respect to ACE in the recognition of the HSM sentences, for any of the three sets of comparisons. However, the number of subjects was quite small for each set, so a statistically significant difference in either direction would be difficult to demonstrate unless the difference in group scores was enormous. In addition, all of the subjects in the study had used the ACE strategy in their daily lives for at least 3.7 years before the acute comparisons with the alternative approaches. Nogueira et al. wisely emphasized the need for further tests and suggested that a longer period of experience with the alternative approaches could lead to better results with one or more of them.

9.5 The Future

Only the first "baby steps" have been taken in moving toward a closer mimicking of normal auditory processing with CIs. A broad plan has been developed (Sect. 9.4.3 and Fig. 9.4), and preliminary studies have been conducted to evaluate various aspects of the plan, mainly the substitution of realistic (nonlinear) filters for the linear filters used in conventional processing strategies for CIs. More studies are needed, with many more subjects and with evaluation of all aspects, both singly and in combinations. (Allowing subjects to gain "take home" experience with the new strategies before testing also might be helpful; this would require portable processors that could be programmed to implement the strategies.)

In addition, advances in stimulus control should be exploited to the fullest as they occur, to represent the details in the outputs of ever more accurate and ever more sophisticated models of auditory function. Further, parts of the normal processing that are not included in the broad plan of Fig. 9.4 should also receive attention, as those parts may be as important or even more important than some or all of the included parts.

Results from the studies reviewed in this chapter show that moving in the direction of a closer mimicking of normal auditory processing with implants is possible. However, it would be utter hubris to suggest by extension that artificial electrical stimulation could ever completely reproduce the normal processing, even with the use of accurate models and even with substantially higher levels of stimulus control. For this reason, all parts of the natural system that remain functional should be preserved and used in conjunction with electrical stimuli whenever possible and to the maximum extent possible.

9.6 Summary

Remarkable progress has been made in the development of implantable auditory prostheses. However, performance is not perfect even for the best patients, and performance is far from perfect for most patients. Performance might be improved by (somehow) increasing the spatial specificity of stimulation; directed training procedures; matching what is presented at the periphery with what the (sometimes compromised) brain can perceive and utilize; incorporation of more accurate auditory models in processing strategies for implants; or combinations of these. Approaches for the incorporation of more accurate (and generally more sophisticated) models are described in this chapter. Studies have been conducted to evaluate in a preliminary way some of the possibilities. The initial results have been encouraging. A more faithful replication of the exquisite and interactive processing that occurs in the normal auditory periphery most likely depends on (1) the use of appropriate models and (2) a requisite level of stimulus control. The latter is limited with present implant systems, but may be improved, perhaps substantially, in the future. Such improvements may well allow an ever closer approximation to the normal patterns of discharge in the auditory nerve, using an auditory prosthesis and electrical stimuli. The closer approximations would require the use of accurate auditory models that reproduce all important details of the processing peripheral to the auditory nerve and perhaps even the efferent control of the cochlea and the stapedius muscle by the brain. A caution is that use of highly sophisticated models in conjunction with limited stimulus control may actually degrade performance, compared to the use of simpler models. Thus, a balance should be sought between model complexity and the available level of stimulus control.

Acknowledgments Initial applications by the authors of relatively sophisticated auditory models in implant design were supported by the United States National Institutes of Health (NIH project N01-DC-2-1002 to Wilson) and by the Spanish Ministry of Science and Technology and IMSERSO (projects CIT-390000-2005-4, BFU2006-07536, and 131/06 to Lopez-Poveda). Material for segments of this chapter was drawn or adapted from several recent publications (Wilson 2006; Wilson and Dorman 2007, 2008a, b, 2009; Wilson et al. 2006).

References

Baumgarte F (1999) A physiological ear model for the emulation of masking. ORL J Otorhinolaryngol Relat Spec 61:294–304.

Bonham BH, Litvak LM (2008) Current focusing and steering: modeling, physiology and psychophysics. Hear Res 242:141–153.

Busby PA, Tong YC, Clark GM (1993) The perception of temporal modulations by cochlear implant patients. J Acoust Soc Am 94:124–131.

Carney LH (1993) A model for the responses of low-frequency auditory-nerve fibers in cat. J Acoust Soc Am 93:401–417.

Delgutte B (1996) Physiological models for basic auditory percepts. In: Hawkins HL, McMullen TA, Popper AN, Fay RR (eds), Auditory Computation. New York: Springer, pp. 157–220.

Deng L, Geisler CD (1987) A composite auditory model for processing speech sounds. J Acoust Soc Am 82:2001–2012.

Dorman, MF, Spahr AJ (2006) Speech perception by adults with multichannel cochlear implants. In: Waltzman SB, Roland JT Jr (eds), Cochlear Implants, 2nd ed. New York: Thieme, pp. 193–204.

Dorman MF, Gifford RH, Spahr AJ, McKarns SA (2007) The benefits of combining acoustic and electric stimulation for the recognition of speech, voice and melodies. Audiol Neurotol 13:105–112.

Favre E, Pelizzone M (1993) Channel interactions in patients using the Ineraid multichannel cochlear implant. Hear Res 66:150–156.

Fishman KE, Shannon RV, Slattery WH (1997) Speech recognition as a function of the number of electrodes used in the SPEAK cochlear implant speech processor. J Speech Lang Hear Res 40:1201–1215.

Friesen LM, Shannon RV, Baskent D, Wang X (2001) Speech recognition in noise as a function of the number of spectral channels: comparison of acoustic hearing and cochlear implants. J Acoust Soc Am 110:1150–1163.

Fu Q-J, Nogaki G (2004) Noise susceptibility of cochlear implant users: the role of spectral resolution and smearing. J Assoc Res Otolaryngol 6:19–27.

Garnham C, O'Driscol M, Ramsden R, Saeed S (2002) Speech understanding in noise with a Med-El COMBI 40+ cochlear implant using reduced channel sets. Ear Hear 23:540–552.

Helms J, Müller J, Schön F, Moser L, Arnold W, et al. (1997) Evaluation of performance with the COMBI 40 cochlear implant in adults: a multicentric clinical study. ORL J Otorhinolaryngol Relat Spec 59:23–35.

Hinojosa R, Marion M (1983) Histopathology of profound sensorineural deafness. Ann N Y Acad Sci 405:459–484.

Kiefer J, von Ilberg C, Hubner-Egener J, Rupprecht V, Knecht R (2000) Optimized speech understanding with the continuous interleaved sampling speech coding strategy in cochlear implants: effect of variations in stimulation rate and number of channels. Ann Otol Rhinol Laryngol 109:1009–1020.

Kim KH, Kim JH, Kim DH (2007) An improved speech processor for cochlear implant based on active nonlinear model of biological cochlea. Conf Proc IEEE Eng Med Biol Soc 1:6352–6359.

Koch DB, Downing M, Osberger MJ, Litvak L (2007) Using current steering to increase spectral resolution in CII and HiRes 90 K users. Ear Hear 28:39S–41S.

Lawson DT, Wilson BS, Zerbi M, Finley CC (1996) Speech processors for auditory prostheses: 22 electrode percutaneous study – results for the first five subjects. Third Quarterly Progress Report, NIH project N01-DC-5-2103. Bethesda, MD: Neural Prosthesis Program, National Institutes of Health.

Leake PA, Rebscher SJ (2004) Anatomical considerations and long-term effects of electrical stimulation. In: Zeng F-G, Popper AN, Fay RR (eds), Auditory Prostheses: Cochlear Implants and Beyond. New York: Springer, pp. 101–148.

Lim HH, Lenarz T, Anderson DJ, Lenarz M (2008) The auditory midbrain implant: effects of electrode location. Hear Res 242:74–85.

Loeb GE, White MW, Merzenich MM (1983) Spatial cross correlation: a proposed mechanism for acoustic pitch perception. Biol Cybern 47:149–163.

Lopez-Poveda EA, Meddis R (2001) A human nonlinear cochlear filterbank. J Acoust Soc Am 110:3107–3118.

Lopez-Poveda EA, Plack CJ, Meddis R (2003) Cochlear nonlinearity between 500 and 8000 Hz in listeners with normal hearing. J Acoust Soc Am 113:951–960.

McCreery DB (2008) Cochlear nucleus auditory prostheses. Hear Res 242:64–73.

Meddis R (1986) Simulation of mechanical to neural transduction in the auditory receptor. J Acoust Soc Am 79:702–711.

Meddis R (1988) Simulation of auditory-neural transduction: further studies. J Acoust Soc Am 83:1056–1063.

Meddis R, O'Mard LP, Lopez-Poveda EA (2001) A computational algorithm for computing nonlinear auditory frequency selectivity. J Acoust Soc Am 109:2852–2861.

Miura M, Sando I, Hirsch BE, Orita Y (2002) Analysis of spiral ganglion cell populations in children with normal and pathological ears. Ann Otol Rhinol Laryngol 111:1059–1065.

National Institutes of Health (1995) Cochlear implants in adults and children. NIH Consensus Statement 13(2):1–30. (This statement also is available in JAMA 274:1955–1961.)

Nogueira W, Kátai A, Harczos T, Klefenz F, Buechner A, Edler B (2007) An auditory model based strategy for cochlear implants. Conf Proc IEEE Eng Med Biol Soc 1:4127–4130.

Otto SR, Brackmann DE, Hitselberger WE, Shannon RV, Kuchta J (2002) Multichannel auditory brainstem implant: update on performance in 61 patients. J Neurosurg 96:1063–1071.

Oxenham AJ, Bernstein JGW, Penagos H (2004) Correct tonotopic representation is necessary for complex pitch perception. Proc Natl Acad Sci U S A 101:1421–1425.

Parnas BR (1996) Noise and neuronal populations conspire to encode simple waveforms reliably. IEEE Trans Biomed Eng 43:313–318.

Qin MK, Oxenham AJ (2006) Effects of introducing unprocessed low-frequency information on the reception of envelope-vocoder processed speech. J Acoust Soc Am 119:2417–2426.

Robert A, Eriksson JL (1999) A composite model of the auditory periphery for simulating responses to complex tones. J Acoust Soc Am 106:1852–1864.

Rubinstein JT, Wilson BS, Finley CC, Abbas PJ (1999) Pseudospontaneous activity: stochastic independence of auditory nerve fibers with electrical stimulation. Hear Res 127:108–118.

Schatzer R, Wilson BS, Wolford RD, Lawson DT (2003) Speech processors for auditory prostheses: signal processing strategy for a closer mimicking of normal auditory functions. Sixth Quarterly Progress Report, NIH project N01-DC-2-1002. Bethesda, MD: Neural Prosthesis Program, National Institutes of Health.

Spahr A, Dorman M, Loiselle L (2007) Performance of patients fit with different cochlear implant systems: effect of input dynamic range. Ear Hear 28:260–275.

Tchorz J, Kollmeier B (1999) A model of auditory perception as a front end for automatic speech recognition. J Acoust Soc Am 106:2040–2050.

Turner CW, Reiss LAJ, Gantz BJ (2008) Combined acoustic and electric hearing: preserving residual acoustic hearing. Hear Res 242:164–171.

Tyler RS, Preece JP, Lansing CR, Otto SR, Gantz BJ (1986) Previous experience as a confounding factor in comparing cochlear-implant processing schemes. J Speech Hear Res 29:282–287.

Williams EJ, Bacon SP (2005) Compression estimates using behavioral and otoacoustic emission measures. Hear Res 201:44–54.

Wilson BS (1997) The future of cochlear implants. Br J Audiol 31:205–225.

Wilson BS (2004) Engineering design of cochlear implant systems. In: Zeng F-G, Popper AN, Fay RR (eds), Auditory Prostheses: Cochlear Implants and Beyond. New York: Springer, pp. 14–52.

Wilson BS (2006) Speech processing strategies. In: Cooper HR, Craddock LC (eds), Cochlear Implants: A Practical Guide, 2nd ed. Hoboken, NJ: Wiley, pp. 21–69.

Wilson BS, Dorman MF (2007) The surprising performance of present-day cochlear implants. IEEE Trans Biomed Eng 54:969–972.

Wilson BS, Dorman MF (2008a) Interfacing sensors with the nervous system: lessons from the development and success of the cochlear implant. IEEE Sensors J 8:131–147.

Wilson BS, Dorman MF (2008b) Cochlear implants: a remarkable past and a brilliant future. Hear Res 242:3–21.

Wilson BS, Dorman MF (2009) The design of cochlear implants. In: Niparko JK, Kirk KI, Mellon NK, Robbins AM, Tucci DL, Wilson BS (eds), Cochlear Implants: Principles & Practices, 2nd ed. Philadelphia: Lippincott Williams & Wilkins, pp. 95–135.

Wilson BS, Finley CC, Lawson DT, Wolford RD, Eddington DK, Rabinowitz WM (1991) Better speech recognition with cochlear implants. Nature 352:236–238.

Wilson BS, Lawson DT, Zerbi M, Finley CC (1994) Recent developments with the CIS strategies. In: Hochmair-Desoyer IJ, Hochmair ES (eds), Advances in Cochlear Implants. Vienna: Manz, pp. 103–112.

Wilson BS, Finley CC, Lawson DT, Zerbi M (1997) Temporal representations with cochlear implants. Am J Otol 18:S30–S34.

Wilson BS, Lawson DT, Müller JM, Tyler RS, Kiefer J, et al. (2003) Cochlear implants: some likely next steps. Annu Rev Biomed Eng 5:207–249.

Wilson BS, Schatzer R, Lopez-Poveda EA (2006) Possibilities for a closer mimicking of normal auditory functions with cochlear implants. In: Waltzman SB, Roland JT Jr (eds), Cochlear Implants, 2nd ed. New York: Thieme, pp. 48–56.

Zhang X, Heinz MG, Bruce IC, Carney LH (2001) A phenomenological model for the responses of auditory-nerve fibers: I. Nonlinear tuning with compression and suppression. J Acoust Soc Am 109:648–670.

Zwicker E (1985) A hardware cochlear nonlinear pre-processing model with active feedback. J Acoust Soc Am 80:154–162.

Zwicker E, Peisl W (1990) Cochlear preprocessing in analog models, in digital models and in human inner ear. Hear Res 44:209–216.

Chapter 10
Silicon Models of the Auditory Pathway

André van Schaik, Tara Julia Hamilton, and Craig Jin

10.1 Introduction

Neuromorphic engineering is a discipline of electrical engineering that aims to develop signal processing systems using biological neural systems as inspiration. As such it has the dual goal of building models of neural systems in order to better understand them, and of building systems to perform useful functions for particular applications. While neuromorphic engineers extensively develop and use computer models of neural systems, which are generally less expensive and faster to create, the main goal is to implement the models in electronic hardware. The greatest advantage of a hardware model is that it gives solutions in real time. Where it may take hours and even days to simulate a small time period in a complex software model, results from a hardware model are obtained instantly. As computational power increases, neural models are also becoming more and more complex, so that for the foreseeable future there is a need for hardware implementations to allow models to interact with the world in real time. An often underestimated benefit of real-time operation is that in watching the behavior of the system change *during* the tuning of parameters, the researcher develops a much better intuition about the influence of these parameters. In addition to real-time operation, hardware models suffer from unavoidable mismatch and noise, just like neural systems. This forces the models to be robust to noise and mismatch, which is not the case for computer models, where noise and mismatch would have to be specifically added.

Neurons communicate using nerve impulses, or spikes. It is generally considered that the exact shape of the spike is not used to transfer information, and as such a spike may be considered a binary event: a spike is either present or absent. On the other hand, a spike may occur at any moment in time, so that time may be considered a continuous variable. Figure 10.1 shows some typical engineering

A. van Schaik (✉)
School of Electrical and Information Engineering, The University of Sydney,
Sydney, NSW 2006, Australia
e-mail: andre@ee.usyd.edu.au

R. Meddis et al. (eds.), *Computational Models of the Auditory System*,
Springer Handbook of Auditory Research 35, DOI 10.1007/978-1-4419-5934-8_10,

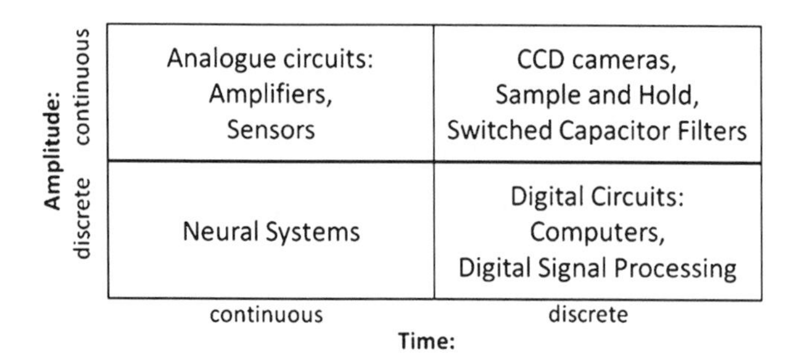

Fig. 10.1 Signal processing domains

applications classified by whether time and amplitude are continuous or discrete variables. In analog systems, such as, for example, most audio amplifiers, both time and amplitude are continuous variables. The camera in a mobile phone handset uses a charge-coupled device (CCD) imager, in which light is converted into a continuous variable – charge – but the readout is clocked and time is thus discretized. In a digital computer, both amplitude and time are discretized. It is interesting to see that there are currently no engineering applications in the quadrant in which neural systems operate. This is mainly due to a current lack of understanding of the advantages of operating in this mode, and this should provide a solid motivation to engineers for studying neural systems.

Models of the auditory pathway typically start with the development of a model of the cochlea. The first electronic cochlea was published by Jont Allen (1985), while electronic models of spiking neurons first appeared in the 1940s. These electronic models, however, used discrete electronic components, which made them large and ill suited for the creation of large scale complex models. In the 1980s, very large scale integration (VLSI) technology became available and affordable, allowing designers to put many components on a small piece of silicon, measuring only a few millimeters on a side. This opened up the way for the implementation of much more complex models. These VLSI electronic systems are the focus of this chapter.

10.2 Silicon Cochleae

Lyon and Mead (1988) built the first silicon cochlea, and there have been many variations and improvements since this initial design. Figure 10.2 shows a tree diagram illustrating the progression of silicon cochlea modeling since Lyon and Mead's first silicon cochlea.

All silicon cochleae discretize the basilar membrane (BM) using a number of filters or resonators with an exponentially decreasing fundamental frequency (from base to apex) to mimic the frequency distribution along the BM. The systematic

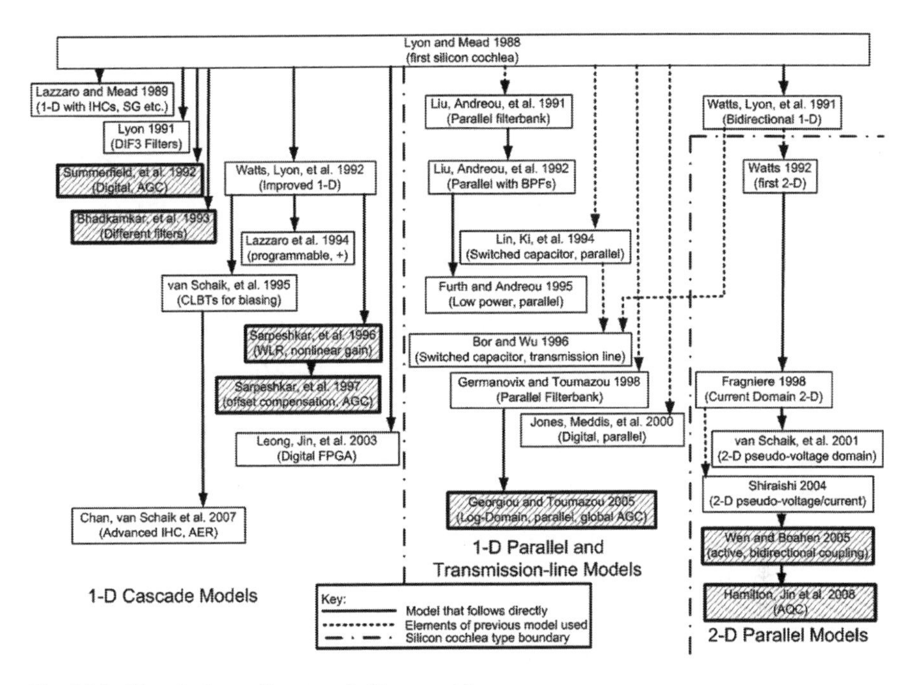

Fig. 10.2 Historical tree diagram of silicon cochleae

changes in the properties of the BM with longitudinal position, such as stiffness and width, are generally modeled by systematic changes in the parameters of the cochlear filter elements.

Figure 10.3 classifies silicon cochleae as one of three types: one-dimensional cascade (Fig. 10.3b), 1-D parallel (Fig. 10.3c), or two-dimensional (Fig. 10.3d), based on the coupling between the cochlear elements. The 1-D cascade silicon cochleae model longitudinal wave propagation of the BM from base to apex (the x-direction in Fig. 10.3a), and each successive segment adds its filtering to that of the segments before it. The combined effect creates the steep high-frequency slope so typical of the frequency tuning of BM sections, while the individual filters are typically only of order two. This allows for compact implementations using simple filters. The first silicon cochlea was of this type and the 1-D cascade silicon models have since had many generations of improvements. A disadvantage of the 1-D cascade model is that it has poor fault tolerance which is inherent in the cascade structure – if one of the early sections fails, all the sections that follow will have the wrong input signal. Also, each section adds a delay to the signal that is inversely proportional to the center frequency of that filter, which can lead to unrealistically long delays, particularly at the low-frequency end of the cochlea. Noise, generated internally by each filter, also accumulates along the cascade, which reduces the dynamic range of the system.

The parallel filter bank cochlea model of Fig. 10.3c is often used as the front-end in software auditory pathway models, because of its ease of implementation.

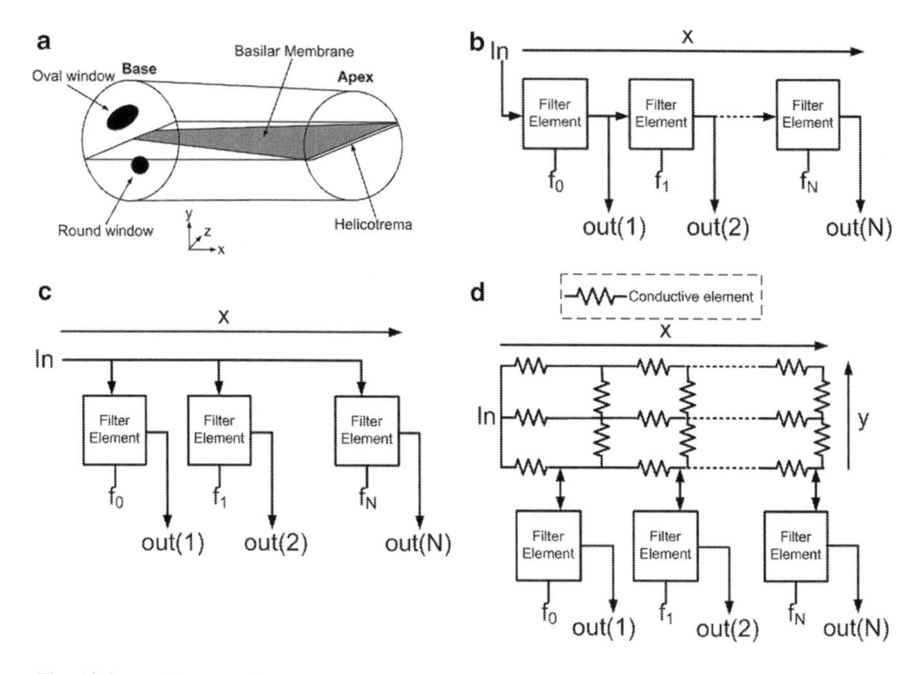

Fig. 10.3 (**a**) The uncoiled cochlea. (**b**) The 1-D cochlea cascade structure. (**c**) The 1-D parallel structure. (**d**) The 2-D structure

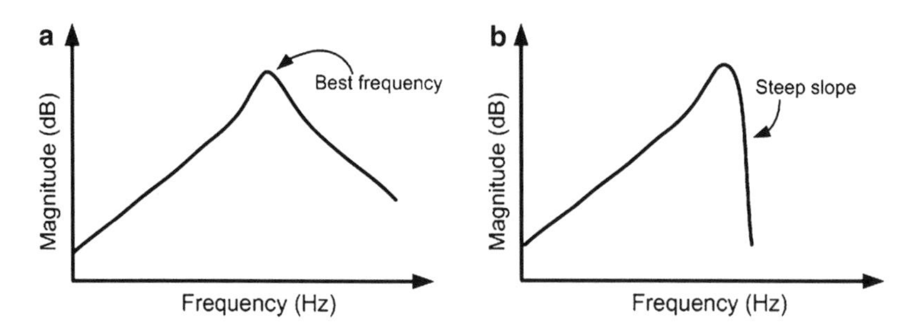

Fig. 10.4 (**a**) Output from a second-order filter element in the parallel filter model. (**b**) Output from a second-order filter element in the 1-D cascade model

Although this type of model does not suffer from the disadvantages of the 1-D cascade model, it is not preferred for silicon implementations because each filter acts independently. To create the same steep high-frequency slope, filters of much higher order are needed which use up significantly more area and power. A comparison of the output of a single second-order section for both a 1-D cascade (a) and a 1-D parallel model (b) is shown in Fig. 10.4.

The 2-D silicon cochleae model the fluid within the cochlea duct as well as the BM, taking both the longitudinal and vertical wave propagation into account. This design couples neighboring filters through the conductive elements modeling the cochlear fluid. These systems can be seen as combining the advantages of the two 1-D structures: coupling of the parallel filters allows the steep high-frequency slope to be generated with second-order filters, while fault tolerance is much improved over the cascade structure and delay accumulation is avoided. Implementations of this type are much more recent and as such many implementation issues are still being sorted out.

Silicon cochleae are classified as "active" when the gain and/or selectivity of the cochlear filter elements changes dynamically based on the intensity of the input, essentially increasing the gain and frequency tuning at low intensities and reducing these at high intensities. The silicon cochleae that can be considered active are shaded in Fig. 10.2. Of these, the majority of models use either a fixed nonlinearity or automatic gain control (AGC). Conventionally, AGC simply means changing the gain in response to changes in the input signal. In this case, however, AGC often incorporates the idea of automatic quality factor control (AQC). Hence, not only does the gain change in response to the input but so too does the bandwidth of the filter.

In the 20+ years since the development of the first silicon cochlea there has been a steady improvement in both the modeling complexity and the results from silicon cochleae. The quality of the output of a silicon cochlea is based on how well it matches certain key biological features. Specifically, these features normally include the steepness of the frequency response curve at high frequencies above the characteristic (or best) frequency of a filter element, the highest achievable quality factor of the cochlea filter elements, and the silicon cochlea's ability to demonstrate characteristic nonlinearities such as large-signal compression, two-tone suppression, combinational tones, and so on. Figure 10.5 shows the results from selected silicon cochleae that demonstrate some of these features. The interested reader should examine the references in Fig. 10.2 for a more thorough exploration of silicon cochlea results.

Figure 10.5 shows a comparison between measured results of the mammalian cochlea and those from silicon cochleae. Figure 10.5a shows that the steep slope above the best frequency is present in both the measured results and those from the silicon cochleae. The results in Fig. 10.5b–d illustrate the nonlinear behavior of the cochlea: compressive gain (b), two-tone suppression (c), and combinational tones (d). The silicon cochleae for these three experiments are all 2-D. Two tone suppression and combinational tones are dependent on bidirectional interactions between filter elements and would not be seen in 1-D cascade or parallel models.

The silicon cochlea, although improving substantially over the last two decades, still falls well short of the performance of the biological cochlea in terms of frequency selectivity, dynamic range, gain (and active gain control), and power consumption. The human cochlea has 3,000 inner hair cells that sense the vibration of the BM whereas the silicon cochlea has had, at most, several hundred filter elements. The designers of silicon cochleae are restricted by the area available on a chip and the mismatch of elements across a single chip. In most silicon cochleae it is very difficult to have matching frequency and gain characteristics between two individual filter elements, let alone all the filter elements on a single chip. Matching can be improved by using higher power and larger devices but this then restricts the

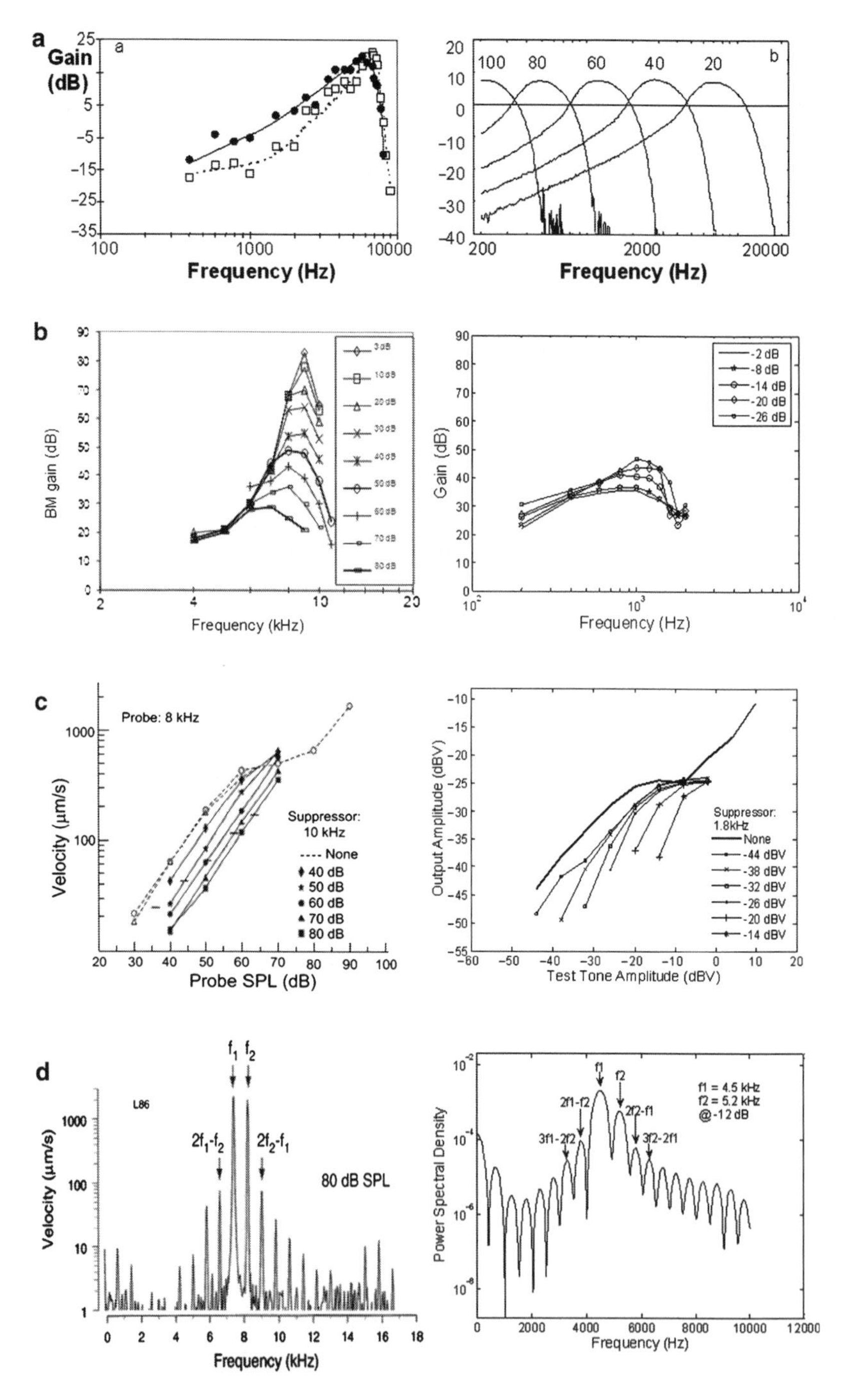

Fig. 10.5 (**a**) BM frequency response measurements for the squirrel monkey (used with permission from Rhode 1971) (*left*) and from the 1-D cascade silicon cochlea in (adapted from van Schaik et al. 1995) (*right*). (**b**) Gain at a place along the BM of a chinchilla in response to changes in input

number of elements that can be integrated on a single chip. Designing, fabricating, and testing a silicon cochlea is a process that takes at least 6 months, so that there are often years between successive generations of a particular model. Creating a silicon cochlea is an extremely challenging engineering task requiring not only biologically sound cochlea models but also good circuit design skills. Improvements are slow but steady and the authors hope to be able to have a silicon cochlea available in the next few years that implements a reasonable model of biological cochlear filtering in real time and that is robust enough to be made available to other researchers.

10.3 Silicon Models of the Inner Hair Cell

The silicon cochleae discussed in the previous section model the vibration of the BM. In the case of the active silicon cochlea, this includes some model of the function of the outer hair cells (OHCs). The next step in modeling auditory processing is to add a model of the inner hair cells (IHCs) which convert the BM vibration into a neural signal. This transduction occurs in two steps. First, vibration of the cilia on the IHC opens ion channels in the cell, causing a change in intracellular voltage. Second, this change in membrane voltage causes neurotransmitter release into the synaptic cleft between the IHC and the spiral ganglion cells (SGCs) that innervate it.

At rest, the mechanically controlled conductance of the IHC ion channels is around 20% of the maximum conductance. Flexing the hairs in one direction closes the channels, saturating at zero conductance, whereas flexing in the other direction opens the channels, saturating at the maximum conductance. Hudspeth and Corey (1977) have shown that conductance as a function of hair bundle deflection has a sigmoidal form. Because the intracellular voltage changes only by about 8 mV between these two extremes, the voltage difference across the conductance changes little from the 125 mV at rest. Therefore, the current entering into the cell through this conductance can be thought of as directly proportional to the change in conductance.

The cell membrane of the IHC has associated with it a membrane capacitance as well as a leakage conductance. The overall result of this is that the IHC membrane low-pass filters the stimulus current resulting from hair bundle motion. This may be appropriately modeled by a first-order low-pass filter with a 1 kHz corner

intensity (*left*, adapted from Ruggero 1992) and BM gain at the output of a filter element in the 2-D silicon cochlea described in (adapted from Hamilton et al. 2008b) (*right*). (**c**) Two-tone suppression from chinchilla data (used with permission from Ruggero et al. 1992) (*left*) and in the 2-D silicon cochlea described in Hamilton et al. (2008b) (*right*). (**d**) Combinational tones from experiments with a chinchilla (used with permission from Robles et al. 1997) (*left*) and in the 2-D silicon cochlea described in Hamilton et al. 2008 (©2008 IEEE) (*right*)

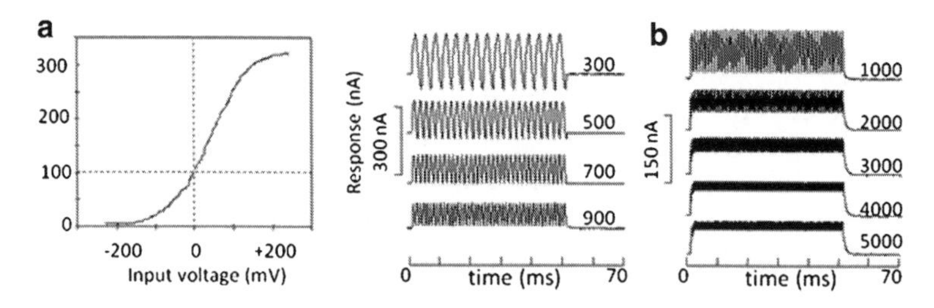

Fig. 10.6 Results from a silicon IHC (van Schaik 2003, ©2003 IEEE). (**a**) Input current (in nanoamperes) as a function of silicon cochlea output (in millivolts). (**b**) Intracellular voltage of the IHC represented as a current signal in response to a 50-ms pure tone. The tone had a fixed amplitude and its frequency was varied as indicated (in hertz) along each trace

frequency. For electronic models, it is convenient to represent the intracellular voltage of the IHC by a current signal, which can then be used directly to stimulate a spiking neuron model. Figure 10.6 shows results from such a silicon IHC (van Schaik and Meddis 1999). The asymmetry of the conductance function, together with the low-pass filtering by the cell membrane, causes the AC component of the response to decrease with frequency, so that at frequencies above a few kilohertz only the DC component remains.

Combining the silicon cochlea with the model for the intracellular IHC voltage results in the response shown in Fig. 10.7 when a 110-Hz harmonic complex (Fig. 10.7a) is used as the input. The response of the silicon cochlea (Fig. 10.7b) shows that the first three harmonics are resolved for the low-frequency sections, while harmonics four to six display amplitude modulation of the partial, indicating the presence of a few harmonics in the output of each section. From the seventh harmonic onward many harmonics are present in the output of each filter, resulting in the characteristic wave packets seen in the top ten traces in Fig. 10.7b. The output of the IHC circuit (Fig. 10.7c) shows some distortion of the low-frequency wave forms due to the nonlinearity of Fig. 10.6a. At higher frequencies the AC component of the output decreases, until for the 3-kHz filter, it is mainly the envelope of the wave packet that is represented by the IHC output.

To complete the electronic IHC, it is important to also model the neurotransmitter release, or at least to model the adaptation resulting from the fact that neurotransmitter is used up as it is released into the cleft. This leads to adaptation so that the response of the auditory nerve (AN) is strongest at the onset of a sound. A computer model that has been widely used to simulate the IHC is the Meddis IHC model (Meddis 1986), and an electronic version of this model was developed by McEwan and van Schaik (2003). The output of this model (Fig. 10.8) shows adaptation to the presentation of a 500-Hz pure tone of 200 ms duration at various stimulus levels. This also shows that the onset response grows more quickly with stimulus level than the sustained response.

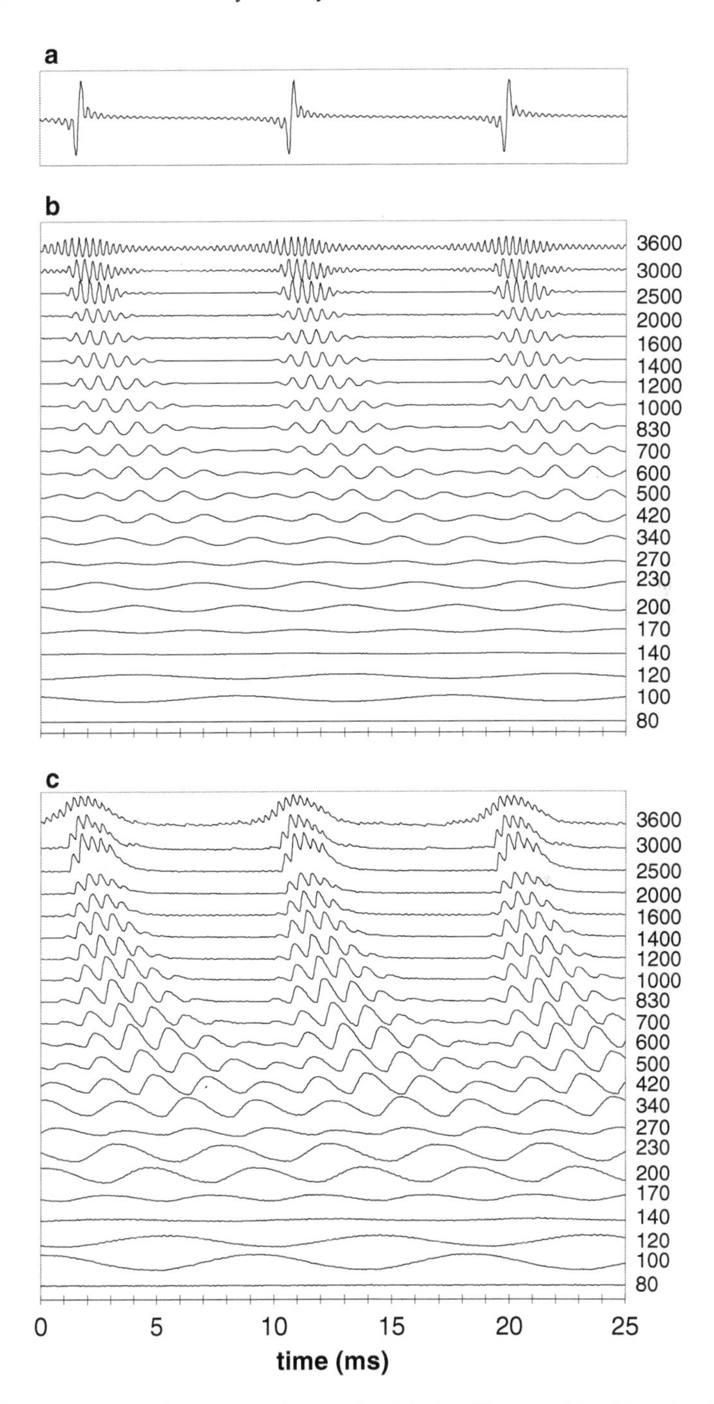

Fig. 10.7 Response a 110-Hz harmonic complex (**a**) of a silicon cochlea (**b**) and an IHC (**c**). The best frequency of each cochlear output is listed (in hertz) along the *right-hand side* of (**b**) and (**c**)

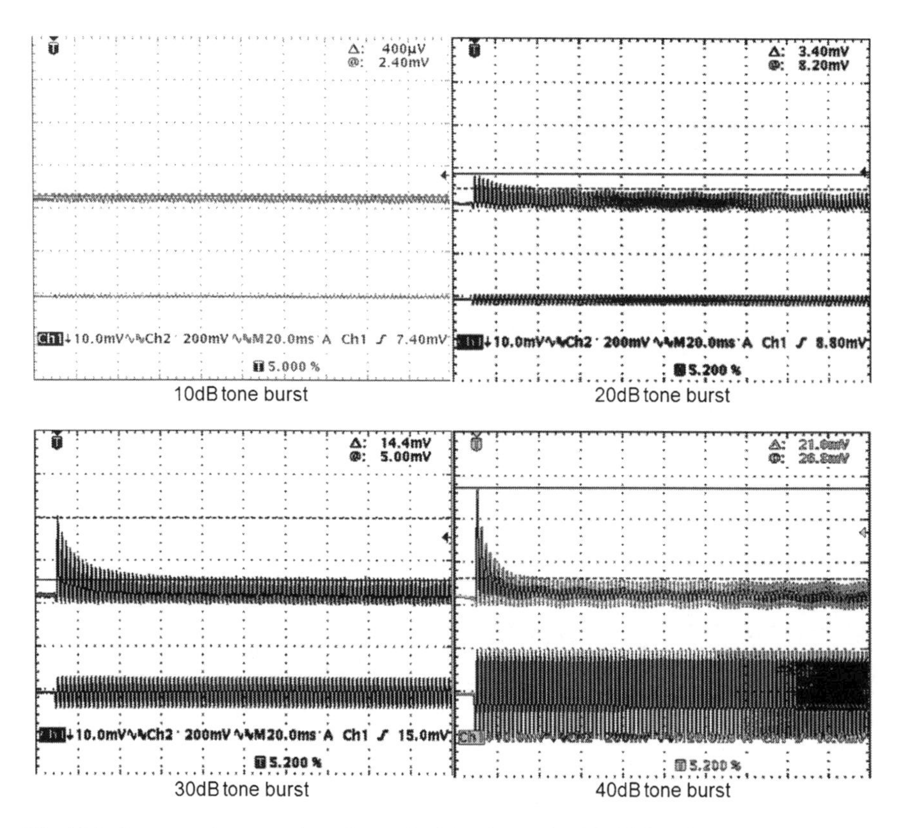

Fig. 10.8 Response (*top trace* in each panel) of the electronic "Meddis" IHC to tone burst (*bottom trace* in each panel) at varying input levels (From McEwan and van Schaik 2003.)

10.4 Silicon Models of Early Auditory Processing

The next step in creating an electronic model of auditory processing is to implement an SGC. In fact, all auditory processing after the IHC is performed by various neurons and rather than building very detailed models of the many different types of neurons in the brain, it is preferable to build only one model (or a few at most) that can be made to behave like various different types of neuron by changing the bias parameters of the circuit. The simplest neuron model that can be made to exhibit various behaviors is the leaky integrate-and-fire (LIF) model (Lapicque 1907), and many silicon neurons are based on this.

The LIF neuron model requires a membrane capacitance, a conductive leak, a firing threshold, and a reset mechanism. All these elements are relatively easily incorporated onto a silicon circuit. To create more realistic behaviors a refractory period, that is, the period after a neuron spikes during which it is unable to spike again, can easily be added to an LIF model. This was done in the simple 10-transistor circuit presented by van Schaik et al. (1996) to obtain a neuron that could model

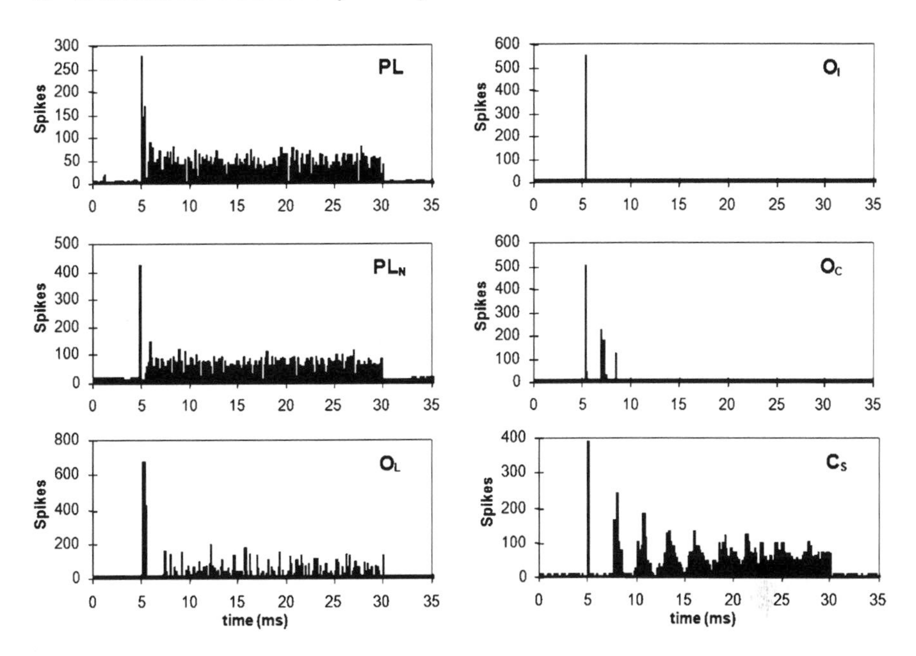

Fig. 10.9 Poststimulus time histogram of an electronic neuron modeling various types of neurons found in the ventral cochlear nucleus. (From van Schaik et al. 1996, ©1996 IEEE.) See text for details

various types of neurons found in the ventral cochlear nucleus, which is one of the sites of innervation by the AN. Figure 10.9 shows the results from this circuit as a poststimulus time histogram (PSTH) using repeated stimulations with a 25-ms 5-kHz pure tone burst. The burst was first filtered by a silicon cochlea and silicon IHC and the section that responded best to the 5-kHz tone was used as input to the neuron. For these experiments the analog output of the IHC was used directly to stimulate the neuron, without first modeling the conversion into spikes by the SGC. However, the output of the primary-like neuron (PL) shown in Fig. 10.9 is identical in this experiment to what the output of the SGC would be. The PL response is typical of a spherical bushy cell, mainly found in the anteroventral cochlear nucleus (AVCN). This cell receives input from only one AN fiber through a large synaptic contact on its soma, called the end bulb of Held, and as a result outputs one spike for every AN input spike. It shows a high spiking rate at the onset of the tone and a much lower sustained rate. The response was obtained from the circuit by giving the neurons a medium level membrane leakage current – to avoid long-term integration of the input signal – and a short refractory period – so that the neurons are ready to fire when the next input arrives.

The primary-like with notch PSTH (PL_N) is very similar to the primary-like PSTH and is the typical response of a globular bushy cell, which is also mainly found in the AVCN. The globular busy cell receives synaptic contacts from a few AN fibers that originate in the same region of the cochlea. Therefore, its PSTH is still similar to the PSTH of an AN fiber, but the probability of the cell firing at the onset of the stimulus is almost one, because at least some of the AN input fibers will carry a spike at the onset of the stimulus. If the cell always spikes at the onset of the stimulus, it will always be in its refractory period just after it spikes. This creates the notch in the PSTH. The globular bushy cell thus enhances the onset of a stimulus. Furthermore, the cell typically needs a few simultaneous AN spikes in order to generate an output spike. This means that the cell only responds to correlated action potentials of several fibers, thereby suppressing the random activity present on the AN. The globular bushy cell therefore can be said to improve the signal-to-noise ratio. A more detailed silicon model of the globular bushy was proposed by Wittig and Boahen (2006). This model demonstrates both the phase-locking and enhanced signal-to-noise response when compared to silicon models of AN fibers. In Wittig (2007) the biophysical mechanisms that produce the precise timing behavior of the globular bushy cell are examined with the aid of the silicon model.

The primary-like with notch (PL_N) response is obtained from the circuit using a relatively high membrane leakage current, to avoid long-term integration of the input signal, a threshold voltage that ensures that the neuron will always fire at the onset of the stimulus, and a refractory period that matches the duration of the notch.

The onset locker PSTH (O_L) is similar to the PL_N, but with a lower and less constant sustained rate. This PSTH is the typical response of the large octopus cells found in the posteroventral cochlear nucleus (PVCN). This response is obtained by the same high probability of spiking at the onset, which creates the onset peak and the notch in the PSTH. Having a higher leakage current than the globular bushy cell, and more synaptic inputs, it needs even more synchronous action potentials in order to generate a spike. Therefore, its signal to noise ratio will be even higher than the signal-to-noise ratio of the PL_N response.

The onset inhibitory PSTH (O_I) is a less common response of the octopus cells. It shows only an initial peak and hardly any activity after. This response can be obtained by increasing the neuron's leak and threshold, so that only at the onset of the stimulus is the output of the IHC large enough to cause the neuron to spike.

The onset chopper PSTH (O_C) is the response of a large multipolar stellate cell, which is mainly found in the PVCN, but also relatively frequent in the AVCN. This type of stellate cell has its soma largely covered by synaptic contacts, and also has very short onset latency. A spike arriving at a synapse on the cell soma creates a relatively large membrane voltage variation. The cell also receives many synapses from AN fibers, and has a high threshold of firing, but has a smaller leakage current than the other two onset cells. This neuron is also assumed to produce the transient chopper response, which is like the sustained chopper response (C_S), but with a less regular form of chopping.

Chopping is the standard behavior of a spiking neuron that completely resets after spiking. After an action potential is generated, such a cell cannot be activated during its refractory period. Once its refractory period is over, the cell starts to integrate the input spikes over a certain time, controlled by the leakage current of this leaky integrator. This yields a fairly constant firing rate of such a cell as response to an input signal with a constant mean spike rate, thus generating the typical chopping behavior. If the sustained input level (from the AN) is too low to reach the firing threshold, the cell will chop only at the stimulus onset.

The sustained chopper PSTH (C_S) is the typical response of the other type of stellate cell. This stellate cell hardly has any synapses on its soma and thus receives most of its inputs on its dendrites. Because the dendrites low-pass filter the incoming spikes, the membrane potential will rise only smoothly. This explains the regularity of the sustained chopper cell. Regularity is slowly destroyed due to the integration of noise over time, as can be seen in the last 10 ms of the PSTH in Fig. 10.9. In van Schaik and Meddis (1999) this setting of the circuit was used to implement a model of amplitude modulation sensitivity in the auditory brain stem. This model used the silicon cochlea, IHC, and C_S neuron as well as the same neuron circuit configured to model coincidence detectors in the inferior colliculus. The coincidence detecting behavior was obtained by giving the circuit a high leak, so that spikes were not integrated over time, and a high threshold, so that several spikes would have to arrive simultaneously from a number of choppers to cause the cell to fire.

Although the above simple neuron circuit could already be used to model a large variety of neurons, one important feature was missing from the circuit. Just like the IHC, many neurons show adaptation in their firing rate with a higher onset rate and a lower sustained rate when stimulated with a constant input current. This adaptation was included by Indiveri (2003) through the slow increase of a membrane leakage current when the neuron spikes. The Izhikevich neuron model (Izhikevich 2003) also uses a slow adapting variable with negative feedback to adapt the firing rate of the neuron. This model further includes a quadratic positive feedback term of the neuron's membrane voltage to its input, which can create many complex behaviors, such as bursting in cortical neurons. This neuron model was recently implemented by Wijekoon and Dudek (2008). Another recent neuron model, the Mihalas and Niebur neuron model, instead slowly adapts the neuron's firing threshold when the neuron spikes (Mihalas and Niebur 2009). This model also allows for the modular addition of other spike induced currents to obtain cortical bursting neurons and has recently been implemented by Folowosele et al. (2009).

10.5 Summary

This chapter introduced neuromorphic engineering, and a number of different silicon models that represent various functions in the auditory pathway have been presented. As understanding of the auditory pathway has improved, the VLSI designs have utilized models that have gradually included more and more of the impressive signal processing characteristics that are found in biology. These silicon models are used in a variety of applications, from a tool for researchers to validate mathematical models in real time to audio front-ends in speech recognition systems.

The complexity of a model used will depend on its end application, however, the spirit of neuromorphic engineering remains the same no matter what the application: to leverage biological functionality in modern technology.

The quest to build better and more realistic silicon auditory pathway models is being confronted with many of the challenges that biology has had to overcome. While there has not been a silicon model built to date that rivals biology, the situation is improving, with better models and better technology. In just over 20 years, implementations have improved from simple 1-D filter cascades in silicon cochlea modeling to more biologically plausible 2-D structures that include outer hair cell functionality. There have also been improvements in circuit design and as miniaturization of CMOS fabrication technology continues, we can include more circuits (filters, cell structures, etc.) on a single integrated circuit.

Even after two decades since Lyon and Mead's (1988) first silicon cochlea, designers are still faced with many of the same design challenges: noise, dynamic range and mismatch to name but a few. This chapter has presented some examples, however, to show that silicon designs have improved significantly in 20 years and by overcoming some of the design challenges that still face the neuromorphic engineer.

References

Allen J (1985) Cochlear modeling. IEEE ASSP Mag 2:3–29.

Bhadkamkar N, Fowler B (1993) A sound localization system based on biological analogy. In Proceedings of the IEEE International Conference on Neural Networks, 28 March–1 April 1993, San Francisco, CA. 1902–1907.

Chan V, Liu S-C, van Schaik A (2007) AER EAR: a matched silicon cochlea pair with address event representation interface. IEEE Trans Circuits Syst I 54:48–59.

Folowosele F, Hamilton TJ, Harrison A, Mihalas S, Niebur E, Cassidy A, Andreou A, Etienne-Cummings R (2009) A switched capacitor implementation of the generalized linear integrate-and-fire neuron. 24–27 May 2009, Proc IEEE Int Symp Circuits Syst 2149–2152.

Fragniere E (1998) Analogue VLSI emulation of the cochlea. Département d'électricité Lausanne: EPFL; 221.

Furth PM, Andreou AG (1995) A design framework for low power analog filter banks. IEEE Trans Circuits Syst I 42:966–971.

Georgiou J, Toumazou C (2005) A 126-μW cochlear chip for a totally implantable system. IEEE J Solid State Circuits 40:430–443.

Germanovix W, Toumazou C (1998) Towards a fully implantable analogue cochlear prosthesis. IEE Colloq Analog Signal Processing 10:1–1011.

Hamilton TJ, Jin C, Tapson J, van Schaik A (2008a) An active 2-D silicon cochlea. IEEE Trans Biomed Circuits Syst 2:30–43.

Hamilton TJ, Tapson J, Jin C, van Schaik A (2008b) Analogue VLSI implementations of two dimensional, nonlinear, active cochlea models. 20–22 November 2008, Proc IEEE Biomed Circuits Syst 153–156.

Hudspeth AJ, Corey DP (1977) Sensitivity, polarity, and conductance change in the response of vertebrate hair cells to controlled mechanical stimuli. Proc Natl Acad Sci U S A 74:2407–2411.

Indiveri G (2003) A low-power adaptive integrate-and-fire neuron circuit. Proc IEEE Int Symp Circuits Syst 4:820–823.

Izhikevich EM (2003) Simple model of spiking neurons. IEEE Trans Neural Netw 14:1569–1572.

Jenn-Chyou B, Chung-Yu W (1996) Analog electronic cochlea design using multiplexing switched-capacitor circuits. IEEE Trans Neural Netw 7:155–166.

Jones S, Meddis R, Lim SC, Temple AR (2000) Toward a digital neuromorphic pitch extraction system. IEEE Trans Neural Netw 11:978–987.

Jyhfong L, Wing-Hung K, Edwards T, Shamma S (1994) Analog VLSI implementations of auditory wavelet transforms using switched-capacitor circuits. IEEE Trans Circuits Syst I 41:572–583.

Lapicque L (1907) Recherches quantitatifs sur l'excitation électrique des nerfs traitée comme une polarisation. J Physiol (Paris) 9620–635.

Lazzaro J, Mead C (1989) Circuit models of sensory transduction in the cochlea. In: Mead C, Ismail M (eds). Analog VLSI Implementations of Neural Networks. Norwell, MA: Kluwer, pp 85–101.

Lazzaro J, Wawrzynek J, Kramer A (1994) Systems technologies for silicon auditory models. IEEE Micro 14:7–15.

Leong MP, Jin CT, Leong PHW (2003) An FPGA-based electronic cochlea. EURASIP J Appl Signal Processing 7:629–638.

Liu W, Andreou AG, Goldstein MH (1991) An analog integrated speech front-end based on the auditory periphery. Proc Int Joint Conf Neural Netw 2:861–864.

Liu W, Andreou AG, Goldstein MH (1992) Voiced-speech representation by an analog silicon model of the auditory periphery. IEEE Trans Neural Netw 3:477–487.

Lyon RF (1991) Analog implementations of auditory models. In: Human Language Technology Conference Proceedings of Workshop on Speech and Natural Language, 19–22 February 1991, Pacific Grove, CA. Morristown, NJ: Assoc Comp Ling 212–216.

Lyon RF, Mead C (1988) An analog electronic cochlea. IEEE Trans Acoust Speech Signal Processing 36:1119–1134.

McEwan A, van Schaik A (2003) An analogue VLSI implementation of the Meddis inner hair cell model. EURASIP J Appl Signal Processing 7:639–648.

Meddis R (1986) Simulation of mechanical to neural transduction in the auditory receptor. J Acoust Soc Am 79:702–711.

Mihalas S, Niebur E (2009) A generalized linear integrate-and-fire neural model produces diverse spiking behaviors. Neural Comput 21:704–718.

Rhode WS (1971) Observations of the vibration of the basilar membrane in squirrel monkeys using the Mossbauer technique. J Acoust Soc Am 49:1218–1231.

Robles L, Ruggero MA, Rich NC (1997) Two-tone distortion on the basilar membrane of the chinchilla cochlea. J Neurophysiol 77:2385–2399.

Ruggero MA (1992) Responses to sound of the basilar membrane of the mammalian cochlea. Curr Opin Neurobiol 2:449–456.

Ruggero MA, Robles L, Rich NC (1992) Two-tone suppression in the basilar membrane of the cochlea: mechanical basis of auditory-nerve rate suppression. J Neurophysiol 68:1087–1099.

Sarpeshkar R, Lyon RF, Mead CA (1996) An analog VLSI cochlea with new transconductance amplifiers and nonlinear gain control. Proc IEEE Int Symp Circuits Syst, Atlanta, USA, 3:292–296.

Sarpeshkar R, Lyon RF, Mead C (1998) A low-power wide-dynamic-range analog VLSI cochlea. Analog Integr Circuits Signal Processing 16:245–274.

Shiraishi H (2004) Design of an Analog VLSI Cochlea. Master's Thesis, Electrical and Information Engineering, The University of Sydney, Sydney, NSW: 90.

Summerfield CD, Lyon RF (1992) ASIC implementation of the Lyon cochlea model. IEEE Proc Int Conf Acoustics Speech Signal Process, San Francisco, USA, 5:673–676.

van Schaik A (2003) A small analog VLSI inner hair cell model. Proc IEEE Int Symp Circuits Syst, Bangkok, Thailand, pp 17–20.

van Schaik A, Fragniere E (2001) Pseudo-voltage domain implementation of a 2-dimensional silicon cochlea. Proc IEEE Int Symp Circuits Syst, Sydney, Australia, 2:185–188.

van Schaik A, Meddis R (1999) Analog very large-scale integrated (VLSI) implementation of a model of amplitude-modulation sensitivity in the auditory brainstem. J Acoust Soc Am 105:811–821.

van Schaik A, Fragniere E, Vittoz EA (1995) Improved Silicon Cochlea using Compatible Lateral Bipolar Transistors. Advances in Neural Information Processing Systems, The MIT Press 8:671–677.

van Schaik A, Fragniere E, Vittoz E (1996) An analogue electronic model of ventral cochlear nucleus neurons. Proc Fifth Int Conf Microelect Neural Netw, Lausanne, Switzerland, 1996, 52–59.

Watts L (1992) Cochlear Mechanics: Analysis and Analog VLSI. Ph.D. Thesis, California Institute of Technology, Pasadena, CA: 173.

Watts L, Lyon RF, Mead C (1991) A bidirectional analog VLSI cochlear model. Advanced research in VLSI, Santa Cruz, MIT Press, Cambridge, MA, 153–163.

Watts L, Kerns DA, Lyon RF, Mead CA (1992) Improved implementation of the silicon cochlea. IEEE J Solid State Circuits 27:692–700.

Wen B, Boahen K (2005) Active Bidirectional Coupling in a Cochlear Chip. Advances in Neural Information Processing Systems 17, The MIT Press.

Wijekoon JHB, Dudek P (2008) Compact silicon neuron circuit with spiking and bursting behaviour. Neural Networks 21:524–534.

Wittig JH (2007) Biophysical Mechanisms for Precise Temporal Signaling in the Auditory System. Ph.D. Thesis, Bioengineering University of Pennsylvania, Philadelphia, PA.

Wittig JH, Boahen K (2006) Silicon neurons that phase-lock. Proc IEEE Int Symp Circuits Syst, Kos, Greece.

Index